T0229593

DESIGN OF EARTH DAMS

GEOTECHNIKA 2

Selected translations of Russian
geotechnical literature

DESIGN OF EARTH DAMS

A. L. GOLDIN & L. N. RASSKAZOV
Energoatomizdat, Moscow

Translated from Russian and edited by
R. B. ZEIDLER
H*T*S, Gdańsk

A. A. BALKEMA / ROTTERDAM / BROOKFIELD / 1992

Originally published as: Proyektirovaniye gruntovykh plotin

© 1987 Energoatomizdat, Moscow

English translation:
© 1992 A.A. Balkema, P.O. Box 1675, 3000 BR Rotterdam, Netherlands

ISBN 90 6191 173 7

Distributed in USA and Canada by:
A.A. Balkema Publishers, Old Post Road, Brookfield, VT 05036, USA

Contents

Preface

Earth dams are the most common impoundment structures. Their construction can be well mechanized, and they consume relatively little labour, energy, and cement, compared with concrete dams. Contemporary earth dams have reached heights of 300 m (Nurek Dam, USSR) and a volume of 130 million cubic metres (Tarbela, Pakistan).

The Soviet earth dams have been constructed under various climatic conditions. Viluy and Ust'-Khantay Dams have been erected in permafrost zones, where Kolyma Dam is also under construction. Nurek, Charvak, Gissarak, and Papan Dams have been built in the Soviet Middle Asia (henceforth referred to as the five union republics in the south-west of Northern Asia), where the highest world dam (Rogun, 335 m) is constructed as well. The Caucasus has its dams of Sarsang, Mingechaur and Zhinval, while Izobilnen Dam and others have been built in the Crimea. A number of large earth dams are under construction in the Soviet Far East, southern Siberia, the Urals, and the Carpathian mountains.

A variety of climatic, geological, topographic, hydrologic and seismic conditions determine for earth dams the technology of construction, design, foundation conditions, etc.

Many earth dams impound huge quantities of water, so their failure might have calamitous consequences.

Stringent requirements are imposed on the design and construction of such responsible structures. Modern design requires accurate static and dynamic computations basing on thorough analysis of stress-strain conditions. Manuals on hydraulic engineering seldom cover the realm of these topics.

The first Soviet handbook on the design of earth dams was published by Nichiporovich in 1973. It has been very important for training of specialists. Many problems have since been explored better, for instance the properties of soil in earth dams, stress-strain relationships, consolidation, and transitions between dams and their foundations, not to mention the computer-aided design (CAD). Optimization methods have been elaborated, and field data have been collected for Charvak and Nurek Dams.

Derivation of mathematical models is based on the theories of elasticity and plasticity. It is assumed here that courses on the above theories are prerequisites for the hydraulic engineering course, although some problems in comprehension may always arise. Similar gaps can appear in numerical modelling techniques.

Optimization of earth dams involves unfamiliar notions but in view of dramatic progress in this area the topic must not be skipped. By and large, a better understanding of the material presented herein is guaranteed through reference to the cited studies.

The book consists of nine chapters. Chapter One describes basic properties of soil in earth dams and provides guidelines on the prediction of their strength, deformability and permeability. Suitability of soil for placement in the body of an earth dam is discussed.

Chapters Two and Three present methods of seepage and consolidation computations. Chapter Four deals with the stress-strain conditions, displacement and settlement, while Chapter Five dwells on seismic stability. Chapter Six concentrates on the working capacity of earth dams, along with their stability on circular cylindric slip surfaces. Chapters Seven and Eight discuss different designs of earth dams and their constituents and foundations. Chapter Nine is devoted to the systems analysis of economic and technological factors.

Some problems dealt with in technical manuals such as hydraulic computation of seepage, certain consolidation topics, slope revetments and stability, quality control and particular design of hydraulic-fill dams, etc are not covered in this handbook.

Numerous materials produced in great a many Soviet scientific, teaching and consulting institutions are quoted.

The handbook has been written for students and postgraduates of hydraulic engineering institutes, or some other civil engineering branches. It will also be very helpful to designers and consultants tackling the questions of earth dams; many problems summarized in this book are dispersed in periodicals, thus less handy if at all available. The Authors hope their book will lay sound foundations for training of hydraulic engineers.

The Authors express their sincere gratitude to all who contributed to the production of this handbook.

Introduction

Performance of earth dams. Modern design of earth dams involves an analysis of static and dynamic conditions. Phenomenological schemes (basing on experimental findings) of soil mechanics are useful in this respect. Methods derived in the mechanics of continuous media for the movement of real soil are solved in terms of mathematical models embodying the most important properties of soil.

Mathematical models basing on the theory of plasticity, for irreversible deformations, have been put forth owing to the progress in computer technology. Irreversible deformations depend not only on the magnitude of loading but also on its sequence. The theory of plasticity does not deal with the duration and intensity of loading (this is analysed in rheology), although real deformability does depend on these factors, which is particularly important for static loading. In many cases rheological processes must therefore be taken into account in the design of earth dams.

Seepage plays an important role in the operation of earth dams. It can be unsteady during construction and longer initial stages. The interaction of earth dams and groundwater flow is reflected in theories of consolidation and seepage. Seepage computations for earth dams are complex because of undefined boundary conditions and seepage anisotropy of soil.

Analysis of the operation of an earth dam under seismic loading embodies various computational methods accounting for earthquakes, propagation of waves in the foundations of an earth dam, etc.

The performance, that is primarily the slope stability of an earth dam can be assessed by simplified engineering methods. This is facilitated by analysis of stress-strain conditions, although the transition to slope stability is by no means definite. The methods based on the concept of circular cylindric slip surfaces have therefore lost their initial value.

The interaction of earth dam and its foundations is one of the primary design factors. Dissolution of easily soluble components of soils containing gypsum or other salts becomes of particular interest. The prediction of leaching may be based on a theory of seepage with dissolution of salts.

Design of an earth dam is thus possible for known soil properties of the nearest

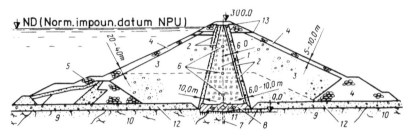

Figure I.1. Nurek Dam. (1) Saphedobian sandy loam core; (2) filters; (3) pebble fills, upstream and downstream fills being henceforth referred to as *thrust prisms*; (4) rock armour and cover layer (being also included in the category of *thrust prisms*); (5) upstream construction cofferdam; (6) inspection galleries; (7) seepage cut-off ('counterseepage curtain'); (8) concrete plug; (9) sandstones; (10) sandstones and aleurolites; (11) aleurolites; (12) alluvium; (13) antiseismic measures.

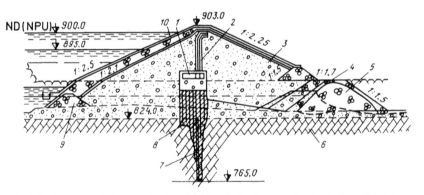

Figure I.2. Atbashin Dam. (1) concrete plug (slab); (2) diaphragm of polyethylene foil between two layers of sand; (3) pebble soil; (4) transitions with inverse filter; (5) river closure dyke; (6) limestones; (7) grout curtain in limestones; (8) seepage prevention curtain in pebbles; (9) rubble dyke ('thrust prism'); (10) coarse stone cover layer.

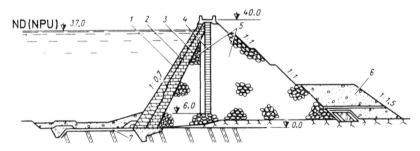

Figure I.3. Shirokov rockfill dam. (1) timber facing; (2) masonry on mortar; (3) dry pitching; (4) concrete-block hole for settlement inspection; (5) rock fill; (6) gravel and sand fill for better stability; (7) upstream blanket.

Table I.1. Classification of dams by material and design.

	Dam type by design				
Dam type by material	Homogeneous Design	Constr. method	Central core Design	Constr. method	
Earth		Fill Hydr.fill Expl.fill		Fill Hydr.fill	
Earth-rockfill	–	–		Fill Hydr.fill Fill/comp.	
Rock		Fill Expl.fill	–	–	

	Dam type by design				
Dam type by material	Sloping core Design	Constr. method	With screen Design	Constr. method	
Earth	–	–		Fill Hydr.fill	
Earth-rockfill		Fill Hydr.fill		Fill Fill/comp.	
Rock	–	–		Fill Fill/comp. Expl.fill	

	Dam type by design	
Dam type by material	With diaphragm Design	Constr. method
Earth		Fill Hydr.fill
Earth-rockfill	–	–
Rock		Fill Fill/comp. Expl.fill

(1) Earth dam is made of fine soil. (2) Earth-rockfill dam incorporates thrust prisms of coarse grained soil, providing principal stability of dam, and seepage prevention constituents (core or screen) of fine soil (clayey or sandy). (3) Rock dam consists primarily of coarse grained soil (rock rubble or gravelly pebbles) in the dam body and a nonsoil material to prevent seepage. Such dams are often referred to as rockfill dams. (4) Fill/comp. denotes filling of soil upon its compaction in layers.

quarries, with proper account for the objectives of the given hydraulic engineering scheme, the climatic, geologic, and hydrological conditions, availability of machinery and equipment, construction practice and experience, etc. The variety of earth dam types and design criteria make the selection a formidable and responsible task.

Optimization of earth dam design relies on technological and economic factors. In the systems analysis one considers the factors controlling the capacity and cost of structure, and selects optimum solutions.

All the aforementioned factors should be tackled in the design of earth dams.

Classification of earth dams. Earth dams are grouped by various criteria — materials, design, construction technology, height, and seepage prevention measures; other are also possible. The classification in Table I.1 is based on materials and design, along with the construction technology applicable to each type of earth dam. The classification by seepage prevention measures is presented in Chapter 7.

The classification by dam height provides low ($H < 25$ m), medium ($25 < H < 75$ m), high ($75 < H < 150$ m) and very high ($H > 150$ m) dams. The division is fairly conditional but makes possible a quite complete description of dam type. For instance, Nurek Dam (Fig.I.1) is a very high earth-rockfill dam with central core, Atbashin Dam (Fig.I.2) a medium-height rockfill dam with a polyethylene diaphragm, while Shirokov Dam (Fig.I.3) is a medium rockfill dam with a timber screen.

CHAPTER 1

Soil as a material of dam body

1.1 CLASSIFICATION OF SOIL, GRAIN-SIZE DISTRIBUTION AND BASIC PHYSICO-MECHANICAL PROPERTIES

Soil is an aggregate of solid particles between which water fills up voids, partly or completely. One of the basic soil characteristics is its grain-size distribution. Lower limits of grain-size distribution curves are shown in Figure 1.1. Denomination of soil fractions also depends on the condition of particle surface. Grains having diameters smaller than 2 mm are referred to as fine, while those with $d > 2$ mm are called coarse. Coarse sediment with round surface is classified as gravel, pebbles, cobbles and boulders, while soil with sharp edges constitutes respectively druss or rubble.

Dam body is often composed of inhomogeneous soil being a mixture of fine and coarse fractions. Moraine or alluvial soil, shown in Figure 1.1 between the dashed lines, contains clayey fractions. In the absence of the latter one has loose coarse grained soil such as mixtures of sand and gravel; gravel and pebbles; sand, gravel and pebbles; or rock, often a man-made product.

The grain size distribution curves depicted in Figure 1.1 provide a certain approximate description. In reality soil can occupy any area in Figure 1.1.

Inhomogeneity of grain sizes is given in terms of the coefficient of uniformity

$$\eta = \frac{d_{60}}{d_{10}} \tag{1.1}$$

in which

d_{60}, d_{10} = size (diameter) of grains not exceeded by 60% and 10% of sample mass, respectively.

The coefficient is of particular importance to loose soil, of which most earth dams are made. Soil non-uniformity can also be described through some other coefficients employing various diameters.

Soil with $\eta > 3$ is regarded as uniform, while for $\eta \leq 3$ one has non-uniform soil. Any loose soil stratifies upon dumping but this effect is negligible for $\eta \leq 10$. Soil with $\eta \leq 3$ also stratifies upon hydraulic filling, although this is often neglected in

1

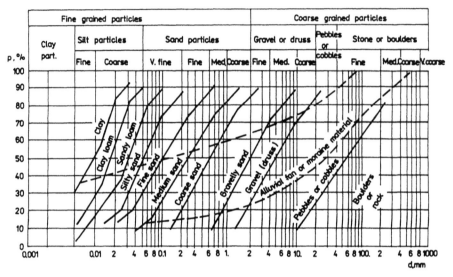

Figure 1.1. Grain-size distributions for various soils.

computations. The quarry soil used for construction of hydraulic-fill dams on the Rivers Volga and Dneper has been uniform, with $\eta = 2-3$. Gravel and pebble soil and rock rubble are usually non-uniform, with $\eta \geq 100-200$.

Coarse grained soil is often described by its mean weighed grain diameter

$$d_{mw} = \frac{\sum_{i=1}^{n} d_i \Delta P_i}{100} \tag{1.2}$$

in which

ΔP_i = percentage of fraction with diameter d_i, %.

The above diameter is often about d_{50}.

Fine soil, with clay particles ($d \leq 0.005$ mm) is referred to as clayey soil; it may be clay, sandy loam or clay loam. The kind of clayey soil may be defined by reference to the textural triangle (Fig.1.2), after Okhotin, but this method is fairly inaccurate. One usually resorts to the plasticity index I_p, being the difference between the water contents of a remolded soil at the transition from liquid to plastic or plastic to semi-solid state, respectively referred to as liquid limit and plastic limit, $w_L - w_P$.

Some authors propose mere use of w_L which seems to be the most precise quantity, e.g. $I_p = 0.75 w_L - 11.0$ (Rozanov 1983). This relationship is recommended if data on w_P is unavailable or if a range of experimental error is to be found.

Atterberg in 1911 classified soil by its plasticity index — sandy loam with $I_p < 7$, clay loam for $7 \leq I_p \leq 17$, lean clay with $17 < I_p \leq 30$, normal clay with $30 < I_p \leq 60$, fat clay for $60 < I_p \leq 100$, and very fat clay for $I_p \geq 100$. Soil with low $I_p \leq 1$ is not plastic, and belongs to sand.

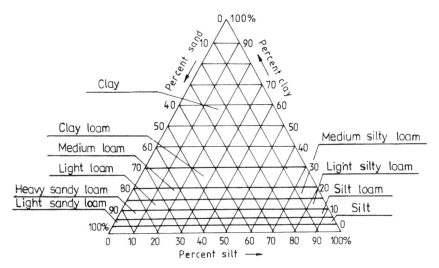

Figure 1.2. Textural triangle for clayey soils.

Goldshteyn (1973) argues that I_p is insufficient to provide complete description. Liquid limit is a very important parameter, a single value of which can be attributed to a few values of the plasticity index, and vice versa.

Casagrande in 1948 provided the well-known classification, improved later by Seed et al. (1964). The improvement is based on the content of organic matter which provides a sound distinction. For

$$I_p > 0.98(w_L - 27.5) \tag{1.3}$$

soil does not contain organic matter, and it does otherwise. Some qualitative parameters are identified for plausible analysis of quarry materials and their zoning in dam body, such as brittle failure of soil at the water content below I_p if $I_p < 0.98(w_L\text{-}27.5)$. For $w_L > 50\%$ soil is stronger at low water contents.

The effect of soil structure and consistency on physico-mechanical properties of undisturbed soil has also been investigated (by Terzaghi and Peck). Each of the proposals on the classification has some shortcomings. The classification by grain-size distribution is fast, while inclusion of liquid limit by Equation 1.3 accounts for organic matter even prior to testing.

The liquidity index I_L has been introduced in the following form for numerical assessment of clayey soil

$$I_L = \frac{w - w_P}{w_L - w_P} \tag{1.4}$$

Clayey soil in its initial stage of formation is often called mud. It usually contains organic matter. Plasticity index provides for mud a classification similar to that for

Classification by I_L

Sandy loam		Clay loam and clay	
Hard	0	Hard	0
plastic	0–1	semihard (stiff)	0–0.25
liquid	1	stiff plastic (firm)	0.25–0.50
		plastic	0.50–0.75
		liquid plastic (soft)	0.75–1
		liquid	1

clay. Hence mud is arranged in classes of sandy, loam clayey and clayey. Mud is characterized by low compaction and water contents exceeding w_L, i.e. $I_L > 1$, cf. Equation 1.4. Muds possess structural strength, so that they transform into liquid if the internal bonds are broken.

Loess is another particular variety of clayey soil; it is of aeolian origin, and contains high quantities of silt. Its degree of saturation (Eq.1.10) is much higher than one, and porosity — above 50%. A characteristic feature of natural loess consists in anisotropic seepage (much greater vertically than horizontally) and wet settlement, in particular under loading exceeding the self-weight.

If natural loess is washed out and transported to a new destination, many of its properties change, and one refers to loessial clayey soil under such natural conditions. Similar transformation of properties will occur under hydraulic filling: the anisotropy as to seepage and deformation will decrease or disappear at all.

Grain shape controls many properties—the interaction of minerals and water and density in fine soil, and porosity and strength in coarse grained soil. Soil grain forms may be given in terms of characteristic diameters. Quantitative estimates are provided for coarse grained soil only. For $d_3 < d_2 < d_1$ one has the following definition of shape factor S_F:

$$S_F = \frac{d_3}{d_1} \qquad (1.5)$$

in which
$$d_1 = \text{maximum grain diameter,}$$
$$d_3 = \text{minimum grain diameter.}$$

For $S_F = 1$ the form is close to spheroidal or cubic. In construction of earth dams one should prefer $S_F > 0.3$.

Mineralogical composition of soil, along with the grain size distribution, determines soil properties. The primary compounds of minerals are SiO_2, Al_2O_3, Fe_2O_3 and the oxides MgO, CaO, Na_2O and K_2O.

Fractions with primary and secondary minerals are distinguished with respect to the mineralogical composition of soil. The former are coarse products of rock disintegration ($d > 2$ mm). Sand (0.05 to 2 mm) and silt (0.005 to 0.005 mm) also contain primary minerals, firstly quartz. Clay minerals belong to secondary, as products of rock weathering and thermodynamical transformations, reflected in the chemical and crystalline structure.

Secondary minerals, which form clay particles, belong to several groups of kaolinite, montmorillonite, etc. Each group determines particular properties of grains. Clayey soil consists of clay and coarser ($d > 0.005$ mm) grains. Properties of clayey soils are determined by the properties of clay, sand and silt fractions, their percentage and interactions. The properties of secondary mineral groups are displayed chiefly through their interaction with water. For instance, montmorillonite is responsible for swelling, to which kaolinite is not susceptible.

In most cases clay particles of soil are mixtures of different components.

Mineralogical composition of coarse grained soils is also very important to the properties of the latter, primarily to the strength of grains.

Easily soluble salts such as $NaCl$, KCl, $CaCl_2$, $NaHCO_3$, $Ca(HCO_3)$, $MgSO_4$ and Na_2SO_4, and less soluble such as gypsum ($CaO_4 \cdot 2H_2O$) are encountered among soil minerals. Soviet standards put a limit on easily soluble and soluble salts, 0.5 to 5%, depending on type of soil; dissolution of salts due to seepage can bring about serious changes in soil properties.

Density of soil particles ρ_p is one of the most important properties. It is the ratio of dry soil mass to its volume. It depends only on mineralogical composition and usually varies in the narrow range from 2650 to 2770 kg/m³. In the presence of metal oxides, density may increase to 3500 kg/m³ and more; for organic matter it decreases under the above figures.

Unit weight of soil particles reads

$$\gamma_p = \rho_p \cdot g$$

in which

g = acceleration due to gravity.

Dry density of soil in situ ρ_{dry} is the ratio of mass of dry soil particles to their volume. Together with this one also distinguishes dry unit weight $\gamma_{dry} = \rho_{dry} \cdot g$.

Soil density ρ_s is the ratio of the mass of soil and water in voids to the respective volume. One often uses unit weight of soil $\gamma_s = \rho_s \cdot g$.

Porosity is the most important property of both loose and cohesive soil; it is the volume of voids in a unit volume of soil, in relative terms:

$$n = \frac{\rho_p - \rho_{dry}}{\rho_p} \tag{1.6}$$

Equation 1.6 can also be written as

$$n = \frac{\gamma_p - \gamma_{dry}}{\gamma_p} \tag{1.6a}$$

Soil porosity is often characterized by the void ratio ϵ, being the ratio of the volume of voids to the volume of solid particles in a soil sample:

$$\epsilon = \frac{n}{1-n}; \quad \text{or} \quad \epsilon = \frac{\rho_p - \rho_{dry}}{\rho_{dry}} \tag{1.7}$$

For $\epsilon \leq 0.5$ soil is considered dense while for $\epsilon \geq 1$ it is regarded as very loose.

Mass water content is the ratio of the mass of water in voids of unit volume of soil M_w to the density of solid particles ρ_{dry} in this volume

$$w = \frac{M_w}{\rho_{dry}} \cdot 100 \tag{1.8}$$

If soil is fully saturated with water one has

$$M_w = \rho_w n = \frac{\rho_w \epsilon}{1+\epsilon}; \quad w = w_{sat} = \frac{\rho_w \epsilon}{(1+\epsilon)\rho_{dry}} = \frac{\rho_w \epsilon}{\rho_p} \tag{1.9}$$

in which
ρ_w = density of water.

The degree of saturation determines the amount of water in voids:

$$D_S = \frac{w}{w_{sat}} = \frac{w\rho_p}{\epsilon \rho_w} \tag{1.10}$$

For $D_S = 0$ soil is regarded as a single phase, although it contains soil particles and gas; in the absence of water the gaseous phase hardly affects soil properties. For $D_S = 1$ soil is considered two-phase because it contains solid particles and water. In all remaining cases soil is a three-phase medium. In practical engineering computations soil can be regarded as a quasi-single phase if the ultimate value of D_S upon loading is below 0.7–0.8.

Unit weight of soil γ_s can be expressed through dry unit weight of soil and weight of water:

$$\gamma_s = \gamma_{dry}(1 + w) \tag{1.11}$$

Water in voids of clayey soil may appear in different states — free water, with usual physical properties of ordinary water, and adsorbed water, the molecules of which are attracted to soil particles and are maintained by forces much greater than the forces of gravity. The adsorbed water creates a continuous film about particles so that it is often referred to as skin water.

Free water contains gravity water, on which the forces of gravity act, and capillary water, which is also affected by capillarity.

The adsorbed water in soil differs from ordinary water by such properties as boiling and freezing temperatures, heat content etc. The adsorbed water can be separated from soil under pressure of tens or hundreds MPa, with simultaneous increase in temperature.

The differences in properties of adsorbed water and free water are displayed, for instance, in dissolving capacity. The adsorbed water makes seepage difficult as it reduces soil voids. The adsorbed water affects considerably the compressibility

Table 1.1.

Property	Formulae
Dry unit weight of soil	$\gamma_{dry} = \frac{\gamma_s}{1+w}$
Volume of gas in unit volume of soil	$V_g = (\frac{\epsilon}{\gamma_p} - \frac{w}{\gamma_w})\gamma_{dry}$
Submerged unit weight of soil	$\gamma_{sub} = \frac{\gamma_p - \gamma_w}{1+\epsilon} = \gamma_{dry} - \frac{\gamma_w}{1+\epsilon}$
Unit weight of saturated soil	$\gamma_{sat} = \gamma_{dry} + n\gamma_w$

γ_w = unit weight of water, approximately 10 kN/m^3.

of soil (reduces it). Rheological properties of fine soil (soil deformability in time) may be explained in terms of a higher viscosity of the adsorbed water.

Some other common physico-mechanical properties of soil are given in Table 1.1.

Porosity and void ratio are quantities which do not fully describe the degree of soil compaction. For sandy and coarse granular soil one often uses the following density index:

$$I_D = \frac{\epsilon_{max} - \epsilon}{\epsilon_{max} - \epsilon_{min}} \qquad (1.12)$$

in which

ϵ_{max} and ϵ_{mean} = void ratios of extremely loose and compacted soil, respectively,

ϵ = real void ratio.

The quantity γ_{dry} for coarse grained soil assumes high values. Gravelly pebble soil (basically an aggregate of round coarse particles) can have γ_{dry} from 21 to 24 kN/m^3, with the fraction $d < 5$ mm of 20 and 30%, respectively; this depends on grain size distribution.

Rock rubble (being primarily an aggregate of badly rounded or unrounded coarse unsorted fragments, mostly man-made) can have a wider range of dry unit weight of soil, depending on the degree of grain uniformity; it varies from 18 to 22 kN/m^3.

Upon filling of clayey soil with coarse grained inclusions (material from alluvial fans or moraine deposits) γ_s can reach 23–24 kN/m^3. Dramatic decrease in total porosity in this soil can be caused by clay particles in voids between coarse particles.

Thermophysical parameters of freezing and thawing soils are required in computations of thermal conditions within dams and their foundations under severe climatic conditions. One usually distinguishes the following three thermophysical quantities: thermal diffusivity α, thermal conductivity λ, and thermal content C (C_F for frozen soil and C_T for thawing soil). These quantities depend on unit weight and water content (Table 1.2).

In some computations one must know thermal contents of water and ice, which are approximately 4.2 and 2.1 kJ/(kg°C). These quantities (hence the soil counterpart as well) depend on temperature, but in the range from 0 to 29°C the heat

Table 1.2.

γ_{dry} kN/m^3	w %	Thermal conductivity, kW/(m·^{o}C)						Volum.heat kJ/(m^3·^{o}C)	
		Sand		Sandy loam		Clay/clay lm.		C_T	C_F
		λ_T	λ_F	λ_T	λ_F	λ_T	λ_F		
12	5	0.466	0.605	-	-	-	-	1190	1085
14	5	0.664	-	-	-	-	-	1381	1260
14	25	-	-	1.07	1.35	0.837	1.163	2050	1510
16	5	0.872	1.06	-	-	-	-	1585	1423
18	35	-	-	1.76	2.24	1.548	1.935	3100	2100
20	15	2.037	2.56	1.63	1.74	-	-	2469	1967

Thermal diffussivity can be found for a known thermal content and thermal conductivity: $\alpha = \lambda/C\gamma_0$.

content of water varies by not more than 1% and that of ice by 10%.

1.2 SOIL PERMEABILITY

Soil permeability, expressed in terms of the coefficient of permeability is one of the most important properties in the design of dams, seepage-confinement units, drainage and other structural constituents.

Seepage is possible within any structural component of an earth dam, including foundations and banks. It is known that the rate of seepage in soil, the latter being a porous medium, is generally described by the Smreker equation (Khristianovich 1940)

$$v = kI^{1/m} \tag{1.13}$$

in which
k = coefficient of permeability,
I = hydraulic gradient,
m = factor from 1 to 2 which determines the conditions of seepage as a function of soil type.

Some other equations are available but are less convenient. For $m=1-1.1$ seepage can be considered laminar, and Equation 1.13 transforms into the Darcy equation

$$v = kI \tag{1.14}$$

The above regime of seepage takes place in fine (clayey and sandy) soils, and also in some coarse grained ones (gravelly-sandy or even compacted gravelly pebbles).

In rock, the factor m is usually between 1 and 2, which generates specific conditions of seepage in a dam made of such soil. This must be taken into account in dams with non-uniform material produced by controlled explosion methods, when the compactness of rock varies with the height of dam and the grain size distribution also varies, thus affecting k and m. For $m \approx 2$ one encounters turbulent seepage, and transition is distinguished in the range $m = 1.1 - 1.85$.

Approximate values of the coefficient of permeability, in m/s, satisfying the Darcy equation (Eq.1.14) can be found for cohesionless soils by the empirical formulae derived by Pavchich

$$k = \frac{4S_F}{\nu} \sqrt[3]{\eta} \frac{n^3}{(1-n)^2} d_{17}^2 \frac{g}{1000} \tag{1.15}$$

in which

S_F = 1 for sand and gravel and 0.35–0.4 for pebbles,
ν = kinematic viscosity of water,
d_{17} = 17% grain diameter,
η = coefficient of uniformity,
n = porosity,
g = acceleration due to gravity.

Despite the incoherence of Equation 1.15 due to increasing k with increasing η and usage of d_{17} instead of d_{10}, which would be in accordance with Hazen (1895), Equation 1.15 has been found satisfactory for a wide range of soils. Its experimental verification with sand and gravel or gravel and pebbles shows a scatter of 8%. Equations 1.14 and 1.15 determine some fictitious velocities of seepage taking place in the entire space. In reality, flow occurs in voids, which in clayey soil are contracted by adsorbed water, and the degree of that contraction depends in general on the hydraulic gradient (Burenkova & Mokran 1983).

The real seepage rate v_r depends generally on active porosity n_a, that is on the porosity reduced because of the presence of adsorbed water:

$$v_r = \frac{v}{n_a} \tag{1.16}$$

with $n_a \approx n$ in sandy soil.

For nonlinear seepage, a rock of non-uniform grain-size distribution may have k varying from 1 to 10 cm/s, and m may vary from 1.05 to 1.72. Greater values of k and m correspond to rock with $\eta \approx 20$–30 and a relatively loose structure ($\rho_{dry} \cong 1750$–1800 kg/m³); lower values having $\eta \cong 100$–200 and $\rho_{dry} = 2000$–2200 kg/m³.

Hence if soil is fairly non-uniform and compacted in a dam constructed by controlled explosions, one may use the laminar relationship (Darcy law), while in other cases one encounters transition conditions; it is only in coarse uniform rock that turbulent seepage will ocur (these recommendations are based on findings by Korchevskiy & Pokrovskiy for the dam on River Burlykiya).

A preliminary assessment of permeability in cohesive soil may be given by the empirical formula of Zhilenkov

$$k = 4 \cdot 10^{-11} exp[\frac{\epsilon}{0.17\epsilon_L - 0.048}] \tag{1.17}$$

in which

ϵ = void ratio

ϵ_L = void ratio at liquid limit:

$$\epsilon_L \approx 1.06 \frac{\gamma_{p,d<1mm}}{\gamma_w} w_L \qquad (1.18)$$

The quantity 1.06 in this equation is the correction factor due to the presence of air; $\gamma_{p,d<1\ mm}$ stands for the unit weight of particles below 1 mm, as these particles in clayey soil are used in determination of w_L and w_P.

Equation 1.17 holds true if the contents of particles finer than 1 mm are above 30–40%; otherwise the fine particles are insufficient to fill up the voids and the quantity k in Equation 1.17 will be underestimated. From analysis of Equation 1.17 it follows that for $\epsilon \to 0$ k goes to 4×10^{-11} cm/s.

It should be noted that most soils are composed of elliptical or plate-type grains, that is with dominance of one or two dimensions. Upon rolling soil particles are oriented horizontally, which brings about anisotropy in seepage properties, that is various k in different directions. Pronounced anisotropic properties are displayed by loess. One also encounters anisotropy in clayey seil (usually with coarse inclusions) and in gravelly pebbles.

Coefficient of seepage anisotropy k_x/k_y is used to quantify the seepage anisotropy of soil (see Chapter 2), in which k_x = the largest coefficient of permeability (usually in the direction of the x-axis) and k_y = permeability in the direction perpendicular to the x-axis.

1.3 VOLUME CHANGE OF SOIL

In clayey soil, the changes under static load are primarily controlled by mutual dislocation of solid particles.

In coarse grained soil, the major factors of deformation are the crushing and breaking of solid particles under loading, and much less by reorientation of grains. In sandy soil (quartz particles) one encounters processes of reorientation and mutual disclocation of grains, together with the crushing effects.

One distinguishes volumetric change and form change of soil.

Volume change of saturated (two-phase) clayey soil is possible only if water is squeezed out of soil. Since soil voids are small and partly occupied by adsorbed water, the squeezing takes place slowly, and the condition of soil deformation is spread over a long span of time. Higher viscosity of adsorbed water also slows down the process of deformation (both volumetric and form type). Deformation of clayey soil in time is described in terms of the theory of soil consolidation and creep (Chapter 3).

Deformation in sandy soil under static load is fairly fast so that the temporal variability is not considered.

In coarse grained soil the processes of buckling, crushing and structural reorientation due to destruction of grains and redistribution of load between them take long time although the mechanism is different than that in clayey soil; it is controlled by creep of material at contacts. This process also appears at the contacts subject to compressive stress, and particularly at the contacts where compressive and displacement stresses occur.

Although the deformation is a long process in most soils, the major deformations (elastic and plastic, about 50–60% of ultimate deformation of loose soil and slightly less in three-phase clayey soil, in which deformation is linked to consolidation) take place at a rate of propagation of elastic and plastic waves, respectively, so they are practically 'instantaneous' even under dynamic loading, for instance seismic. About 20% of change occurs after load is applied, which is a very short period compared with the duration of operational loads measured in minutes, hours and days. Therefore the short-term deformation alone is often analysed, i.e. without inclusion of creep properties of soil.

In clayey soil, 'fast' damping of deformation occurs for $D_s \leq 0.8$–0.85, when the process is mostly controlled by the gaseous phase.

Soil is a definite discrete body for which the mathematical apparatus of the mechanics of continuous media has been applied. The criterion of application has been formulated by Gersevanov. This criterion is usually obeyed by all soils used in dam construction.

The mechanics of granular media is presently in its initial stage of development (Deresevich 1961), but some elements and notions have already been adopted in the domain of soil, although still in terms of continuum mechanics.

The mechanics of continuous media makes use of stress tensors which can be split up into two tensors, spherical and deviatoric

$$\begin{vmatrix} \sigma_{xx} & \sigma_{xy} & \sigma_{xz} \\ \sigma_{yx} & \sigma_{yy} & \sigma_{yz} \\ \sigma_{zx} & \sigma_{zy} & \sigma_{zz} \end{vmatrix} = \begin{vmatrix} \sigma & 0 & 0 \\ 0 & \sigma & 0 \\ 0 & 0 & \sigma \end{vmatrix} + \begin{vmatrix} \sigma_{xx} - \sigma & \sigma_{xy} & \sigma_{xz} \\ \sigma_{yx} & \sigma_{yy} - \sigma & \sigma_{yz} \\ \sigma_{zx} & \sigma_{zy} & \sigma_{zz} - \sigma \end{vmatrix} \qquad (1.19)$$

in which
$$\sigma = \frac{\sigma_{xx} + \sigma_{yy} + \sigma_{zz}}{3} = \text{mean stress}.$$

In the tensor form Equation 1.19 reads

$$\sigma_{ij} = \sigma + S_{ij} \qquad (1.19a)$$

in which
$$S_{ij} = \sigma_{ij} - \sigma = \text{components of stress deviator tensor}.$$

If the relationship between stresses and strains is linear, the symmetric tensor describes the volume change and the deviator represents variations of form. These particular relationships between stress and strain are analysed in the linear theory of elasticity.

If the relationships are nonlinear one encounters the so-called secondary effects

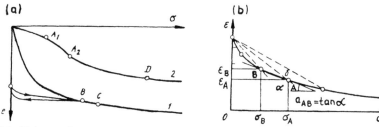

Figure 1.3. Volume change of soil.

— the impact of the stress deviator in volume change. This effect often brings about additional expansion or compression of soil upon displacement. It has been referred to as dilatancy.

The strain tensor can also be decomposed into the spherical and deviatoric parts:

$$
\begin{vmatrix}
e_{xx} & e_{xy} & e_{xz} \\
e_{yx} & e_{yy} & e_{yz} \\
e_{zx} & e_{zy} & e_{zz}
\end{vmatrix}
=
\begin{vmatrix}
e & 0 & 0 \\
0 & e & 0 \\
0 & 0 & e
\end{vmatrix}
+
\begin{vmatrix}
e_{xx} - e & e_{xy} & e_{xz} \\
e_{yx} & e_{yy} - e & e_{yz} \\
e_{zx} & e_{zy} & e_{zz} - e
\end{vmatrix}
\tag{1.20}
$$

in which

$$e = \frac{e_{xx} + e_{yy} + e_{zz}}{3}, \text{ i.e. } e = \text{one third of volume change.}$$

By analogy to Equation 1.19a the tensor form of Equation 1.20 reads

$$e_{ij} = e + \epsilon_{ij} \tag{1.20a}$$

in which

$$\epsilon_{ij} = e_{ij} - e = \text{components of strain deviator.}$$

Since the total volume change is a convenient notion one usually takes $e = e_{xx} + e_{yy} + e_{zz}$. This notation will be used henceforth in the analysis of stress-strain relationships for the spherical tensor.

Volume change of soil. For cubic compression, the deformation is illustrated in Figure 1.3a, curve 1 versus mean stress for loose soil; it has curvature of one sign. For a preconsolidated soil (curve 2) the character of the relationship is different. The segment from O to A_1 describes linear deformation of soil under loading — segment of repeating loading with A_1 corresponding to the maximum value of σ for preconsolidation. The section of curve 2 between A_1 and A_2 depicts the transition from preconsolidation to primary deformation analogous to curve 1.

One may also observe an additional characteristic point D corresponding to the transition of soil from a three-phase to two-phase status (Chapter 3), that is the transition to full saturation (with water).

Apparatus for soil deformation tests. The relationship $\sigma = f(e)$ may be found with the apparatus for which all components of stress and strain tensors are known for a sample. This apparatus includes departure gauges. The testing can be twofold:

1. Soil sample is cylindric, and tests are carried out under condition $\sigma_1 > \sigma_2 = \sigma_3$, in which σ_1, σ_2, σ_3 = principal stresses, σ_1 stress due to load applied to sample phases, $\sigma_2 = \sigma_3$ = stresses on sides of sample. In this case the sample is squashed. A particular case can constitute simple squashing under vertical load $\sigma_2 = \sigma_3 = 0$.

2. Test condition with $\sigma_1 = \sigma_2 > \sigma_3$. The maximum stresses ($\sigma_1 = \sigma_2$) are applied to sides of sample and the minimum stresses to its faces. Sample failure occurs upon elongation of sample and decrease in its diameter.

The first method is standard—the stresses $\sigma_2 = \sigma_3$ remain constant during growth of σ_1, after a certain value is reached along the loading path for $\sigma_1 = \sigma_2 = \sigma_3$.

Triaxial apparatus is a term often used as a counterpart of departure meter. All present soil tests are also conducted in more complex apparatus in which samples are cubic or tubular; these instruments may permit testing of soil properties in more sophisticated stress-strain situations.

Oedometer is a simple device employed for testing of deformation properties of soil. A soil sample is cylindric but has proportions different from those used in triaxial apparatus, in which one usually has $H = 2$–$2.5d$, in which H = sample height and d = sample diameter. A high soil sample can reduce the squeezing effect at ends of samples, thus distorting the homogeneity of stress conditions. In oedometer one has $H = 1$–$0.33d$, depending on absolute dimensions of the instrument and type of soil, so as to diminish the effect of soil friction on walls.

Soil in oedometer is contained in a metallic housing so that one has $\sigma_2 = \sigma_3$, but these quantities are unknown, and one component of the stress tensor is known only. Because σ_2 and σ_3 are unknown and $\sigma_2 = \sigma_3 < \sigma_1$, there is no possibility of finding $\sigma = f(e)$, as it is the case in triaxial apparatus, despite the fact that all strain components are known (vertical displacements are measured so that one has ϵ_1 for $\epsilon_2 = \epsilon_3 = 0$). Accordingly, the oedometer tests yield $\sigma_y = \sigma_1 = f(e_1 = e_y)$. Oedometer simulates the process of one-dimensional deformation, in the direction of sample height. The relationship $\sigma_y = f(e_y)$ is referred to as compressive but recently one also refers to compressive as the relationship $\sigma = f(e)$ (although there is a principal difference between these two cases). The deformation of soil under compression measured in oedometer is controlled not only by the mean stress σ (component of the spherical tensor stress) but also by the components of stress deviator (numerically unknown); the latter can increase or decrease the strain (compare the remark on dilatancy); since they are unknown, the sign of error also remains unknown a priori.

The relationship $e = f(\sigma)$ may be obtained if all components of measured stresses are predetermined and respective components of strain tensor are found. Several types of instruments are available to comply with these requirements; the widest application has been found by the triaxial apparatus, which is relatively the simplest and most accurate device.

Soil compressibility. It can be found in oedometer or under cubic compression in

triaxial apparatus. The following relationships are found to be generally applicable.
   1. Exponential relationship, being most widespread in computations

$$e = \left(\frac{\sigma}{E_0}\right)^n \tag{1.21}$$

in which
   E_0 = modulus of oedometric deformation,
   $\sigma = 1/3(\sigma_1 + \sigma_2 + \sigma_3) = 1/3(\sigma_x + \sigma_y + \sigma_z)$ = mean stress,
   n = curvature factor.

Equation 1.21 is also given as $E_0 e = \sigma^n$, but in this case the modulus of volume change is more complex. Values of E_0 and n for some soils are given in Table 1.3.
   2. Relationship proposed by Grigoryan

$$e = e_\infty[1 - exp(-\lambda\sigma)] \tag{1.22}$$

in which
   e_∞ = reference volume change for $\sigma \to \infty$, i.e. the deformation that would be reached if the dry unit weight of soil approached the unit weight of soil particles,
   λ = empirical coefficient.

Equation 1.22 is closer to reality as the volume change is asymptotic while that in Equation 1.21 increases without limits with growing σ; for real static $\sigma \leq 10$ MPa both relationships are realistic.
   The following factor of volumetric compressibility may be related to the void ratio ϵ

$$\epsilon = -A \log_e(\sigma - \sigma_0) + c \tag{1.23}$$

in which
   A, σ_0 = empirical parameters having complex dimensions (difficult conversion of units).

Equation 1.23 describes well the compressibity of disturbed soil in a small range of pressures or in strongly deformable soil of undisturbed structure. One should remember that Equations 1.21–1.23 have not encompassed the initial (linear) and transient segments of deformation graphs (Fig.1.3a).
   Some other functions may also be used in description of soil deformation under cubic compression.
   Linearization is suggested to simplify the compression curve (Fig.1.3b). An experimental coefficient of compressibity a is introduced for each segment

$$a = \frac{\epsilon_1 - \epsilon_2}{\sigma_2 - \sigma_1} \tag{1.24}$$

The higher the number of rectilinear segments of the curve $\epsilon = f(\sigma)$, the more accurate is the experimental data for a series of $a_{1,2}$. One often uses $\sigma_1 = 0$ for

Table 1.3.

Kind of soil	w_L	γ_p, kN/m^3	γ_{dry}, kN/m^3	E_0, MPa	n
Gravel & pebbles	-	27.2	22.5	39.4	0.8
Rock	-	26.8	19.4	65	0.67
Clay	0.436	27.4	16.1	10	1
Clay	0.436	27.4	17.8	25	1
Skeletal sandy loam*	0.18	27.2	21.1	12.9	0.68

*Fines constitute 50% of soil volume.

simplification, in which case the rectilinear segments in the form of chords (solid lines in Figure 1.3b) are replaced by secants (dashed lines in Figure 1.3b).

Deformation processes in soil may obey various laws. Considered above were obvious cases of loading, that is the growth of one component of the stress tensor. It is only in this case that the process of loading and unloading can be described fairly simply.

Assume that volume change of soil is characterized by curve 1 (Fig.1.3a). If one unloads soil sample from point B (load decreasing to zero) then one finds out that the practically linear character of deformation occurs in the initial stage of unloading (the latter displaying reversible and irreversible deformations); if the loading is repeated the initial branch will be met at point C.

With high accuracy one may assume that deformation on the branch of repeated loading is linear by point B, and the points B and C may be deemed identical.

Soil is dumped in a dam upon compaction, most often by rolling, until a certain compactness is reached. If the rolling machinery is removed, unloading occurs and afterwards dam is loaded by the overlying soil. Analogous processes are repeated in laboratory — soil sample is provided a certain density corresponding to the density in dam body, the consolidating load is relieved thereafter, the sample is again placed in the test apparatus and loaded anew. A character of operation is shown by curve 2 in Figure 1.3a.

Experiments in the above apparatus can be conducted under various conditions of water outflow from soil sample. A notion of undrained system is introduced to specify experimental conditions of impossible outflow of water from soil, in which the degree of saturation D_S is greater than 0.8. In this case a considerable part of load is applied to water so that soil is less deformable than in the drained system, in which water may be removed from sample under the effect of hydraulic gradient generated by the external load. Both systems employ different experimental schemes. Either of these two systems may be chosen, depending on prescribed conditions.

1.4 FORM CHANGE

Referred to as form change is the variation of linear dimensions of a sample without volumetric effects. The analysis and description of soil behaviour under the conditions of form change create most sophisticated problems. It should be noted that oedometer tests are accompanied not only by volume changes but also by form changes as the height of sample varies for a constant radius.

Triaxial apparatus provides information on all components of stress and strain tensors, and the stress condition in this apparatus is relatively homogeneous (discrepancies are observed only close to the press). Owing to this property, a triaxial apparatus is the best device applicable to analysis of form changes.

Results of soil tests along the deviatoric segments (form change segments) are usually given in the invariant form:

$$T = f(\Gamma)$$

where

$$T = \frac{1}{\sqrt{6}}\sqrt{(\sigma_1 - \sigma_3)^2 + (\sigma_1 - \sigma_2)^2 + (\sigma_2 - \sigma_3)^2};$$

$$\Gamma = \sqrt{\frac{2}{3}}\sqrt{(e_1 - e_3)^2 + (e_1 - e_2)^2 + (e_2 - e_3)^2}$$

(1.25)

in which

T = second invariant of stress deviator,
Γ = second invariant of strain deviator,
$\sigma_1, \sigma_2, \sigma_3$ = components of principal stresses,
e_1, e_2, e_3 = components of principal strains.

The use of invariants is purposeful if the material is isotropic and the principal axes of stress and strain tensor coincide (in this case one says that the tensors are coaxial).

In triaxial tests, the condition of coaxiality is complied with, and a certain initial anisotropy may be neglected.

Instead of T and Γ one often uses their proportional counteparts, stress intensity σ_i and strain intensity m_i, or otherwise normal stresses and strains on the octahedral plane σ_{oct} and e_{oct} (that is the plane equally inclined to principal planes).

The relationships $T = f(\Gamma)$ for various initial values of σ are shown in Figure 1.4. A whole family of curves is depicted, while for the relationship $e = f(\gamma)$ one had only one curve (obviously for $T \neq 0$ one would have an analogous family due to dilatancy but this will be shown below).

Lode-Nadai parameter. The relationship $T = f(\Gamma)$ is controlled not only by the initial values of σ but also by other factors. Let $T = T_1 = T_2$ but let T_1 occur under the condition that σ_1 is maximum principal stress and $\sigma_2 = \sigma_3$ is the minimum one

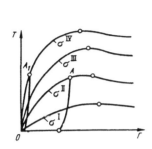

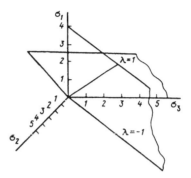

Figure 1.4. Form change of soil (left).
Figure 1.5. Stress condition with planes $\lambda=-1$ and $\lambda=+1$ (right).

(for standard tests in triaxial apparatus) while T_2 occurs for maximum $\sigma_1 = \sigma_2$ and minimum σ_3 (the second possible methodology of triaxial tests mentioned above).

The soil experiments for various modes of T_1 and T_2 provide different Γ. Hence it is not only absolute Γ but also the mode of its interrelations that control the form changes. The character of relationship between components of principal stresses is described by the Lode-Nadai parameter λ_δ. This parameter is also referred to as the parameter of stress condition type; it is the third invariant of stress deviator:

$$\lambda_\delta = \frac{2\sigma_2 - \sigma_1 - \sigma_3}{\sigma_1 - \sigma_3} \tag{1.26}$$

For $\sigma_2 = \sigma_3$ one obtains $\lambda_\delta = -1$ while for $\sigma_2 = \sigma_1$ one has $\lambda_\delta = +1$. If σ_2 is between σ_1 and σ_3 one obtains $-1 \leq \lambda_\delta \leq +1$. The parameter λ_δ is one of the elements of the loading path, that is the ratio between components of principal stresses for invariable configuration of their planes. The planes $\lambda_\delta = -1$ and $\lambda_\delta = +1$ are shown in Figure 1.5 in coordinates of principal stresses. All possible conditions expressed in terms of principal stresses are in between these planes. The condition $\sigma_1 \geq \sigma_2 \geq \sigma_3$ is not satisfied outside the volume bounded by these planes.

The Lode-Nadai parameter for strains λ_e is computed by analogy to Equation 1.26, and uses components of principal strains.

For $|\lambda_\delta - \lambda_e| \neq 0$ the similarity of stress and strain tensors disappears. The same parameters can be constructed with regard to increments in stress components $\Delta\sigma_{ij}$ and strain components Δe_{ij}. Acordingly, $|\lambda_{\Delta\sigma} - \lambda_{\Delta e}| \neq 0$ will denote dissimilarity of tensors of strain and stress rates. In the theory of plasticity one accepts the hypothesis on the similarity of stress and strain tensors for plastic strain. Hence the condition $|\lambda_\delta - \lambda_{\Delta e}| \neq 0$ points to the fact that the hypothesis is not satisfied ($\lambda_\dot{e}$ is most often used instead of $\lambda_{\Delta e}$). Hence the parameters λ_δ, λ_e, $\lambda_{\Delta\sigma}$, and $\lambda_{\Delta e}$ describe the applicability of hypotheses or theories on soil conditions.

Revolution of the planes of principal stresses. If the planes of principal stresses are rotated by angles α_1, α_2, α_3 with regard to their original counterparts (the relationship $\cos^2 \alpha_1 + \cos^2 \alpha_2 + \cos^2 \alpha_3 = 1$ being complied with) then the process in a nonlinearly deformable plastic material (such as soil) can affect deformations. Upon transition to invariant notation for plane deformation problems one should consider the revolution of octahedral planes α_δ and α_e, since in this case they are equal to angles of rotation of the principal axes of stress and strain tensors respectively. One of the angles of rotation can be neglected in problems of plane deformation, while in three-dimensional problems one has three axes of principal stresses and three axes of principal strains.

By analogy to the Lode-Nadai parameter one may find the parameters α_σ, $\alpha_{\Delta\sigma}$, α_e, and $\alpha_{\Delta e}$ (with the equivalent notation $\alpha_{\dot{\sigma}}$ and $\alpha_{\dot{e}}$) and interrelate them. For $|\alpha_\sigma - \alpha_e| = 0$ the coaxiality of stress and strain tensors under complex stress-strain conditions occurs while for $|\alpha_{\Delta\sigma} - \alpha_{\Delta e}| = 0$ one encounters coaxiality of the stress and strain rate tensors, etc.

Finally, if one changes experimentally the relation between σ and T for $\lambda_\sigma = $ constant and $\alpha_\sigma = $ constant, then the deformation will be different although the values σ and T are kept constant. The relations between σ and T are often replaced by relations between σ and σ_i (as σ_i is proportional to T), because non-negative value of σ_i for $\lambda_\delta = -1$ is $\sigma_1 - \sigma_3$ (i.e. if $\sigma_2 = \sigma_3$, $T = \frac{1}{\sqrt{3}}(\sigma_1 - \sigma_3)$ and $\sigma_i = \sqrt{3}T$). Then the path parameter reads

$$\mu = \frac{d\sigma}{d\sigma_i} \tag{1.27}$$

For triaxial tests with $\sigma_2 = \sigma_3 = $ constant (standard method) one may easily obtain $\mu = 0.33$. Hence the three parameters, λ_σ, α_σ and μ, characterize their loading path at a point. The trajectory for $\mu < 0$ is referred to as unloading, although the notion does not describe properly the real process as it can be both loading and unloading. One must note that soil is a very complex medium so that one may have $|\lambda_\sigma - \lambda_e| \neq 0$; $|\lambda_\sigma - \lambda_{\Delta e}| \neq 0$; $|\lambda_{\Delta\sigma} - \lambda_{\Delta e}| \neq 0$; $|\alpha_\sigma - \alpha_e| \neq 0$; $|\alpha_\sigma - \alpha_{\Delta e}| \neq 0$ etc but soil properties are usually idealized mathematical modelling parameters. Depending on the degree of idealization one may describe soil deformability under load with one or another degree of reliability.

Figure 1.4 shows curves with points of limit equilibrium denoted by circles. Before these points are reached soil is in subritical condition, and afterwards soil is in limit condition.

The process of deformation is different along the two segments. The segment of limit state (to the right of the point of limit equilibrium) is characterized by soil flow, that is unlimited deformation of soil at constant or even decreasing stresses. Reduced stress brings about accelerated deformation after the point of limit condition has been reached. This is supercritical condition of soil.

Experiments have shown that the lowest stresses in supercritical state can occur under conditions of controllable strain, for instance vertical, if vertical loading is given by a press at a constant rate of punch motion (electric drive is suitable for this purpose). Under conditions of controlled stresses in supercritical region one observes the progressive destruction of sample. On the other hand, the process of sample deformation occurs in the subcritical stage if tests are conducted at controllable stresses; one may also observe growth of stresses if conditions of controllable deformation are satisfied.

Lomize and his students have investigated the effect of loading pattern on soil deformations. The mechanism of form changes is similar to volume changes but the major role in this process is played by irreversible disclocation of solid grains with regard to one another, or their destruction.

Mathematical models. At present one has a few classes of mathematical models (simply referred to as models) which describe soil deformation along the path of deviatoric loading, that is functions of the type $T = f(\Gamma, \sigma, \lambda_\sigma, \mu, \alpha_\sigma)$.

First of all it is desirable to have the relationship between stresses and strains in a differential form so that one would have a function of the type $dT = f(\alpha, \Gamma, \sigma...)$ with exclusion of $\lambda_\sigma, \mu, \alpha_\sigma$ as the integration will occur along the loading path given in implicit form for the sequence of dam construction and reservoir filling (Chapter 4).

The relationships commonly used do not consider multiple phases of soil (quasi-single phase soil is assumed), skeleton creep, etc that is the process of soil deformation is not considered as time-varying although some more complete solutions are available. This chapter of soil mechanics is still in the stage of intensive growth.

Deformability of soil is often described as the function $T = f(\Gamma, \sigma...)$, i.e. it is assumed that loading path does not affect the character of deformation, that stress and strain are coaxial, similar, etc and there is a straightforward relationship between stresses and plastic (irreversible) strains. Such relationships are referred to as strain ones. The class of strain models is very wide, and at present it is already more or less clear how to incorporate the loading patterns in these models.

The strain theory of plasticity has been used by Ioselevich (1967) for practical applications; he has included loading patterns. Deformation theories have been applied for problems in which loading patterns slightly affect the soil foundation or are similar. In the latter case one is recommended to model the loading pattern if the parameters of strain theory are determined experimentally.

Dolezhalova (1967, 1969, 1983) singled out 15 characteristic strain trajectories in earth dams and determined correspondence between strain trajectories and stress patterns in these soils. She also displayed the effect of loading pattern on strain characteristics of soils; relationships are given which provide background for stress-strain computations for earth dams.

In the strain model of Dolezhalova one uses traditional characteristics E and ν, which are selected as functions of not only stresses but also loading patterns. For instance, trajectory number 3 corresponds to the stress trajectory if soil strength at a given point grows, that is for displacement unloading — for increasing E and decreasing ν. Incremental method of loading, that is stage by stage construction has been used in the model.

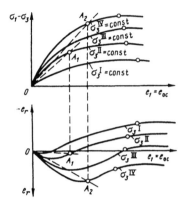

Figure 1.6. Graphical results of triaxial soil tests.

Shortcomings of this model, like in other strain models, include ambiguity of solutions (nonuniqueness).

One of the versions of strain models was suggested by Rasskazov (1969); it is summarized as follows. By constructing usual standard graphical plots of results of triaxial tetsts (Fig.1.6) and by using generalized Hooke's law for each point of the graph, for instance A_1 and A_2 one obtains:

$$E = f_1(\frac{\sigma_1}{\sigma_3},\sigma_3), \quad \nu = f_2(\frac{\sigma_1}{\sigma_3},\sigma_3)$$

in which

E and ν = modulus of linear deformation and Poisson's ratio, respectively.

Since soil is a clearly defined material with plastic properties at any practical stress occurring in earth dams and foundations, E is not the modulus of elasticity but rather characterizes the deformability under load only. The same is true for ν. Values of E and ν for some soils are given in Figure 1.7. They are also used in solutions of soil problems with regard to displacements, stresses and stability of earth-rockfill dams (Rasskazov 1972). In the construction of these graphs one uses the following relationships obtained from the Hooke law:

$$E = \frac{(\sigma_1 + 2\sigma_3)(\sigma_1 - \sigma_3)}{\sigma_3(e_1 - 2e_3) + \sigma_1 e_1}$$

$$\nu = \frac{\sigma_3 e_1 - \sigma_1 e_3}{\sigma_3(e_1 - 2e_3) + \sigma_1 e_1}$$

(1.28)

These relationships are given in the form provided by Goldshtein et al. (1977).

The graphs in Figure 1.7 can be used for approximate estimates of deformations — one must assume the ratio $\sigma_1/\sigma_3 = 2$–3, find the mean values of E and ν for an approximate range of stresses and estimate the possible deformation of structure or foundation.

Aside from the aforementioned assumptions, the strain models have some other shortcomings due to their implementations in two- and three-dimensional problems (see Chapter 4). For instance, from Figure 1.7 it follows that one must resolve strain relationships by iterations, as E and ν are functions of stress condition. Hence one must give E and ν for various elements of dams and foundation, find stress components at various points of dam by any method, find a better approximation for E and ν at these points, and again solve the stress problem. The convergence of iteration is unclear.

Models basing on the theory of plastically hardening body. Drucker (1962) postulated applicability of the associated flow rule for elastoplastic materials. The theory has been worked out for metals with strictly elastic volume change; Drucker assumed that rock could also conform to this assumption.

From the properties of crystals Lin (1976) derived the basic assumptions for loading surface in bodies with volumetric plastic strains.

In passing, with reference to the application of the theory of hardening plastic bodies to soil, one should mention after Drucker that the external apparent similarity of friction and plastic strain is deeply incoherent, particularly if its 'instability', that is friction, weakens with intensifying flow, which occurs in soil in supercitical condition. In the system with deformation controlled by friction one may encounter cases in which infinitesimal perturbation brings about finite displacement of a system, which is impossible in a hardening material or in ultimately plastic material equivalent to such motion, as neither plastic material nor the forces generating the plastic strains are capable to do an infinitesimal work, as indicated by Drucker (1962). Despite these drawbacks, the theory has been widely applied in analysis of soil deformation, which can be attributed to many advantages of this type of description.

The following conclusions can be drawn from soil tests:

1. Full information in infinitisemal neighbourhood of a given point (the soil sample may be regarded as a point) consists of elastic, plastic and viscous parts, the plastic and viscous parts bringing about residual strains under full unloading. The elastic part is small under primary loading, compared with the plastic one. The test procedures are usually so devised as to include a large part of viscous strains into plastic ones (the methodology is based on the criterion of stabilized deformation at any step of loading).

2. A yield limit exist in soil, but prevailing plastic strains are observed within the entire range of primary loading.

3. It can be assumed that material is a linearly deformable body under unloading.

4. Soil is a plastic compressible body which changes its volume nonlinearly under primary loading.

A first major study in the application of the theory of hardening plastic bodies to soil belongs to Ioselevich & Didukh (1970) and Ioselevich et al. (1975). Some investigations were conducted abroad.

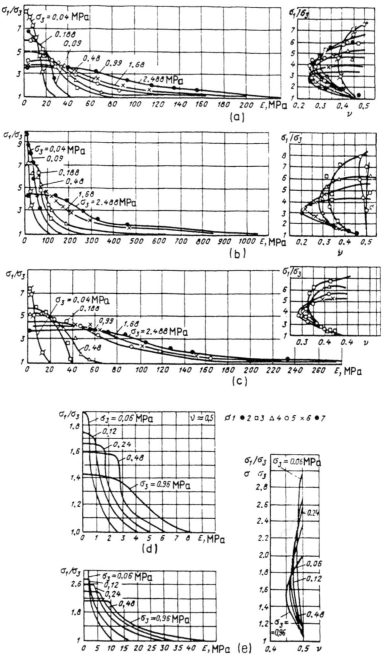

Figure 1.7. Modulus of linear deformation E and Poisson's ratio ν in various soils tested in triaxial apparatus. (a) diorites, $d_{max} = 200$ mm, $I_D \approx 0.9$; (b) shingle, $d_{max} = 200$ mm, $I_D \approx 0.9$; (c) limestone, $d_{max} = 200$ mm, $I_D \approx 0.9$; (d) sandy loam, $\gamma_{dry} = 17$ kN/m³, $w = 13.5\%$, $S_D = 0.8$; (e) sandy loam, $\gamma_{dry} = 17$ kN/m³, $w = 13.5\%$, $S_D = 0.6$.

The loading surface f is taken by Ioselevich & Didukh (1970) as a function of the first and second stress invariants, strength parameters detemined by the Mises-Schleicher strength condition (Section 1.5), and two invariants of the plastic strain tensor. Upon right assumption that elastic strains are small Mises and Schleicher neglected them.

The loading surface is a graphical representation of the function $f=0$ in which f = function of the aforementioned arguments, with the following properties:

a) The vector of plastic strain rate Δe_{ij}^p is orthogonal to the loading surface;

b) The direction of Δe_{ij}^p does not depend on the direction of stress rate vector $\Delta\sigma_{ij}$;

c) The sign of the differential function with regard to stresses $d'f$ is '+' for loading and '-' for unloading, whereupon $d'f = \frac{\partial f}{\partial\sigma}d\sigma + \frac{\partial f}{\partial T}dT$, or in stress components $d'f = \frac{\partial f}{\partial\sigma_{ij}}d\sigma_{ij}$. For $d'f=0$ one has neutral loading $\Delta e_{ij}^p=0$.

In general, the arguments of the loading function can be not only components of the stress tensor and strain tensor but also soil moisture, temperature (for frozen soil), strength parameters and some other factors. The quantity $d'f$ is total differential with regard to stresses only, without inclusion of other arguments of the loading surface. Usual graphical representation of the loading function on the plane in coordinates σ, T or σ, σ_i is the projection of loading surface on the given plane. The plastic strain rate in terms of the associated flow rule reads:

$$d\,e_{ij}^p = \begin{vmatrix} h\frac{fd}{d\sigma_{ij}}d'f & \text{at } f = 0 \text{ and } d'f > 0; \\ 0 & \text{at } d'f \leq 0, \end{vmatrix} \tag{1.29}$$

in which

$$h = \left(\frac{df}{de_{\alpha\beta}^p}\frac{df}{d\sigma_{\alpha\beta}}\right)^{-1};$$

e_{ij}^p = components of plastic strain tensor.

The orthogonality condition for the vector de_p with respect to the loading surface requires that it is colinear with the vector $grad\,f$:

$$d\,e^p = dh\,\frac{\partial f}{\partial\sigma_{ij}} \tag{1.29a}$$

in which

$dh > 0$ as the vector de^p should be oriented outwards from the loading surface.

Hence Equation 1.29 is a form of the Drucker postulate which can be written as

$$dh = h\frac{\partial f}{\partial\sigma_{kl}}d\sigma_{kl} \tag{1.29b}$$

The function h is a particular hardening law, in which it depends only on σ_{ij} and other arguments but does not depend on $d\sigma_{ij}$.

In general Equation 1.29 can be given as

$$de^p_{ij} = h \frac{\partial f}{\partial \sigma_{ij}} \frac{\partial f}{\partial \sigma_{kl}} d\sigma_{kl} \qquad (1.29c)$$

If both sides of Equation 1.29c are divided by dt one gets the flow rule of an elastoplastic medium with hardening. This law is referred to as the associated rule for it is directly linked to the condition of flow (Eq.1.42).

Plastic flow is a process of irreversible deformation. A large part of strain work is converted into heat.

Stresses depend on loading path. The equations describing plastic deformations must not be finite relationships between stresses and strains but should be written down in differential forms.

Plastic changes in volume must not be neglected in soil, although in the theory of flow it is assumed that the relative changes in volume are elastic.

Components of the stress deviator and the components of the deviator of plastic strain rates are proportional (Eq.1.19):

$$d\epsilon^p_{ij} = d\lambda S_{ij} \qquad (1.29d)$$

Elastic deformations can be neglected in soil as they are small under static loading; in this case Equation 1.29d is referred to as the Saint Venant-Mises equation of plastic flow.

Another form of Equation 1.29d is also obtained if its two sides are divided by dt, so that the strain rate appears on the left-hand side and the equation is analogous to the equation of flow of a viscous liquid.

The isotropic strenthening is better applicable for soil in the theory of plastic flow:

$$dA_p = \Phi'(T)dT \qquad (1.29e)$$
in which
  A_p = work of stresses spent on plastic strains,
  $\Phi'(T)$ = hardening function,
  T = shear stress rate in Equation 1.25.

Eventually one obtains

$$d\epsilon^p_{ij} = F(T)dT S_{ij} \qquad (1.29f)$$

Equations 1.29d and 1.29f show that the vector of plastic strain rate is coaxial with the stress vector in six-dimensional space of stress components. Equations 1.29d and 1.29f can be obtained from Equation 1.29a since the Mises flow condition yields

$$\frac{\partial f}{\partial \sigma_{ij}} = \frac{\partial (T^2)}{\partial \sigma_{ij}} = S_{ij}$$

and Equation 1.29a becomes

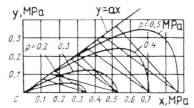

Figure 1.8. Loading surface in sand.

$$de^p_{ij} = d\lambda S_{ij}$$

Hence the plastic strain rate can be determined if the loading function is found experimentally. Since soil displays volumetric plastic strains under isotropic compression, the projection of the loading surface on the plane σ, T should intersect the hydrostatic axis, and this intersection should be at a right angle due to convexity of the loading surface, as expressed by the Drucker condition. The axes $x = \sigma/\sqrt{3}$ and $y = \sqrt{2}T$ were taken by Ioselevich & Didukh (1970) instead of σ and T.

As the result of triaxial tests along unloading trajectories ($\mu < 0$), projections of the loading surface have been found by Ioselevich & Didukh (1970) as functions of hydrostatic compression for sand and Charvak sandy loam (Fig.1.8). Unfortunately, the volume changes were not measured with a sufficient accuracy, and axial deformations were taken as an indicator of incipient plastic deformation (the accuracy of measurements was 0.01 mm). All loading surfaces were stationary and began at the origin of coordinates.

The associated flow rule (Ioselevich & Didukh 1970) was verified by arrangement of a 'fan-like' loading with small increments $|\Delta\sigma| = \sqrt{\Delta x^2 + \Delta y^2} = 0.02$ MPa for $\sigma = 0.7$ MPa, all over the loading surface. Results for sandy loam are given in Figure 1.9a, and for sand in Figure 1.9b. The loading and unloading were arranged in the order of the numbers shown, while the semi-solid circle denotes the axial deformations measured.

The maxima of axial deformations are arranged on the vector corresponding to purely deviatoric loading (direction 24), for which $\Delta x = 0$; since the points were selected on the shallow branch of the loading surface and the normal line was almost vertical, while that of axial deformations was practically identical with the direction of full deformation vector, there should no be doubt about applicability of the associated flow rule. Accordingly, it was shown that the loading surface cannot be stationary at the origin of coordinates x, y and should move along with loading (Ioselevich et al. 1975); an analogous proof is provided by Ivlev & Bykovtsev (1971).

The fan-like loading is aimed at finding the form of the loading surface about a given point and at satisfying the associated flow rule, that is orthogonality of the vector Δe^p_{ij} vs. loading surface. Indeed, if a small value of $|\Delta\sigma|$ applied to the loading surface along one of the 24 trajectories corresponds to a certain Δe^p_{ij}

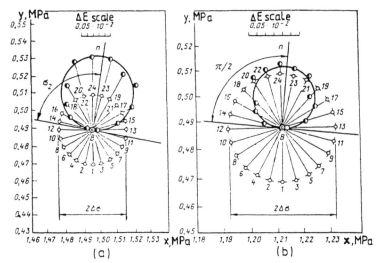

Figure 1.9. Measurement of strain ϵ on loading surface (B) versus direction of loading vector (Ioselevich & Zuyev's data). (a) Charvak sandy loam ($\gamma_{dry}=17.2$ kN/m^3, $w=18.3\%$); (b) Lubertsi sand ($\gamma_{dry}=15.6$ kN/m^3).

upon loading and unloading, the entire loading process goes along this trajectory; in the opposite case unloading occurs. Hence analysis of all 24 trajectories should identify the ones corresponding to loading and unloading. The boundary between the two classes is the projection of the loading surface, while the coincidence of the maximum rate Δe^p_{ij} and the normal to the surface at a given point provides substantiation for the associated flow rule. As shown by Ioselevich & Didukh (1970), the fan-like loading involves certain methodological inaccuracies due to variation of the original loading surface under each cycle of loading and unloading. Upon transition of soil into limit state the loading surface is degenerated into a straight circular cone, the side of which coincides with a conical surface of Mises-Schleicher and the direction of the vector of plastic strain rate is no longer orthogonal to the loading surface but instead depends on the loading history, so that smoothness of the loading surface is violated (a singular point appears).

Experimental evidence on the deviation from the orthogonality of the vector of plastic strain rates in limit state has been provided by Lomize & Sukhanov. The function for which the smoothness is violated is referred to as singular, in contrast to the regular point, at which basic properties of the function are not infringed.

A primary shortcoming cf experimental investigations (Ioselevich & Didukh 1970) consists in low accuracy of measurements for volume change, so that conclusions have been based on axial deformations. Owing to reconstruction of the system of measurements of volume changes it was possible to increase the accuracy for axial and volume changes, up to 0.00002, which has proved very important

(Ioselevich et al. 1979). The intersection of the loading surface and the hydrody-namic axis is of particular interest for the right angle the transition to deviatoric loading along the initial section brings about neutral loading (along tangential line to the loading surface about $T=0$), and there will be no deformation. Tests have however shown that deformations do exist along the initial section of the deviatoric loading, and moreover they are significant; this stems from tests with loading in relatively small steps.

An extensive series of experiments on the deformation process at the intersection with hydrostatic axis was continued. The fan-like loading-unloading scheme was applied. The value of $|\bar{\Delta\sigma}|$ was 2% σ. However, as the result of the fan-like loading the surface was deformed, and one had to select a certain sequence of loading and unloading so that the distortions were as small as possible. In terms of this technique the loading region begins for the highest distortion of loading surface (sequence 13) since it is due to $\bar{\Delta\sigma} = +\sqrt{\Delta\sigma^2}$ that the highest distortion of the loading surface occurs, as proved experimentally, while the lowest ones correspond to gradual increase in the component of the quantity ΔT. The sequences one to twelve correspond to unloading, and since they are inside the loading surface 'no distortion' occurs. As the point was on the hydrostatic axis it has been only a half of the loading fan that was analysed (Fig.1.10a, b).

Shown in Figure 1.10b are the loading sequence and results of measurements for axial (left-hand side) and total (right-hand side) deformations. The point of hydrostatic compression 0.3 MPa in Figure 1.10a is not shown to scale of the drawing. Figure 1.10b indicates that the maximum axial deformation corresponds to purely deviatoric loading as in the tests by Ioselevich & Didukh (1970). The first part of fan-like loading corresponds to underestimated axial deformations (total inclusive), since after the sequence 7 (8 and 13) pronounced dislocations in the position of the loading surface are observed. It is important that already in the sequence 4 axial deformations occur, so that the projection of the surface on the plane σ, T was oriented to the hydrostatic axis at an angle below 45° and above 30°, but not 90°; the singular point corresponding to distortion was generated. Numerous investigations for various σ have shown that the angle of approach of the loading surface to the hydrostatic axis was always in this range. Even more interesting are the results for the vector of total strain rates $\overline{\Delta e} + \bar{\Delta\Gamma}$: the maximum of the vector always corresponded to the sequence 7, and the vector itself in all experiments was practically normal to the loading surface at this point on the plane σ, T. Other loading sequences indicate that total rate vectors were not normal to the loading surface. Deviations were up to 90° (sequence 3, Figure 1.10b). Hence the point on the hydrostatic axis is singular also in terms of the associated flow rule which requires that the vectors of strain rates are located on the normal line of the loading surface (by definition).

Analysis of the configuration of the projection on the plane σ, T shows that the latter is elongated 'missile-type' (Fig.1.10) and moves behind the loading point, practically detaching immediately about the origin of coordinates. The final part

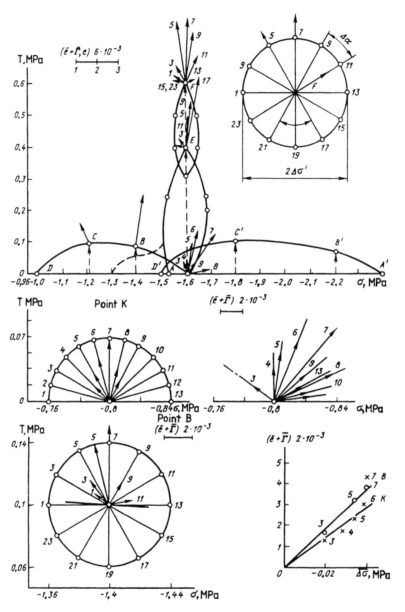

Figure 1.10. Loading surfaces. (a) movement of loading surface; (b) fan-type loading, axial and total strains (with singular point A on the hydrostatic axis); (c) fan-like loading at regular point; (d) total strain and total stress.

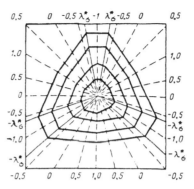

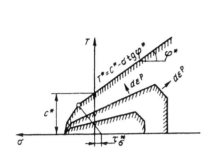

Figure 1.11. Loading surface and its ultimate layout, after Zaretskiy & Lombardo (1973) (left).

Figure 1.12. Development of loading surface on deviator plane in triaxial compression tests (right).

of the surface corresponds to $\sim \frac{1}{2}\sigma$. The type of approach to the hydrostatic axis along the end segment requires further details, but the experiments carried out substantiate the opinion that the angle is about 70–90°, so that smoothness might be violated here.

Basing on the analysis of slip planes in polycrystalline materials Lin has shown that loading surface in volumetrically compressible materials is missile-type. Behaviour of the vector rates for total plastic strain at regular points of loading surface (for instance B, C, B', C') is illustrated in Figure 1.10c for tests about point B. The result is analogous to singular points (K, A, A'), as vectors were not located along the normal line of the loading surface.

The behaviour of the loading surface under hydrostatic loading is depicted in Figure 1.10. The relative length of the loading surface (ratio of σ at origin and end) gradually decreases although the absolute length (difference of the above values) increases.

Subsequent investigations were concentrated about deviatoric loading. As for hydrostatic compression, the surface was missile-type. Respective vectors of total plastic strain rate are shown in Figure 1.10a at points A, E and F. The results indicate that the loading surface detaches from the coordinate axes upon approach of the limit state of soil. Vectors of total plastic strain rate have their maximum for purely deviatoric loading and do not lay on one straight line since the point is singular.

Upon approach of the limit state with increasing T the negative dilatancy decreases (consolidation decreases due to the deviation); in proximity of the limit state it seems to change sign as expansion begins. For instance, dilatancy does not occur at point F, and the total vector of plastic strain rates is vertical. The greatest deviation from the vertical orientation due to strengthening appears about point A. Upon consideration of axial deformations (for point F they are shown in Figure 1.10a), one may basically draw the same conclusions on distortion of smoothness at the origin of the loading surface and its transformation and motion in the planes σ, T upon deviatioric loading. Approaching the limit point on purely deviatoric trajectory of loading the volume of the loading surface decreases due to reduction of its length along both T and σ.

The above experiments have given rise to objections on the applicability of the associated flow rule to soil (note the cited study by Drucker) and permitted verification of the hypothesis on isotropic soil used in construction of the energetics model (described below), i.e. the hypothesis on proportionality of total vector of plastic strain rate to the total vector of stress rate. The hypothesis proves satisfactory (Fig.1.10d). It should be noted that the plausible confirmation is due to the Bushinger effect (hardening effect) which intensifies with growing number of loading and unloading sequences. If one tries to exclude this effect, each loading sequence should have its respective soil sample, which produces errors by two orders of magnitude higher than the quantities measured for the assumed accuracy of measurements and type of experiment.

Analysis of the projection of loading surface on deviatoric plane (Ivashchhenko & Zacharov 1973) has exposed elongation of the surface in the direction of loading point, proportionality of plastic strain rate to stress rate, and deviation of the vector of plastic strain rate from the normal line. Unfortunately, the low accuracy of measurements does not permit conclusions on smoothness of the surface about the loading point.

The above investigations complement each other and support the hypothesis on the proportionality of stress and strain rate tensors.

Analysis of the soil loading surface shows that the associated flow rule for a hardening plastic body is not fully applicable to soil. Hence one must use some additional assumptions; which is also a general requirement for other models of soil.

The loading surfaces obtained can be regarded as envolopes of an infinite number of singular points, while each singular point can be considered an intersection of large (possibly infinite) number of loading surfaces.

Construction of the functions describing the motion of loading surface and its transformation under loading is difficult.

One must bear in mind that the associated flow rule has advantages because even an approximate description of soil behaviour under loading requires a mathematical expression for the loading function. In this case all problems of hydrostatic compression, dilatancy and deviatoric loading, loading vs unloading (particularly important for dynamic effects), subtrical and critical conditions, etc. are solved jointly by a single relationship.

The use of the associated flow rule is very ambiguous and requires estimates of possible errors.

The associated rule for stress-strain problems of earth was employed by Zaretskiy & Lombardo (1983). Assumptions were made which seem to increase the accuracy of solution. The loading surface is taken piecewise-linear (Fig.1.11) in the plane σ, T, consisting of three segments which upon conjugation generate two singular (angular) points. The projection of the loading surface on deviatoric plane (Fig.1.12) also yields piecewise-linear functions with singular points on projections of the axes $\lambda_\sigma = +1$ and λ_σ^*, in which the former stands for the Lode-Nadai parameter and the latter is an experimental soil factor. Moreover, it was assumed that the loading surfaces begin at the origin of coordinates. This means that plastic strains do not arise upon cubic compression and subsequent unloading to zero; this is in a certain contradiction to results of tests. However, since unloading in earth-rockfill dams primarily occurs on the upstream side due to increase in water level, as a result of buoyancy, the above assumption is applicable in practical situations.

Theoretical computations and experimental findings show that upon active loading, no matter how complex, the loading vector in terms of this soil model is at a singular point of loading surface. Upon unloading and repeated loading from the point of regular loading surface the latter is transformed, practically immediately thereafter, and the loading vector appears again at the singular point. Hence the model so formulated is practically close to the theory of sliding in which all loading occurs at the conical point of loading surface, and the direction of vector of plastic strain rates depends on the direction of the vector of full loading but is limited by the 'plastic cone'. This practically confirmes Koiter's hypothesis which postulates that sliding theories can be derived on the assumption that the loading surface is an envelope of finite or infinite numbers of regular surfaces all of which possess all properties of loading surface.

Hence the loading surface consisting of an infinite number of singular points satisfying only one function is considered by Zaretskiy & Lombardo (1983). This function describes the boundary between the elastic and elastoplastic regions, so that the primary properties of loading surface shown by Ioselevich et al. (1979) are not conflicting.

In his fundamental study Rabotnov (1979) poses the question: 'is the motion of loading surface indeed real and can one use it in the theory of plasticity ?'. He himselves answers this question: 'there are versions of the theory of plasticity (Ilyushin) which are not based on the notion of the loading surface but describe directly the components of stress tensor as certain functionals for loading path; one of the fundamental arguments in the construction of such theories is the noted impossibility of making a strict distinction between elastic and plastic strains under experimental conditions. It appears that any of the existing theories of plasticity can be verified experimentally if one considers detailed reasonable averaging of results supporting the experimental findings at least for a certain limited set of experimental programmes'.

Relationships for properties of strain models and associated flow rule. These functions usually determine non-holonomic relationships between components of strain tensor and stress tensor, so that certain restrictions are imposed on strain rates.

Since the loading surfaces play the role of a boundary between elastic and elasto-plastic regions in the stress space or stress-strain space, one may generally refrain from using the loading surface if one constructs a functional (Rasskazov 1973, Rasskazov & Dzhkha 1977):

$$d\sigma_{ij} = f(\sigma, de_g, d\epsilon_{ij}^p) \tag{1.30}$$

in which

σ_{ij} = components of stress tensor,
e_g = volume changes due to cross-correlations, that is the ones depending on displacement strains (dilatancy),
ϵ_{ij}^p = components of plastic strain tensor,
σ = mean stress.

Equation 1.30 implies coaxiality of the tensors of stress and strain rates.

The hypotheses assumed in the associated rule and in Equation 1.30 can be verified if the tensors of stresses and strains are compared as to their axes; this was done by Ryabchenkov who measured strains and stresses in sand under plane deformation on a stand $220 \times 120 \times 75$ cm, under conditions of expansion of a cylindric space having the initial diameter of 12 m. Uniform loading of 0.03 MPa was applied to the surface of sand.

The experiments have shown that the evolution of the stress-strain condition for quasi-equilibrium conditions is accompanied by the resolution of the axes of stress and strain tensors and generation of non-colinear and dissimilar tensors. The comparison of coaxiality is depicted in Figure 1.13. The highest discrepancy at various points of the stand is observed for total tensors of stresses and strains (deflection of diagonals of respective squares with increasing load). The deflections of the axes of the stress rate tensor $\alpha_{d\sigma}$ and the strain rate tensor α_{de} are practically close to the respective deviations between stress tensors and strain rate tensors. The quantities λ_σ and λ_e, etc. did not coincide, either. Hence this points to the approximate status of the assumptions, with the degree of accuracy depending on condition of soil. The associated flow rule in the interpretation provided by Zaretskiy & Lombardo (1983) gives close results in the presence of singular points and relationships such as Equation 1.30 along the segment of active loading.

Complete Equation 1.30 reads

$$
\begin{aligned}
d\,\sigma_{ij} = \ & \frac{E_0^n}{n|\sigma|^{n-1}} \left[de - \text{sign } (\Gamma - \Gamma_0)\frac{M d\Gamma}{|\sigma|} \right] \sigma_{ij} + \\
& + 2|\sigma|^{1-n} \left\{ f(v)\frac{E_0^n}{\exp} \left[-b(1 - \bar{k}) \right] + \right. \\
& + \left. G_0\bar{k} \left[1 - \exp(-b(1 - \bar{k})) \right] \right\} d\epsilon_{ij}
\end{aligned} \tag{1.31}
$$

in which

δ_{ij} = Kronecker delta (1 for $i = j$ and 0 for $i \neq j$),

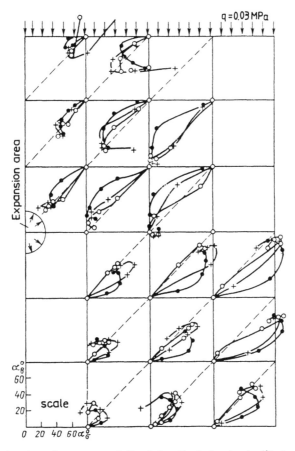

Figure 1.13. Examination of tensor coaxiality (after Ryabchenkov). (t) stress and strain; (d) stress and strain rate; ($+$) stress rate and strain rate.

Table 1.4.

Type of soil	G_0, MPa	U_0, MJ/m^3	b	M, MPa	Γ_0
Gravel and pebbles	21.3	0	30	0.05	0.01
Rock rubble	18.4	0.09	40	0.03	0.015
Clay	2	0.002	20	0.0	0
Clay	3	0.02	30	0.0	-
Skeletal sandy loam*	2.87	0.05	40	0.01	0.017

*Fine particles make up 50% of soil by volume.

M = modulus of dilatancy,

Γ_0 = critical second invariant of the tensor of form change; for $\Gamma < \Gamma_0$ soil is consolidated at shear strain while for $\Gamma > \Gamma_0$ soil expands,

$[\sigma]$ = permissible stress under massive crushing of coarse grained soil when the dilatancy effect is damped.

Under permissible stresses the grains collapse but do not lose mutual contacts, so that expansion under displacement takes place. In gravel and pebbles one has $[\sigma] \cong 2$ MPa, and for rock $[\sigma] \cong 1$ MPa. For $[\sigma < [\sigma]$ one has $Md\Gamma/\sigma > 0$ and for $\sigma \geq [\sigma]$ one obtains $Md\Gamma/\sigma = 0$. The quantity $f(\nu)$ depends on Poisson's ratio of soil at the original point of deviatoric loading for $\sigma = 0.001$ MPa. For $\nu = 0$ (quite often in soil) one has $f(\nu) = 3/2$ and for $\nu = 0.2$ one obtains $f(\nu) = 1.0$ while for $\nu = 0.3$ the function $f(\nu)$ becomes 0.807. Hence $f(\nu)$ expresses the relationship between the moduli of volume and shear change at incipient deformation. E_o and G_0 are moduli of volume change and shear change (bulk and shear moduli of elasticity), respectively, at $\sigma = 0.01$ MPa if $T \rightarrow 0$ while $\bar{k}$ is a special parameter of hardening which acounts for the effect of loading prehistory in generalized sense:

$$\bar{k} = \frac{k_s - 1}{k_s} \tag{1.32}$$

$$k_s = \frac{U_0 + \int\limits_L \sigma de}{\int\limits_L S_{ij} de_{ij}} \tag{1.33}$$

in which

U_0 = energy of preconsolidation which accounts for the loading prehistory.

More details on k_s are given in Section 1.5.

Since k_s describes the ratio of the energy of form changes and the energy of volume changes, the model of soil work in subritical condition is referred to as energetics model (Rasskazov 1973, 1977).

The models presented by Esta & Hayal (1973) are close to the above formulation.

Equation 1.31 describes the tests of active loading. For unloading one assumes Hooke's law which is written down more conveniently as

$$d\sigma_{ij} = E_{0y} de\, \delta_{ij} + 2G_y de_{ij} \tag{1.34}$$

in which

E_{oy} = modulus of elasticity for volume changes,

G_y = shear modulus of elasticity.

The quantities E_{oy} and G_y are determined on the unloading branch.

One of the shortcomings of the above relationships and procedures is the necessity of making special additional assumptions on loading and unloading. An increase in strain energy can be one of such assumptions — loading for $d\mathcal{E} \geq 0$ and unloading for $d\mathcal{E} < 0$, but this requires long computer times so that one rather uses an indirect condition for earth dams (and even more often for earth-rockfill dams) consisting in the unloading primarily within the upstream prism due to increasing water level in reservoir. This assumption is not very rough.

In finding soil characteristics corresponding to Equation 1.31 it is sufficient to conduct one 'ideal' experiment by standard methods in a triaxial apparatus, that is the quantity of experimental data must be sufficient for statistical processing. Characteristics of soil corresponding to Equations 1.31 and 1.33 are given in Tables 1.3 and 1.4.

If one uses the associated rule the parameters must be determined by several experiments which is not only inconvenient but also increases errors.

Hence the problem of soil deformation under loading is not solved definitely, but the postulated model can be used with practically acceptable accuracy in the design of earth dams. More details on soil models in stress-strain computations for dams are given in Chapter 4.

1.5 CONDITION OF SOIL STRENGTH

The primary condition of soil strength is provided by the general rule of Mohr-Coulomb, by which the soil strength is given as a certain relationship between strain tensor components, and not by individual components, as assumed in the Coulomb law.

In Mohr's interpretation, it is only two components of principal stresses, the major σ_1 and the minor σ_3, that determine the soil strength

$$\sin \varphi_M^* = \frac{\sigma_1 - \sigma_3}{\sigma_1 + \sigma_3 + 2c_M^* \cot \varphi_M^*} \tag{1.35}$$

in which

φ_M^* = angle of internal friction,
c_M^* = cohesion in Mohr's strength condition.

The above condition ignores the effect of the intermediate principal stress σ_2 and the loading path, and assumes the existence of a rectilinear envelope of Mohr's circles.

The simplicity and convenience of the above assumption (simple loading trajectory, small range of stresses σ_3 in the construction of low structures) have given rise to a wide spectrum of applications. The condition is very flexible and can indeed describe the strength of various materials if the loading path does not in general affect the strength; otherwise the Mohr-Coulomb condition is insufficient.

Shown in Figure 1.14 are results of tests on the strength of clayey soil for constant $\sigma = \frac{1}{3}(\sigma_1 + \sigma_2 + \sigma_3)$ on the deviatoric plane, for various Lode parameters for stress.

Figures 1.15 and 1.16 show experimental results (Lomize & Fedorov 1975) for clay loam with various Lode parameters and loading trajectories in the plane $\lambda =$ const. The trajectory parameter was selected as set by Equation 1.27:

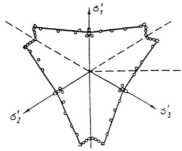

Figure 1.14. Data of triaxial tests on clayey soil samples depicted in the plane σ=const.

$$\mu = \frac{d\sigma}{d\sigma_i}, \quad \sigma_i = \frac{1}{\sqrt{2}}\sqrt{(\sigma_1 - \sigma_2)^2 + (\sigma_1 - \sigma_3)^2 + (\sigma_2 - \sigma_3)^2} \qquad (1.36)$$

Experimental investigations (Lomize & Fedorov 1975) have shown that the trajectory parameter μ also significantly affects the strength (Fig.1.15), while tests reported on by Lomize et al. (1970) show the effect of the rotation of the axes of principal stresses. Hence the loading path is an important element of strength condition.

The strength condition for soil including the effect of intermediate principal stress is known as Mises-Botkin or Botkin condition (Botkin 1940). This condition is applied to soil in combination with the energetics condition of strength by Mises-Huber-Hencky; which for metals means that failure occurs if the energy of form variation reaches a certain limit value. Assuming that the cubic compression must not be ignored in soil (but may be for metals), Botkin considered the strength on an octahedral plane in which the components of normal and tangential stresses are equal to the first and second invariants of stress tensor, with an accuracy to a constant coefficient. It is also assumed that the Coulomb condition is satisfied on the octahedral plane:

$$\tau_{oct} = (\sigma_{oct} + c_{oct}) \tan \varphi_{oct} \qquad (1.37)$$

in which

$$c_{oct} = c \cot \varphi_{oct}$$

An analogous strength condition is referred to as limit deviatoric norm, but that condition cannot account for the effect of loading trajectory parameters, much as the generalized condition of strength by Mohr-Coulomb.

It is seen that components of strain tensor do not intervene. This is justified by the desire to neglect soil strain in the strength condition. The absence of strain tensor components in the energetics strength condition by Huber-Hencky is discussed below after Goldshteyn (1973).

First of all, neglect the energy of volume change; next formulate the strength condition (not for soil) as follows: failure occurs if the energy of form change reaches its limit value for a given material

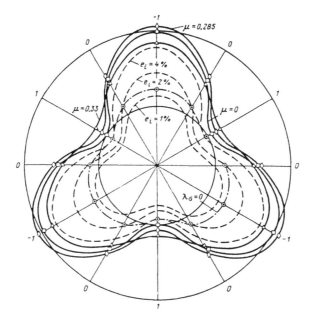

Figure 1.15. Data of clay loam tests for various Lode parameters (after Lomize & Fedorov).

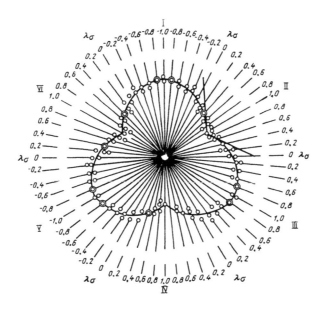

Figure 1.16. Type of limit surface on deviatoric plane $\sigma = 0.5$ MPa.

$$\int T d\Gamma = U_0 \tag{1.38}$$

in which

$\Gamma = \sqrt{\frac{2}{3}}\sqrt{(e_1 - e_2)^2 + (e_1 - e_3)^2 + (e_2 - e_3)^2}$ = second invariant of strain deviator.

For a linear relationship between stresses and strain one has

$$U_0 = \frac{1}{2}T\Gamma \tag{1.39}$$

If strains are given through stresses by the Hook law, and upon subsition in Equation 1.39 one has

$$U_0 = \frac{1 + \nu}{6E}[(\sigma_1 - \sigma_3)^2 + (\sigma_1 - \sigma_2)^2 + (\sigma_2 - \sigma_3)^2] \tag{1.40}$$

in which

E, ν = modulus of elasticity and Poisson's ratio, respectively.

The energy U_a is controlled by cohesion; Huber determined it from the conditions of uniaxial tension:

$$U_0 = \frac{1 + \nu}{E}2\sigma_R^2 \tag{1.41}$$

Mises and Hencky substituted σ_L for σ_R in the case when brittle failure is replaced by plastic flow; for pure shear one has

$$\sigma_L = \sqrt{3}T_L = \sqrt{3}\sigma_1 = \sqrt{3}\tau_L \tag{1.42}$$

Schleicher postulated $T_L = f(\sigma)$.

Pol (Chernousko & Bonichuk 1973) noted that Maxwell in a letter to Thompson (in 1856) drew attention to the fact that 'homogeneous amorphic solid bodies should transform into plastic ones' if the work of 'form changes' reaches a critical state (Pol 1975). From the same position Reiner analysed the strength conditions of Huber. Since one talks about metals it must be assumed that 'hydrostatic pressure does not cause plastic flow or failure' (Reiner 1947).

Numerous soil tests have exposed large plastic volume changes under hydrostatic compression. These strains must be included in the model of soil strength since the presence of volume changes brings about changes in consolidation of soil, and the latter factor obviously controls the strength of soil.

Hencky argues that strength is identified with a certain vessel, and failure occurs if the level of 'liquid energy' reaches edges of this vessel and energy starts 'overflowing' (since one talks about elastic and linearly deforming bodies it is obvious that the energy is the potential energy of elastic strain).

Reiner (1947) assumed that dissipative energy (energy of irreversible dissipation) does not bring about failure — it is neutral. However, volume plastic strains due to compression in soil are identified with the dissipative energy in its considerable part because the elastic energy is much smaller (by a factor of 10 and more), and it is the soil consolidation (due to plastic strain only) which brings about increase in the strength. Assuming plastic failure in soil one must consider not the elastic energy of form changes but the total energy of changes and the total energy of volume changes.

Reiner's model assumes unlimited very slow increase in the 'liquid-energy' in the vessel for some materials caused by relaxation ('opening in the vessel'). However, the possibility of plastic volume change was not considered. Such possibility must include the assumption that slow 'filling of vessel' is accompanied by simultaneous increase in the volume. The assumption of Schleicher reads $T_L = f(\sigma)$, supported by experiments and in agreement with the assumption on the growth of 'vessel-strength'.

From Equation 1.38 it follows that transition from energetics condition to dynamic indicators is possible for linear relationships between stresses and strains.

Computers allow more complex strength conditions with inclusion of strains so that the energetics equations must not be simplified. If one assumes that the energy of volume compression brings about increase in soil strength, the strength condition can read as follows:

$$U_0 + \int \sigma de = \int S_{ij} d\epsilon_{ij} \tag{1.43}$$

in which

U_0 = energy of initial soil strength (due to adhesion, 'interlocking' or preconsolidation).

Upon neglect of the energy of volume change and for linear relationship between stresses and strains one obtains the general condition of Huber-Mises-Hencky. The Botkin condition can be obtained without neglect of volume changes and with a linear relationship between them and strains as well (as for deviators) for the case $U_0 = 0$, that is for the ideal granular medium in loose condition.

Equation 1.43 can be obtained from thermodynamical analysis of soil work. Soil is a plastic medium, the strain energy of which cannot be returned, that is it must be dissipated (Freidental & Geiringer 1962). It can be shown that Equation 1.43 holds true for a dissipative medium if irreversible volume strains and form strains occur, since the energy dissipation under compression at volume change is accompanied by reduction in volume, which is equivalent to increased strength. This is a feature of granular medium in which all deformation processes are accompanied by mutual displacements of grains and the friction opposing this motion is explained by Drucker (1962).

Analysis of Equation 1.43 shows that soil strength is determined by parameters of stress-strain relationships, loading trajectory and a single parameter of strength U_0, which reflects the loading prehistory. Inclusion of the loading trajectory requires an incremental relationship between stresses and strains, in particular Equation 1.37.

From the strength condition in Equation 1.43 one could obtain a new condition in a closed form if integration along the loading trajectory were undertaken. This is, however, impossible in practice because the loading path is given indirectly by the graph of dam construction, and the evolution of loading trajectory at any given point depends on the design of dam and boundary conditions. If one uses an inverse representation of strength utilized by Hencky and later Reiner, in the form of the 'vessel', then in the case of soil the 'vessel' changes volume from 0 (if $U_0=0$) to a certain value $\int_L \sigma\,de$ given at boundaries and by deformation properties of soil. It was already Terzaghi (1933) who pointed to the close relationship between strain parameters and soil strength.

Numerous tests with various soils were conducted to verify Equation 1.31. Along with them it was also limit surfaces on the deviatoric plane ($\mu=0$) which were found for different Lode parameters.

Experimental results illustrated in Figures 1.14–1.16 indicate concavity of the limit surface about $\lambda_d=+1$. The data shows that the limit quantity T_L decreases smoothly from $\lambda_\sigma=-1$ to $\lambda_\sigma=+1$, that is the strength decreases but the surface still remains convex. Due to the sixfold symmetry, the points $\lambda_\sigma=-1$ and $\lambda_\sigma=+1$ are singular, and the directions of the axes of principal stresses change their orientation at these points. Analogous data have been obtained for clay ($w_L=43.6$ and $\gamma_{dry}=17.8$ kN/m^3).

It should be recalled that the limit surface is the ultimate configuration of the loading surface in terms of the theory of hardening plastic bodies. It is obvious that variation of parameters of the loading path will bring about changes in the limit surface although its form will be preserved. There are no constant limit surfaces in the stress space, and the configuration of the surface determines the function of loading path.

It must be noted that the energetics condition holds for compression. For $\sigma > 0$ the character of the relationship between stresses and strains changes and a linear law seems to be valid.

The safety (reliability) factor for energetics condition reads

$$k_s = \frac{U_0 + \int_{L_1} \sigma\,de}{\int_{L_2} S_{ij}\,d\epsilon_{ij}} \tag{1.44}$$

The evolution of design methods for earth dams has given rise to combined strain conditions, but yet they are constructed only in the stress space, with inclusion of some elements of loading patterns (such as the Lode parameter).

A piecewise smooth limit surface has been proposed by Zaretskiy & Lombardo (1983) as follows

$$T = T^* \equiv c_{oct}^* - \sigma \tan \varphi_{oct}^*; \quad -1 \leq \lambda_\sigma < \lambda_\sigma^*$$

i.e. one satisfies the Botkin condition (the minus sign is due to the system of coordinates selected for negative compresssion; the asterisk denotes soil constants).

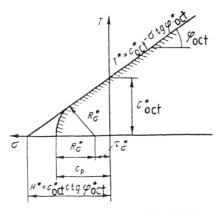

Figure 1.17. Limit surface after Zaretskiy & Lombardo (1983).

For $\lambda_\sigma^* < \lambda_\sigma \leq +1$ one has $\tau_{max} = \tau_{max}^* \equiv c_x^* - s_x \tan \varphi_x^*$, in which c_x^* and φ_x^* are the Hill strength parameters; and $s = \frac{1}{2}(\sigma_1 + \sigma_3)$. The slide plane after Hill is given by the conditions $m = 0$; $l = n = 1/\sqrt{2}$, in which m, n, l = directional cosines; that is the planes pass through the axis σ_2, and the line normal to the plane is inclined to the principal axis σ_3 at an angle $\psi = 45°$ (whereupon $l = n = \cos \psi$). The quantity λ_σ^* is a soil property. For $\lambda_\sigma > \lambda_\sigma^*$ the intermediate principal stress σ_2 does not control the strength parameters, with the exlusion of $\lambda_\sigma = +1$. In approximation one has $\lambda_\sigma^* \approx -0.8$.

The parameters of combined strength condition can be found if one determines the angle of internal friction and cohesion by Mohr (c_M^*, φ_M^*) for $\lambda_\sigma = -1$, that is for standard tests in a triaxial apparatus. One has

$$c_{oct}^* = \frac{2\sqrt{3} c_M^* \cos \varphi_M^*}{3 - \sin \varphi_M^*}$$

$$\tan \varphi_{oct}^* = \frac{2\sqrt{3} \sin \varphi_M^*}{3 - \sin \varphi_M^*}$$

(1.45)

Making use of the experimental relationship between the strength parameter by Mohr (for $\lambda_\sigma = +1$) and by Hill in conditions of planar strain one puts

$$\tan \varphi_M^* = \alpha_1 \tan \varphi_x^*$$

$$c_M^* = \alpha_2 c_x^*$$

(1.46)

From an analysis of experiments with coarse grained soil and sand one obtains $\alpha_1 \cong 1.1$. For cohesionless soil one has $c_M^* = c_x^* = 0$.

In the tension region $(\sigma > 0)$ the truncation of the Mises-Schleicher-Botkin cone takes place over a certain sphere. In the plane σ, T the projection of this surface is given as a circular arc (Fig.1.17) with the radius:

$$R_\sigma^* = (c_{oct}^* \cot \varphi_{oct}^* - c_t) \frac{\sin \varphi_{oct}^*}{1 - \sin \varphi_{oct}^*} \tag{1.47}$$

in which

c_t = tensile cohesion of soil.

The shift from the origin of coordinates is given by τ_σ^*. It is seen that the combined condition is based on the Mohr strength condition.

For first approximation of strength one should know the strength parameters φ_M^* and c_M^* for various soils; they are provided in Table 1.5.

Shear strength of coarse grained soil has been studied recently (since 1960) in connection with the construction of high and superhigh earth-rockfill dams of Nurek, Charvak, El-Infiernillo, Oroville, etc.

First of all one should single out the considerable effect of the density index I_D or the angle of internal friction φ_M^*, which might be 15° and more, and the effect of mean stress σ, the increase of which can bring about reduction of φN^* by 15–20°.

The effect of compaction of coarse grained soil may be estimated as follows (Rasskazov 1968):

$$\Phi_M^* = \theta + \alpha I_D^k \tag{1.48}$$

in which

Φ_M^* = shear angle for $\sigma_3 \to 0$,
τ = shear angle for $I_D \to 0$,
α, k = empirical factors (Table 1.6).

The maximum values of Φ_M^* usually do not exceed 53° for rock rubble and 51° for gravel and pebbles.

Approximate effects of the textural density of coarse grained soil can be found by the rule saying that the increase in soil porosity by 1% brings about reduction in Φ_M^* by 1°.

Decrease in Φ_M^* with increasing σ is caused by the crushing of grains under load. The compressive stresses at interfaces of particles are very high and are proportional to $p^{1/3}$, in which p = contact stress. If soil grains are represented as spheres with the diameter 100 mm, then the load distribution for $\sigma_n = 1$ MPa at the contacts brings about $\sigma_{max} = 250$ MPa. Practically under all real loads one encounters a certain number of broken grains. Increasing σ_n can cause massive crushing of grains. In this case $\varphi_M^* \to \varphi_{M\ min}^*$. Gravel and pebbles consist of rock particles which were rounded by the ambient environment. Individual grains do not have cracks. Grains of gravel and pebbles consist of igneous rock (20–30%) and sedimentary rock (70–80%). Sedimentary rock is primarily limestone, or sandstone in a certain part, so that the soil is gray. The high strength of grains in gravel and pebbles determines the limit of massive failure $[\sigma] \cong 2.0 \pm 0.5$ MPa for σ found by experiments.

The rock used for dam construction is obtained from quarries by explosions so that its edges are sharp and the strength of most soil particles is weaker due to microcracks. Therefore the stresses at incipient massive failure are below 1.2±0.5 MPa. However, for small stresses, $\sigma \leq 1$ MPa, the shear angle of rock is by 2–4° higher than that of pebbles. The difference is caused by the condition of grain surface. The strength of coarse grained soil may be assessed not through φ and c, due to curvilinearity of the envelope, but instead by the shear angle Φ_M^* given by Equation 1.35 under the condition $c=0$. Each value of Φ_M^* is compared with σ_1, σ_n or σ_3. Values Φ_M^* for El-Infiernillo Dam pebbles with $I_D \geq 0.9$ and andesite rock are given in Figure 1.18.

The variation of Φ_M^* with stresses may be approximated as indicated by Gordiyenko

a) for gravel and pebbles:

$$\Phi_{M\sigma}^* = \Phi_M^* - 5 \log \frac{\sigma_n}{\sigma_o} \tag{1.49}$$

b) for rock (rubble):

$$\Phi_{M\sigma}^* = \Phi_M^* - 9 \log \frac{\sigma_n}{\sigma_o} \tag{1.50}$$

in which

$\sigma_n = \frac{2\sigma_1\sigma_3}{(\sigma_1+\sigma_3)}$ = normal stress on shear plane,

$\sigma_0 = 0.2$ MPa.

Equations 1.49 and 1.50 can be used for $\sigma_n > 0.2$ MPa.

Gravel and pebbles are characterized by stable properties which is due to their grain size distribution and mineralogic composition. Gravelly pebbles in Mexico and Irkutsk, on the Island of Taiwan and in Middle Asia have close values of Φ_M^* and deformation properties if their I_D are identical. Some exceptions to this rule are however observed — for instance, for weathered terrace pebbles in which $[\sigma] \approx 0.5$–0.7 MPa and Equation 1.49 does not hold.

From Equation 1.50 it follows that the behaviour of rock for high σ_n is worse than that of gravelly-pebble soil.

Frozen soil occupies a particular position. Its strength depends on the strength of ice, but since it is deformed under loading and loses strength with time, one distinguishes instantaneous and long-term cohesion.

Depending on temperature, the instantaneous cohesion of clay ($w_L=30$–40%) varies from 0.6 MPa ($t=0°$) to 1.6 MPa ($T=$-4°). The respective long-term strength is 1.8 and 0.4 MPa, that is four times less. Silty sand ($w=23\%$) has instantaneous cohesion of 1.1 MPa ($t=0.9°$) and 2.0 MPa ($t=$-4°) with the long-term values of 0.21 and 0.45 MPa.

Soil strength under uniaxial tension (tensile cohesion c_t) is typical of clayey soil. It must be known for assessment of cracking strength in cores and screens (Section 4.5) and in the design of cores and screens subject to contact uplift (Section 2.3). These characteristics are the least explored properties of soil.

The strength of soil at liquid limit, in MPa, has been found experimentally (Istomina 1957) as:

Table 1.5. Strength factors by Mohr-Coulomb condition.

Soil	Porosity 0.41–0.5		0.51–0.6		0.61–0.7		0.71–0.8		0.81–0.95		0.96–1.1	
	φ_M^*	c_M^*	φ_M^*	c_M^*	φ_M^*	c_M^*	φ_M^*	c_M^*	φ_M^*	c_M^*	φ_M^*	c_M^*
Sand:												
grav&												
coarse	43	0.001	40	-	38	-	-	-	-	-	-	-
medium	40	0.002	38	0.001	35	-	-	-	-	-	-	-
fine	38	0.003	36	0.002	32	1	-	-	-	-	-	-
silty	36	0.004	34	0.003	30	2	-	-	-	-	-	-
Clayey; $w_P,\%$:												
9.5–12.4	25	0.006	24	0.004	23	3	-	-	-	-	-	-
12.5–15.4	24	0.021	23	0.01	22	7	21	3	-	-	-	-
15.5–18.4	-	-	22	0.025	21	12	20	9	19	5	18	4
18.5–22.4	-	-	-	-	20	34	19	17	18	14	17	9
22.5–26.4	-	-	-	-	-	-	18	41	17	20	16	18
26.5–30.4	-	-	-	-	-	-	-	-	16	47	15	23

(1) Sandy soil includes quartz of various roundness, not more than 20% of feldspar and not more than 5% of various admixtures (mica, glauconite etc.); (2) Clayey soil belongs to Quaternary residual deposits containing not more than 5% vegetative admixtures at full saturation of voids with water ($S_F > 0$); (3) Data in Table 1.5 does not hold for clayey soil in liquid state (with $I_L > 1$); (4) c_M^* in the last four columns is multiplied by 1000.

Table 1.6.

Parameters	Soil	
	gravelly pebbles	rock (rubble)
θ°	38–41°	40–45°
α°	7–10°	8–12°
k	0.8–1°	0.4–0.7°

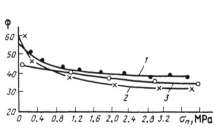 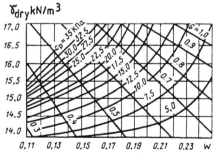

Figure 1.18. Shear strength as a function of normal stress on shear plane ($I_D \geq 0.9$). (1) El-Infiernillo pebbles (40% of grains with $d < 5$ mm); (2) andesite rubble (5–8% of grains with $d < 5$ mm); (3) limestone rubble (12% of grains with $d < 5$ mm) (left).
Figure 1.19. Graph for tensile cohesion c_t of clay loam (right).

$$c_{LL} = 0.14 \cdot 10^{-4} \frac{1}{d_{80}} \tag{1.51}$$

in which

d_{80} = grain diameter, in centimetres, not exceeded by 80% (by mass) of grains.

One must take into account that the maximum diameter of grains in clayey soil should be not more than 1 mm. If there are fractions $d > 1$ mm, the grain size distribution curve should be reconstructed, with identification of the fine fractions $d \leq 1$ mm, and subsequent determination of the respective diameter d_{80}. For other values of water contents the first approximation is provided by Figure 1.19 (Teytelbaum et al. 1975).

1.6 PRIMARY DYNAMIC PROPERTIES OF SOIL

The knowledge of soil characteristics under dynamic loading is a prerequisite in design of dams under seismic loading. As in the case of static loads, all characteristics are contained within the soil models accepted. The basic soil model available for the assessment of seismic effects is the elastic one, although it does not provide the residual deformation of the structure after passage of seismic waves.

In the Hookean model of linearly deforming medium one must know two soil constants E_y and ν_y. It is more convenient to use E_{oy} and G_y which are determined by celerity of propagation of elastic waves:

$$c_p = \sqrt{\frac{(1-\nu)E_y}{\rho(1+\nu)(1-2\nu)}} = \sqrt{\frac{E_{0y} + \frac{4}{3}G_y}{\rho}}$$

$$\tag{1.52}$$

$$c_s = \sqrt{\frac{E_y}{2\rho(1+\nu)}} = \sqrt{\frac{G_y}{\rho}}$$

in which

E_{oy} = modulus of elasticity for volume changes,
G_y = shear modulus of elasticity,
ρ = soil density,
c_p, c_s = celerity of longitudinal and transverse waves, respectively.

The above quantities for various soils are given in Table 1.7 (Krasnikov 1970).

The transition from E_{oy} and G_y to E_y and ν_y is given as follows

$$\nu_y = \frac{3E_{oy} - 2G_y}{6E_{oy} + 2G_y}; \quad E_y = \frac{9E_{oy}G_y}{3E_{oy} + G_y} \tag{1.53}$$

Residual dislocations in dam can be assessed by transition from elastic to deformation characteristics determined upon inclusion of plastic strain. The approximate assessment of residual dislocation in Equation 1.28 encompasses total strain, elastic and plastic.

A more accurate method has been elaborated on the basis of the associated flow rule (Zaretskiy & Lombardo 1983). It is most convenient for dynamic computations because it strictly identifies elastic and elastoplastic regions. This method is discussed in Section 1.4.

The stress-strain problems for earth dams under dynamic loading require the knowledge of damping coefficient related to the critical state (Section 1.4):

$$\xi = \frac{\delta}{\sqrt{(2\pi)^2 + \delta^2}} \qquad (1.54)$$

in which

$\delta = \ln \frac{A_n}{A_{n+1}} = 2\pi\xi\frac{\omega}{\omega_D} =$ logarithmic decrement,

A_n, A_{n+1} = neighbour amplitudes of attenuating oscillations in soil,

ω, ω_D = frequency of eigen oscillations without and with damping, respectively.

In clayey soil the damping factor may be assumed constant, which however varies with density and moisture content, from 0.1 to 0.25 (lower values for $\sigma \geq 0.4$ MPa and higher values for $\sigma \leq 0.1$ MPa) (Rasskazov et al. 1983). In sand under cubic compression for σ above 0.1 MPa one has $\xi = 0.18$–0.2, but for lower loads ξ depends appreciably on soil conditions (in relation to critical state).

The initial value $\xi \approx 0.18$ increases to unity for $\bar{k} \to 0$ (as by Equation 1.32), reaching the value of 0.6 (for $\sigma = 0.06$ MPa) and 0.9 (for $\sigma = 0.3$ MPa), for $\bar{k} \leq 0.8$. Hence if a granular soil approaches the limit state, the damping factor sharply increases at small mean stresses, i.e. near dam slopes.

1.7 SELECTION OF SOIL PROPERTIES FOR DAM BODY

Selection of soil properties for dam body is one of the most responsible stages of design controlling the reliability and economy. The design of dam profile crucially depends on selected characteristics of soil. Design of dam and soil characteristics are considered in Chapter 9. Some ranges of possible properties are discussed in this section. The primary property of soil is its density, plus moisture contents for cohesive soil. These factors control the strength and strain of soil—the higher the density of (compacted) soil the greater its strength and the smaller the permeability.

The best effect upon consolidation of clayey soil is reached if soil has the so-called optimum moisture content w_{opt}. In general, w_{opt} depends on type of soil, type of rolling, number of rollings, etc. In preliminary design one may assume $w_{opt} = w_P$ - 2–4%. Reduction of moisture content by 2% below the plastic limit gives fatter soil, while that reduced by 4% corresponds to leaner soil. Upon rolling by heavy machinery the optimum moisture content is less important than if lightweight machinery is used. In early design stages the parametrization by w_P is doubtful

Table 1.7.

Soil and rock	Unit weight kN/m^3	Celerity of elastic waves, km/s		Ratio c_s/c_p
		long.c_p	transv.c_s	
Loose unsaturated fill soil (sand, sandy loam, clay loam)	14–17	0.1–0.3	0.07–0.15	0.62–0.4
Gravel and sand	16–19	0.2–0.5	0.1–0.25	0.62–0.4
Sand:				
dry	14–17	0.15–0.9	0.13–0.5	0.62–0.55
medium moisture content	16–19	0.25–1.3	0.16–0.6	0.55–0.4
saturated	17–22	0.3–1.6	0.2–0.8	0.4–0.1
Sandy and clay loam	16–21	0.3–1.4	0.12–0.7	0.6–0.3
Clay:				
wet plastic	17–22	0.5–2.8	0.13–0.2	0.4–0.1
dense stiff& hard	19–26	2.0–3.5	1.1–2.0	0.62–0.4
Loess	13–16	0.38–0.4	0.13–0.14	0.35
Semi-rock and rock soil:				
marl 18–26	1.4–3.5	0.8–2.0	0.62–0.4	
sandstone:				
loose weathered:	17–20	1.0–3.0	0.6–1.8	0.55–0.3
dense	20–26	2.0–4.3	1.1–2.5	0.62–0.55
strong limestone	20–30	3.0–6.5	1.5–3.7	0.62–0.55
clayey shale	20–28	2.0–5.0	1.2–3.0	0.62–0.2
Igneous and metamorphic rock:				
cracked (gneiss, basalt, diabase, etc.)	24–30	3.0–5.0	1.7–3.0	0.58–0.48
uncracked (granite, gneiss, etc.)	27–33	4.0–6.5	2.7–4.3	0.65–0.58

so that one may use the relationship $w_P = 0.25w_L + 11$ (in percent, see Section 1.1), which is usually true for clay and clay loam.

Pore pressure (see Chapter 3) may be reduced if one tends to have possibly low moisture content upon dam filling. Sometimes it is recommended that the moisture content be lower by 1% than the optimum one.

The optimum moisture content can be determined in laboratory by the Proctor method, by which soil is tested for compactability by shock loading. Tests are performed in a cylinder having a diameter of 10 cm, in which 3–5 layers of soil, having a total weight of 21 N are arranged and consolidated with the falling weight. The mass of weight, number of strikes and the height of fall are so selected as to obtain the consolidation effect equivalent to one or another type of rolling. For instance, a 2.5-kg weight for 25 strikes on each of three layers, falling from 30 cm, corresponds to consolidation by pneumatic rolling with a machinery having 30–50 tons (medium rolling). The consolidation effect by heavy machinery is equivalent to 25 strikes of a 45-kg weight falling from 45 cm on each of five layers, that is 125

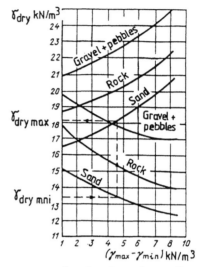

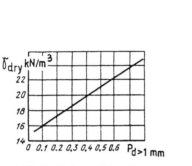

Figure 1.20. Unit dry weight of soil γ_{dry} versus percentage of grains above 1-mm diameter, after Istomina (left).

Figure 1.21. Nomograph for γ_{dry} in critically loose granular soil (right).

strikes in total. The consolidation is carried out for various moisture contents, and the one giving the maximum consolidation is considered the optimum.

Under prototype conditions the optimum moisture content is determined more accurately by soil tests employing industrial machinery.

The degree of consolidation is characterized by the textural density of dry soil. The permissible value of unit weight of dry soil γ_{dry} upon rolling can be determined as follows:

$$\gamma_{dry} = \frac{\gamma_p\gamma_w(1-V)}{\gamma_w + w\gamma_p} \tag{1.55}$$

in which

γ = unit weight of soil particles,
γ_w = unit weight of water,
V = volume of entrapped air, in fractions of unity; it is 0.03–0.05 for clay, and 0.07–0.08 for loess,
w = moisture content in fractions of unity.

For $w \geq w_{opt}$, the value determined by Equation 1.55 may be easily obtained under rolling. In the opposite situation the value determined by Equation 1.55 is not obtained without difficulty (heavy rolling and tamping to be employed) so that accurate γ_{dry} must be determined by test rolling.

It should be noted that Equation 1.55 describes the transition of soil into the condition in which γ_{dry} includes water in soil voids, so correction for the presence of air must be introduced.

If the moisture content of clayey soil is above optimum, determination of the necessary water content under soil rolling in dam body is a complex task. First of all, one must employ technological and economical factors to substantiate the reduction of moisture content. Principles are given in Chapter 9. If the moisture content is to be reduced to the optimum value, than one uses $w = w_{opt}$ in Equation 1.55; and in all other cases one must take for granted the original moisture content.

If the moisture content of soil in quarry is much below the optimum w_{opt}, then the technological and economic advantages of additional moistening are obvious as the increase in the cost of soil is insignificant.

In earth dam construction one often uses pebbles and druss or gravel and pebbles, with clayey or sandy fillers above 50%. Moraine soil and mountainous rock belong to this category, the so-called alluvial materials. If the aggregate is clayey, the unit weight of dry mixture can be found as follows:

$$\gamma_{dry,mix} = 0.96 \frac{\gamma_{dry,\ d<1mm}\gamma_{p,\ d<1mm}}{\gamma_{dry,\ d<1mm}P_{d>1mm} + \gamma_{p,\ d>1mm}P_{d<1mm}} \qquad (1.56)$$

in which

$\gamma_{p,\ d>1mm}$ = unit weight of coarse particles ($d > 1$ mm),

$P_{d>1\ mm}$ = percentage of coarse particles above 1 mm,

$P_{d<1mm}$ = percentage of fine particles below 1 mm.

Equation 1.56 is derived on the assumption that coarse fractions do not create skeleton, that is they are not in mutual contact (nonetheless such soil is referred to as skeletal). The textural density is determined for the moisture content of the coarse fractions taken as 2–3%.

Figure 1.20 shows the curve determined from field investigations on constructed dams; it gives γ_{dry} for known contents of the fraction $d > 1$ mm.

In determination of unit weight of dry coarse grained soil one should be guided not by the absolute textural density but rather by the density index (Eq.1.12), usually assumed as $I_D \geq 0.9$. Moreover, like for clayey soil, the density should be substantiated by consideration of technological and ecomonic factors (Chapter 9)— it might be purposeful to use soil with $I_D = 0.8$ or even 0.7 (see Chapter 8). These problems are solved upon consideration of specific construction factors but usually one assumes $I_D > 0.9$, which reduces the volume of dam. It must be remembered that the index reaches 0.7–0.75 if soil is dumped in dam from trucks.

In order to find ϵ and γ_{dry} by Equation 1.12 one must know I_D and find ϵ_{max} and ϵ_{min} for a given soil. The approximate method proposed by Maslov may be used.

The unit weight of dry soil corresponding to the limit loose state of material can be found from the condition:

$$\gamma_{dry\ min} = \frac{A}{(\Delta P\sqrt{K})^{0.05}} \frac{\gamma_p}{26.5} \qquad (1.57)$$

in which

A = empirical coefficient equal 17.5 kN/m^3 for sand, 18.6 kN/m^3 for rock, and 22.8 kN/m^3 for gravel and pebbles,

ΔP = percentage of reliably defined fractions in grain size distribution.

If the granulometric distribution is found accurately, one assumes $\Delta P = 100$. Difficulties arise in analysis of the contents of fine and coarse fractions. If the fractions d_{95} and d_5 are known one has $\Delta P = 95 - 5 = 90$ etc, in which K = coefficient of grain composition

$$K = \frac{D_2 - D_1}{\Delta P \log \frac{D_2}{D_1}} \sum_{i=1}^{i=n} \frac{\Delta q_i}{d_{2i} - d_{1i}} \log \frac{d_{2i}}{d_{1i}} \tag{1.58}$$

in which

$\quad\quad$ D_2 and D_1 = maximum and minimum reliable fraction, respectively,
$\quad\quad$ Δq_i = percentage of i-th segment of grain size curve,
$\quad\quad$ d_{2i} and d_{1i} = maximum and minimum diameter of i-th segment of grain size curve, respectively.

It is assumed that the curve is split up into n segments.

The unit weight of dry soil corresponding to the ultimate compacted condition $\gamma_{dry, \, max}$ can be determined from the nomograph given in Figure 1.21.

Equation 1.12 requires the following coefficients

$$\epsilon_{max} = \frac{\gamma_p - \gamma_{dry, \, min}}{\gamma_{dry, \, min}}; \quad \epsilon_{min} = \frac{\gamma_p - \gamma_{dry, \, max}}{\gamma_{dry, \, max}} \tag{1.59}$$

From Equation 1.12 for a certain design value I_D one determines ϵ and afterwards γ_{dry} for the dam body which must be achieved upon rolling.

The unit weight of dry sand in hydraulic fill of dams can be determined by the method proposed by Maslov for $I_D = 0.3$–0.33. A simpler estimate of γ_{dry} is given by Antipkin

$$\gamma_{dry} = 1.82 \frac{\gamma_{dry, \, max} \gamma_{dry, \, min}}{\gamma_{dry, \, max} + \gamma_{dry, \, min}} + 1.1 \tag{1.60}$$

Once the density in dam body is known one finds k, by Equation 1.17 for clayey soil and by Equation 1.15 for cohesionless soil. Additional consolidation of soil due to overlying layers should be taken into account (see Section 1.3).

Deformations should be determined by one of the methods discussed in Chapter 4, and variation of k over the height of dam is also to be accounted for (Chapter 2). In preliminary design the strength characteristics can be found from Table 1.5, satisfying the Mohr-Coulomb condition for fine soil, or by Table 1.6 and Equations 1.48, 1.49 or 1.50 for coarse (coarse grained) soil. Transition to other strength conditions is secured by φ_M and c_M known or is found in special experiments.

Deformability of soil can be assessed in approximation by reference to Equation 1.7. Special experiments are required for more accurate design.

It must be remembered that gravelly pebbles have stable permeability, strength and deformation, so that the computations by the above methods are sufficiently accurate.

CHAPTER 2

Seepage in earth dams

2.1 FUNDAMENTALS OF SEEPAGE THEORY

Soviet construction standards in force (SNiP 2.04.05-84) require that seepage characteristics be determined in earth dams and their foundations, i.e. the following activities are stipulated:

a) construction of a hydrodynamic grid (flow net) for groundwater flow, including depression surface as the upper boundary of the flow in the dam body;

b) determination of seepage parameters such as groundwater velocity and gradients;

c) computation of flow rates within the dam body or through the foundation and banks (if necessary).

Construction of depression surfaces and determination of seepage parameters are necessary for proper design of dam, particularly its tailwater part. They are also required for computation of seepage forces in dam, intervening in static computations of slope stability, distribution of soil in dam, seepage strength etc. Knowledge of flow rates in the dam and its foundation is required for computation of water losses and design of special drainage facilities (for instance, hydraulic computations of tubular drainage etc).

Seepage investigations can be theoretical and experimental.

Distinguished in the theory of seepage is the motion of the so-called gravity water, which is driven by the force of gravity. It is the hydrodynamical pressure only that acts in this case. The gravity water moving in a porous soil medium is referred to as groundwater, which also includes capillary water.

Seepage in dam is an open channel-type flow as there is a free surface of the flowing water at which pressure is atmospheric. The upper boundary of groundwater flow is given by the depression surface (or depression curve in plane problems). The water flowing at each point is characterized by hydrodynamic pressure (or head) and velocity.

Groundwater flow can be steady and unsteady. Steady flow is encountered in most seepage problems. In some particular cases, for instance fast moving upstream water, unsteady seepage is dealt with, occurring with varying depression surface.

Incompressibility of groundwater and non-deformability of soil skeleton are taken as assumptions in the theory of seepage.

Laminar (irrotational) flow is assumed in such soils as clay, clay loam, sandy loam, sand and rock with fine cracks. This flow occurs at low velocities, so the inertia forces can be neglected. Flow of groundwater in large size rubble fill and in rock with large cracks is turbulent.

Unlike the motion of water in open channels, the flow of groundwater involves two phases — liquid (water) and solid (soil skeleton). The velocity of water and its flow rate are controlled by permeability of soil, and seepage forces acting on soil skeleton are accompanied by the force of buoyancy (uplift).

The physical properties of soil considered in Section 1.1 have included porosity n. It is customary to assume that the coefficient of area porosity α, characterizing the clearances in the planar cross-section of soil, is equal to void ratio ϵ (for volume porosity).

Moving groundwater has velocity vectors at every point. This velocity varies as to the direction and magnitude. In a small unit plane orthogonal to streamline one may identify a certain average velocity $\bar{v}$ oriented along this streamline and equal to the discharge passing through the voids. This velocity is the real velocity of groundwater, referred to as seepage velocity. On the other hand, if the discharge is related not to the area of voids but instead to the entire area of a given cross-section then one obtains the so called discharge velocity v of a certain fictitious continuous body of water spreading over both voids and soil skeleton. This velocity is in fact dealt with in the theory of seepage, in terms of continuum models. Obviously, the relationship between the seepage velocity and the discharge velocity is controlled by porosity n: $\bar{v} = v/n$.

The Bernoulli equation holds true for steady flow of water in soil:

$$\frac{p}{\rho g} + z + \frac{v^2}{2g} = const \tag{2.1}$$

in which
> $p=$ pressure in water,
> $v=$ velocity of groundwater,
> $g=$ acceleration due to gravity,
> $\rho=$ density of fluid,
> $z=$ geometrical altitude above datum.

It is known that the first term in the Bernoulli equation denotes the piezometric height and the last one is the velocity head. The sum of the first two terms is referred to as piezometric head or simply head:

$$H = \frac{p}{\rho g} + z \quad \text{or} \quad H = \frac{p}{\rho_w g} + z \tag{2.2}$$

in which
> $\rho_w =$ density of water; $\rho_w g = \gamma_w$.

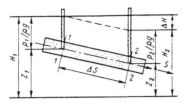 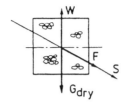

Figure 2.1. Elementary stream tube along line S. Dashed line depicts piezometric head from 1 to 2 (left).
Figure 2.2. Forces acting on unit soil volume in groundwater flow (right).

The last term is usually neglected because of small velocities.

The motion of water in an elementary stream tube (Fig.2.1) involves a hydraulic gradient between points 1 and 2 spaced by the distance ΔS, in which $I = \Delta H / \Delta S$ for $\Delta H = H_1 - H_2$ (heads at points 1 and 2). Hence the seepage discharge can be characterized by the gradient $I = -dH/dS$. In many types of soil the velocity of groundwater is linearly related to hydraulic gradient through Darcy's law, cf. Equation 1.14.

The factor of proportionality between v and I in Darcy's formula is referred to as the coefficient of permeability, having the dimension of velocity. The coefficient of permeability is the velocity of seepage per unit hydraulic gradient. It is the primary physical parameter characterizing the seepage peoperties of soil. In dense clay one has the so-called incipient hydraulic gradient I_0, whereupon it is assumed that seepage through soil begins only at gradients higher than the incipient gradient. In this case Darcy's law reads

$$v = k(I - I_0) \tag{2.3}$$

Many specialists object the existence of the incipient hydraulic gradient and assume that a nonlinear relationship intervenes between v and I in the range of low gradients, so that Darcy's law is not obeyed there. In our theory of seepage applied to clayey soil we are using the linear law derived by Darcy, and follow the concept of incipient hydraulic gradient. The linear law by Darcy has also limitations at high velocities in large voids. In computations for uniform soil (sand, gravel) Pavlovskiy considers applicability of Darcy's law for Reynolds numbers $Rc = \frac{v d_{10}}{\nu}$ below 10. For coarse soil at high speed the relationship between v and I is nonlinear, for instance $I = av + bv^2$ or $v = c \cdot I^n$. The parameters a, b, c and n are found experimentally. If one neglects the linear term compared with $b \cdot v^2$, then the turbulent seepage will be described by Equation 1.13 with head losses proportional to quadratic velocity:

$$v = k\sqrt{I} \tag{2.4}$$

In a uniform (homogeneous) soil medium the seepage properties are identical at all points, and permeability is also identical everywhere. In inhomogeneous soil permeability depends on coordinates. In isotropic soil, seepage properties do not depend on the direction of seepage, while permeability in anisotropic soil depends on the orientation of flow. Permeabilities of clayey soil can vary in time as the result of aggregate structure of soil and chemical suffosion. In general, since permeability depends on porosity, this in turn being computed from the stress condition of soil, one must tackle joint problems of seepage and stress-strain conditions in dams.

Soil porosity or void ratio intervene in Equations 1.15 and 1.18; they can depend on stress-strain relationships, particularly in very high dams. Using the solution for stress-strain conditions (see Chapter 4) one may provide more accurate parameters of groundwater flow.

Analysis of hydraulic action of seepage on soil skeleton is necessary for identification of forces and the depression curve.

A moving liquid exerts hydraulic effects on soil skeleton, bringing about a resulting force composed of the elementary force of water pressure on soil grains and the elementary force of friction, which can be determined by hydraulic gradient and viscocity of water (Chugayev 1967). It is more convenient to determine the resulting force as a sum of a vertical lift force acting on soil skeleton and the so-called seepage force f oriented tangentially to streamline:

$$f = \rho_w g I \tag{2.5}$$

The resistance force equal to f is transferred from skeleton onto the moving water. Hence the soil skeleton in a unit volume of seepage is subject to three volume forces (Fig.2.2): weight G_{dry}, bouyancy force W and seepage force f.

Soil in a seepage region is submerged (bouyant), and its weigth reads

$$G_{sub} = G_{dry} - W = G_{dry} - (1 - n)\gamma_w \tag{2.6}$$

from which it is seen that the bouyancy force is solely applied to soil skeleton, which occupies the volume $1-n=m$.

Euler's equations of steady flow for heavy incompressible liquid (upon neglect of inertia forces) can be written down as follows (Polubarinova-Kochina 1975):

$$
\begin{aligned}
\frac{1}{\rho}\frac{\partial p}{\partial x} + f_x &= 0 \\
\frac{1}{\rho}\frac{\partial p}{\partial y} + f_y &= 0 \\
\frac{1}{\rho}\frac{\partial p}{\partial z} + f_z + g &= 0
\end{aligned}
\tag{2.7}
$$

in which

f_x, f_y, f_z = projections of resistance forces on coordinate axes (z-axis being vertical).

With inclusion of the relationship between velocity of liquid and hydraulic gradient, in the form of Darcy's law, together with the relationships between head and pressure in water (Eq.2.2), and the one between head and seepage force in Equation 2.5, the equation of motion (attributed to Zhukovskiy) can be given as follows:

$$v_x = -k_x \frac{\partial H}{\partial x}; \quad v_y = -k_y \frac{\partial H}{\partial y}; \quad v_z = -k_z \frac{\partial H}{\partial z} \tag{2.8}$$

Equations of motion are supplemented by the continuity equation, which is derived from the budget of inflow and outflow of incompressible liquid (for details see Chapter 3):

$$\frac{\partial}{\partial x}\left(k_x \frac{\partial H}{\partial x}\right) + \frac{\partial}{\partial y}\left(k_y \frac{\partial H}{\partial y}\right) + \frac{\partial}{\partial z}\left(k_z \frac{\partial H}{\partial z}\right) = 0 \tag{2.9}$$

For isotropic and homogeneous soil ($k_x = k_y = k_z = $ const) the continuity equation with inclusion of Equation 2.8 reads

$$\frac{\partial v_x}{\partial x} + \frac{\partial v_y}{\partial y} + \frac{\partial v_z}{\partial z} = 0 \tag{2.9a}$$

The three equations of motion (Eq.2.8) and continuity equation (Eq.2.9a) contain the unknown functions v_x, v_y, v_2 and H. They are principal differential equations of steady groundwater flow. Hence the seepage theory is actually a set of equations of mathematical physics. This set may be solved if boundary conditions are given about the borders of the seepage area. The hydraulic gradient satisfies the following Laplace equation if soil is isotropic and continuously homogeneous:

$$\frac{\partial^2 H}{\partial x^2} + \frac{\partial^2 H}{\partial y^2} + \frac{\partial^2 H}{\partial z^2} = 0 \text{ or } \Delta H = 0 \tag{2.10}$$

From Equation 2.8 it follows that discharge velocity has a potential which is referred to as velocity potential $\varphi = -kH$. This notion is required in solution of seepage problems by the method of conformal mapping. The following equation also holds:

$$\Delta \varphi = 0 \tag{2.10a}$$

The function satisfying the Laplace equation is called harmonic. Integration of the Laplace equation for respective boundary conditions yields H at any point of seepage region. The surfaces $H(x, y, z) = $ const are loci of equal head referred to as equipotential surfaces. The condition $\varphi = const$ is satisfied on those surfaces.

For plane problem, most typical of seepage computations, for example in consideration of the cross-section of an earth dam, Equations 2.8 and 2.9a read:

$$v_x = -k \frac{\partial H}{\partial x}; \quad v_y = -k \frac{\partial H}{\partial y} \tag{2.11}$$

$$\frac{\partial v_x}{\partial x} + \frac{\partial v_y}{\partial y} = 0 \tag{2.12}$$

From continuity equation (Eq.2.12) it also follows that there is a certain function $\psi(x, y)$, for which one has $v_x = \frac{\partial \psi}{\partial y}; v_y = \frac{\partial \psi}{\partial x}$. This function is called stream function, and is related to φ by the Cauchy -Riemann equations:

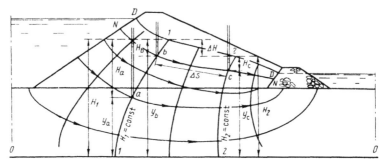

Figure 2.3. Groundwater flow in earth dam. (DD) depression curve; (NN) streamline; (1-1) (2-2) potential isolines.

$$\frac{\partial \varphi}{\partial x} = \frac{\partial \psi}{\partial y}; \quad \frac{\partial \varphi}{\partial y} = -\frac{\partial \psi}{\partial x}$$

The stream function also satisfies the Laplace equation and is conjugated with φ; it is another harmonic function.

The function $W = \varphi + i\psi$ is a function of complex variable $z = x + iy$ and is called complex potential. Solution of the equation $\Delta \Psi = 0$ with respective boundary conditions gives a system of streamlines. The potential isolines $\varphi =$ const and streamlines $\psi =$ const are mutually orthogonal (by virtue of the equation $\frac{\partial \varphi}{\partial x}\frac{\partial \psi}{\partial x} + \frac{\partial \varphi}{\partial y}\frac{\partial \psi}{\partial y} = 0$) and generate the so-called flow net (or hydrodynamical grid).

The flow net in an earth dam is depicted in Figure 2.3. A better explanation is provided by analysis of the heads at points a, b and c, at which piezometers are located. The line D-D denotes the depression curve; H_a, H_b and H_c are piezometric heads at the marked points, and y_a, y_b y_c are coordinates of these points. It is seen that H_a and H_b are identical, since the points are on the same equipotential line. Hydraulic gradients are determined by the ratio of head increments to the distance between respective points. For instance, for the points b and c on the line N-N one has $I = \frac{(H_1 - H_2)}{\Delta S} = \frac{\Delta H}{\Delta S}$. The flow velocity along the segment b-c is given by the formula $v = k \cdot I = k\Delta H/\Delta S$. The discharge between the two streamlines is q $= \psi_1(x, y) - \psi_2(x, y) = c_1 - c_2$.

As mentioned above, while solving differential equations one must know boundary conditions for $H, \varphi,$ and ψ. Let us illustrate this task in Figure 2.4a. The earth dam is founded on many layers. Two layers 1 and 2, having different seepage properties are underlain by the water confining stratum 3, which is practically impermeable. The seepage area is bounded by the lines MM_1ABCNN_1 and LL_1, for which boundary conditions must be found. along the following lines:

1. The permeable segments, in particular on upstream and tailwater sides, given by the segments MM_1 and NN_1 have the hydrostatic pressure distribution, so that $H =$ const or $\psi =$ const (if the segment LL_1 coincides with the x-axis then one has H_{MM_1} corresponding to coordinate of upstream side, and H_{NN_1} corresponds to the tailwater coordinate).

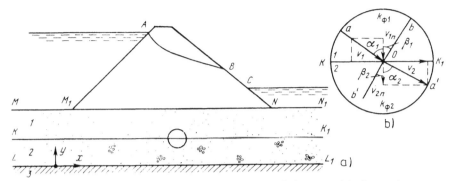

Figure 2.4. Seepage in dam underlain by multiple soil layers. (a) depression curve in homogeneous dam on double-layer foundation; (b) detail of interface at KK_1.

2. Impermeable segments such as LL_1 at the interface of the layers 1 and 3, of which 3 is the water-confining stratum, convey flow along the water-confining stratum, so that the line LL_1 is a streamline on which one has ψ=const.

3. Pressure is atmospheric and head has coordinates of the curve, $H = y$ ($\varphi = -k \cdot y$) on the depression curve (segment AB), being a surface of groundwater flow. If the infiltration from dam surface and evaporation are neglected then one may assume that the depression curve is a streamline so that one has ψ=const along the segment AB. One should note that location of depression curve is unknown a priori and becomes one of the objectives of the computations.

4. At exits, the pressure is atmospheric, so H and φ vary ($H = y$ and $\varphi = k \cdot y$) along the outer segments (segment BC on dam slope).

5. At the interface of layers with different seepage properties ($k_1 \neq k_2$) one has different directions of flow. In Figure 2.4b the rays OA and OA' correspond to the directions of streamline tangents at point O on the side of layers 1 and 2. Assuming that pressure should be continuous at the interface one obtains $\varphi_1/k_1 = \varphi_2/k_2$ or $H_1/k_1 = H_2/k_2$.

From the orthogonality condition of velocity at interface the velocity component should be continuous: $v_{1n}=v_{2n}$, from which one has $\psi_1=\psi_2$ for the so-called seepage law: $\tan \alpha_1/\tan \alpha_2 = k_1/k_2$. In view of the orthogonality of equipotential lines and streamlines one also has $\tan \beta_1/\tan \beta_2 = k_1/k_2$.

Seepage problems become much more complex for anisotropic soil. Such problems often arise if a dam is constructed in dry, layer by layer, and rolling is applied to non-uniform soil (as to grain size). The filling of this kind enhances the differences in the shape of grains which can be elliptical, scally, plate-type, etc. The seepage heterogeneity of soil due to its layered structure should be reduced to a certain seepage anisotropy, characterized by different permeability properties in the vertical and horizontal directions. In a number of cases the ratio of permeabilities in both directions is 10. If the axes of anisotropy coincide with the coordinate

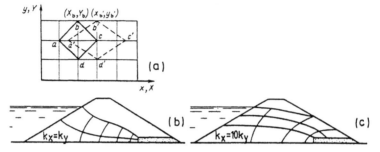

Figure 2.5.   Seepage in anisotropic medium.   (a) transformation of flow net cell for anisotropic medium; (b) flow nets in dam with flat drainage, for isotropic permeability; (c) as (b), for anisotropic permeability.

axes, the equations of motion are

$$v_x = -k_x \frac{\partial H}{\partial x}; \quad v_y = -k_y \frac{\partial H}{\partial y} \tag{2.13}$$

Continuity equations remain unchanged, in the form of Equation 2.12. Equations 2.12 and 2.13 yield the following heads:

$$k_x \frac{\partial^2 H}{\partial x^2} + k_y \frac{\partial^2 H}{\partial y^2} = 0 \tag{2.14}$$

For the transformation of coordinates suggested by Polubarinova-Kochina (1977) one has

$$X = \frac{k_y}{\sqrt{k_x k_y}}; \quad Y = y \tag{2.15}$$

so that Equation 2.14 becomes the Laplace equation in coordinates X, Y:

$$\frac{\partial^2 H}{\partial X^2} + \frac{\partial^2 H}{\partial Y^2} = 0$$

Hence one considers a certain fictitious isotropic region in coordinates x, y with the permeability $k = \sqrt{k_x k_y}$. Flow of groundwater occurs with a certain velocity having the components $v_x = -k\frac{\partial H}{\partial X}$ and $v_y = -k\frac{\partial H}{\partial Y}$. For the head one has the equation $H(X, Y) = \frac{p}{\gamma_0} + Y$. The problem is solved in the fictitious area X, Y and afterwards one transforms it into the x, y problem using the relationship in Equation 2.15. Introduction of the coordinates X, Y means a certain stretching or compression of the area of motion along the x-axis (depending on the ratio of the coefficients of permeability). This means that the flow net obtained must be transformed accordingly upon transition from X, Y to x, y (Fig.2.5a). In more complex cases, when the anisotropy axes do not coincide with coordinate axes the task is more sophisticated, and revolution of the anisotropy axes must be taken into account in the transformation.

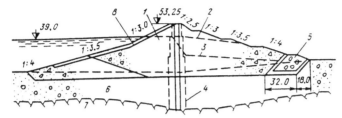

Figure 2.6. Ortotokoy Dam. (1) dam body soil; alluvial fan sediment with 24.5% of the fraction $d < 2$ mm; (2) depression curve in dam prior to construction of clay-cement curtain; (3) as (2), after construction; (4) curtain; (5) drainage prism; (6) alluvium; (7) bedrock; (8) upstream stone revetment.

Figure 2.5b, c illustrates the flow nets for a homogeneous dam having isotropic and anisotroipic soils with tailwater drainage. It is seen that anisotropy brings about considerable change in the depression curve and flow exit on the tailwater side. The role of flat drainage becomes dramatically less important, and the design of a dam must be changed.

A similar situation arose during construction of Ortotokoy Dam on the Chu River; seepage exit on the tailwater side was particular at the tailwater level, so that the stability of the slope was endangered. The seepage conditions had to be changed by construction of a counterseepage diaphragm (core), implemented by grouting of clay and cement, so that the depression curve was lowered on the tailwater side (Fig.2.6).

Nonlinear seepage problems arise if the motion of groundwater is more complex, for instance turbulent seepage in rubble fill, rock with large cracks, etc. A theory has been put forth by Khristianovich (1940). Nonlinear problems must be considered if free surface of water varies with time. However, they are not discussed in this book.

2.2 DETERMINATION OF SEEPAGE PARAMETERS IN EARTH DAMS

General. Seepage parameters in earth dams and foundations must be determined experimentally or analytically in solution of the following design problems:
1. Stability of dam slopes;
2. Stability of soil in terms of internal suffosion;
3. Stability of soil in view of seepage uplift;
4. Design of counterseepage components of dam and its foundation;
5. Design of drainage;
6. Selection of transition zones and inverse filters.

The first problem is solved in Chapter 6, and the remaining objectives are considered in this chapter.

As a rule, seepage computations are conducted for the dam design selected in

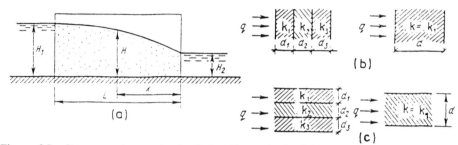

Figure 2.7. Seepage schemes in the hydraulic method. (a) rectangular prism on water-confining stratum; (b) transformation of vertical stratification into uniform layer; (c) transformation of horizontal stratification into uniform layer.

the preliminary stage of studies. This selection is based on analogy to similar situations. Dam design, overall dimensions, components, drainage, etc are postulated afterwards in a more rigorous manner.

Computational and experimental methods for seepage computations in earth dams are available at present. Computational methods may be classified into hydraulic and hydromechanical. Experimental methods include the method of electro-hydrodynamic analogy (EHDA). Physical Hele-Shaw models (slot type) are also possible. Complex problems often arise which require both computational and experimental methods.

One must first select a design scheme, including a variety of factors, accompanied by a design method, which permits solution of a given problem and provides necessary accuracy. Selection of the design scheme should be based on consideration of engineering, geologic, and hydrological factors about the dam, and should also account for seepage properties of soil. The design scheme should reflect the structural properties of the dam and its upstream and tailwater sides.

Selection of the method depends on the complexity of the problem. For a dam on a plain river one may employ a planar scheme, which is however unaccceptable for high dams in narrow canyons. Spatial seepage is a complex mathematical and physical problem. Therefore planar schemes are preferred for a number of vertical cross-sections x,y, coupled with horizontal sections x, z aimed at adequate bank transition (they are so-called plane seepage problems). It is obvious that the predicted seepage conditions will differ from reality.

Hence selection of a design scheme is closely associated with selection of an adequate method of solution. The design scheme may be simplified in terms of the geological structure of the facilities and foundations. Curvilinear boundaries are made rectilinear, and seepage properties are schematized and attributed to a certain mean coefficient of permeability.

The description of seepage computations in dams and foundations given below is merely an outline; details may be found in the cited references.

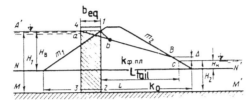

Figure 2.8. Scheme of seepage in homogeneous dam if upstream wedge is replaced by equivalent rectangular body and fixed exit length.

Hydraulic methods. They are most common in the practice of seepage computations. They provide the depression curve, seepage discharge, mean velocity and hydraulic gradient. Simple hydraulic methods are more justified if soil properties are determined with small accuracy, up to an order of magnitude. Hydraulic methods were formulated by Pavlovskiy (1956) and others. Description of the background cases has been provided by Chugayev (1967). Design is facilitated by handbooks (Nedrigin 1983).

The hydraulic methods include a number of simplifying assumptions (Chugayev 1967):

1. Seepage flow varies smoothly, and the curvilinear potential isolines (Fig.2.3) on the tailwater side of a homogeneous dam can be replaced by vertical lines. For flow between two vertical cross-sections one uses the Dupuit formula. For the simplest rectangular cross-section (Fig.2.7a) the depression curve is a descending curve, and the flow rate reads:

$$q = k\frac{H_1^2 - H_2^2}{2L} \tag{2.16}$$

while the thickness of seepage layer becomes

$$H_x = \sqrt{H_2^2 + (H_1^2 - H_2^2)x/L} \tag{2.17}$$

in which

L = distance between cross-sections considered.

2. In groundwater flow through layers of different permeability one must introduce certain fictitious (virtual) zones of seepage. For instance, a series of vertical layers having different thicknesses d_i and permeabilities k_i (Fig.2.7b) can be replaced by a single layer with the permeability for one of the layers (in this case $k = k_1$) having the thickness $d=d_1+\frac{k_1}{k_2}d_2+\frac{k_1}{k_3}d_3+...$, and at the same time the series of horizontal layers (Fig.2.7c) is replaced by one layer having the thickness $d=d_1+\frac{k_2}{k_1}d_2 + \frac{k_3}{k_1}d_3 + ...$ It should be noted that the method of virtual length may be applied if the ratio of permeabilities for adjacent layers is not greater than 10.

3. The upstream wedge of dam can be replaced by an equivalent rectangular massive (Fig.2.9). It is shown (Chugayev 1967) that in the case of a dam with

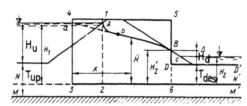

Figure 2.9. Homogeneous dam on permeable foundation.

identical coefficients of permeability of dam and its foundation, a sufficient accuracy is provided if the equivalent width b_{eq} can be taken as $0.4\,H_1$. For a steep upstream slope ($m_1 < 2$) one has

$$b_{eq} = \frac{m_1}{1 + m_1} H_1 \tag{2.18}$$

in which
m_1 = slope factor (cotangent) of the upstream side.

4. The exit segment Δ on the downstream side (for $k_{dam} = k_f$) may be approximated as follows:

$$\Delta = 1.2[A + \sqrt{A^2 + 0.4 g H_2}] \tag{2.19}$$

in which
$A = 0.5[q m_2 - (1 + 0.4/m_2)h_2]$,

$q = \dfrac{H_1^2 - H_2^2}{2(L_{tail} + 0.4 H_1)}$ = unit discharge,

m_2 = tailwater slope factor; the remaining notation being given in Figure 2.8.

If a dam is not situated on a water-confining stratum, the seepage conditions embody flow of water in the foundation. The design depth T_{des} for an infinite water-confining stratum (WCS) is taken as a half-width of dam foot. If WCS is $0.5\,L$ above dam foot, the real distance between WCS and dam foot is deemed the design depth. If the coefficients of permeability of dam soil and foundation soil, k_{dam} and k_f, are different, the preliminary design with T yields a fictitious thickness with k_{dam}, as recommended under item (2), and subsequently the design zone is selected within the fictitious zone.

The above considerations are utilized by Grishin (1979), Nedriga (1983), Pavlovskiy (1956), Rozanov (1983) and Chugayev (1967) where computations are provided for the most typical cases, yielding analogs for a wide class of problems. The most common seepage problems for homogeneous dams on permeable and impervious foundations, dam with cores and screens, variable permeabilities etc are dealt with. Formulae are also given for depression curves, groundwater flow rate, location of exit, etc.

For a homogeneous dam on permeable bed (Fig.2.9) the depression curve and

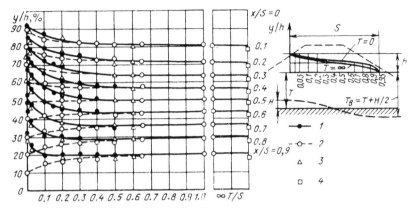

Figure 2.10. Graph for depression curve ordinates in homogeneous earth dams. (1) horizontal water-confining stratum (Numerov's data); (2) sloping water-confining stratum (Voshchinin's data); (3) EHDA (NIS Gidroproyekt's data) for horizontal water-confining stratum ($S=5H$); (4) theoretical solution by Nelson-Skornyakov ($T \rightarrow \infty$).

seepage discharge are found in the following order:

1. For $k_{dam} \neq k_f$ the foundation is reduced to the fictitious region with k_{dam} (line MM') and depth T_{up};

2. Design region with T_{des} (line NN') is taken in the fictitious zone T_{up};

3. Cross-section $1 - 2$ is arranged at point A of upstream water datum, the cross-section $3 - 4$ is constructed by $b_{eq}=0.4H_1$, and point a is found;

4. Elevation Δ of tailwater exit is found, together with B, and $H'_2 = T_{des} + H_H + \Delta$ is computed, with approximate H'_2 as a rough q;

5. Seepage discharge is found by Equation 2.16 for the design profile $N'345DD'$ (solid line), with H'_1 and H'_2, and the depression curve is constructed by Equation 2.17. More accurate versions are possible, with nonlinear equations, but the essence of the method remains identical. The accuracy of the solution is acceptable;

6. The depression curve is corrected by visual construction of Δb, which should be orthogonal to slope line at A. If a drainage prism or internal drainage is arranged on the tailwater side, the section 5–6 passes through the intersection of the internal boundary of drainage facility and tailwater level.

Hydromechanical methods. These methods of linear seepage are based on the Laplace equation. They are more complex than hydraulic methods, but do provide a flow net and seepage velocities at any point. Analytical solutions give insight into the effect of various factors (geometry, etc). Unfortunately, they deal with idealized cases of homogeneous dams (Aravin & Numerov 1955, Nelson-Skornyakov 1949).

The method of complex functions (Nelson-Skornyakov 1949, Pavlovskiy 1956) has been most widespread. Differential equations (Polubarinova-Kochina 1977), integral approaches, etc are also employed. Hydraulic and hydromechanical methods provide comparable results for the simplest cases, e.g. for homogeneous earth dams (Fig.2.10).

Numerical methods are effective, in particular finite differences (FDM) and finite-element method (FEM) for the Laplace equation. Difference equations for partial derivatives, e.g. elliptical are well elaborated (Vazov & Forsyth 1963, Tikhonov & Samarskiy 1966) and are employed in seepage problems in complicated, inhomogeneous, etc bodies.

Since the free surface is not known a priori, the numerical methods require more than one boundary condition, and the problem must be solved in two stages. The first stage consists in approximate solution giving a depression curve, and FDM is employed thereafter (cf. description in Section 3.3); and iterations follow.

FEM for seepage problems is dealt with by Vovkushevskiy & Zeiliger (1980) and Connor & Brebbia (1979). The method is described in Section 4.2 with reference to the stress-strain condition of dams. The Laplace equation (in general, Poisson's equation) may be solved, and then head distribution in dam and foundation is defined. Boundary conditions must be selected. They are of two types: head prescribed in one segment of the boundary and impermeability in the other part. The computational cross-section encompasses the dam and a part of its foundation. Moreover, potentials at the boundary should also be given. The method accounts for heterogeneity of dam and foundation, as to the coefficient of permeability, and simply resolves a complex geometry of the analysed area. The latter is split up into homogeneous subzones with constant permeability. The interface of subzones requires that the following condition be obeyed:

$$v^+ = v^-; \quad k^+ \left(\frac{\partial v}{\partial n}\right)^+ = k \left(\frac{\partial v}{\partial n}\right)^- \tag{2.20}$$

in which

'+' and '-' denote adjacent subzones.

FEM yields the function $H(x, y)$, for a plane problem, which minimizes the functional:

$$\Pi(H) = \frac{1}{2} \int_\Omega k \left[\left(\frac{\partial H}{\partial x}\right)^2 + \left(\frac{\partial H}{\partial y}\right)^2\right] d\omega - \int_{S_F} fv \, dS \tag{2.21}$$

in which

Ω = seepage area,
S_F = boundary of Ω,
$f = k\frac{\partial H}{\partial n}$ = given flux through S_F.

An example for hydraulic gradient in a dam with an asphalt-concrete diaphragm is illustrated in Figure 2.11. The problem is solved on a BESM-6 computer using TUOS algorithm (Vovkushevskiy & Zeiliger 1980); the Laplace equation is solved by piecewise-smooth interpolation of H within a finite element.

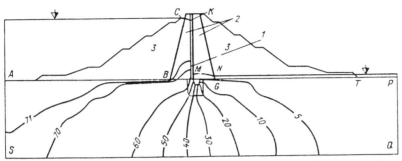

Figure 2.11. Seepage in dam with diaphragm. (1) asphalt-concrete diaphragm; (2) transition; (3) rubble fill (thrust prisms).

The dam and a rectangular element of foundation are included in the scheme. Compartments with constant seepage properties are singled out — foundation, grouting curtain, rubble prisms, saturated and dry zones of soil, and diaphragm, for which permeabilities are given. The depression curves CK and MN are found by simplified methods; more accurate data is given by iterations. A part of soil above the depression curves is in dry condition and is characterized by zero heads, which can be achieved by attributing zero to the coefficient of permeability in the above zones. Rubble fill it entirely permeable ($k \to \infty$), so it is excluded from computations. If it is not (e.g. FEM for both seepage and elasticity problems) one assumes $k = 0$ in it. Zero head field generates zero hydrodynamic forces, which is realistic. A head of H_0 (for upstream SWL) is assumed along ABC, and H_0 of tailwater datum is taken along GTP. The head equal to the ordinates of nodes is given along CK and MN. Impermeability is assumed along $ASQP$. Head isolines computed are depicted in Figure 2.11.

The seepage problem is more complex if permeability depends on stress condition; it then becomes conjugated with the stress-strain problem.

Method of electrohydrodynamic analogy (EHDA). It has been applied to seepage problems by Pavlovskiy (1956) who devised the experimental techniques and apparatus for problems of confined flow; it is now widespread (Filchakov 1960).

The analogy is taken between steady flow of groundwater and stationary electric current, which for plane problems reads as follows.

Flow in soil:

$$v_x = -k\frac{\partial H}{\partial x}; \quad v_y = -k\frac{\partial H}{\partial y} \tag{2.22}$$

while for electric current :

$$i_x = -\sigma\frac{\partial U}{\partial x}; \quad i_y = -\sigma\frac{\partial U}{\partial y}$$

in which

i_x, i_y = components of specific current,

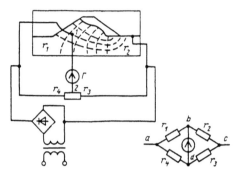

Figure 2.12. Circuit diagram of EHDA.

Groundwater flow	Electric current
Head H	Voltage U
Coefficient of permeability $k(x,y)$	Specific conductivity $\sigma(x,y)$
Flow rate $Q = \int_L v_S dL = - \int_L k \frac{\partial H}{\partial S} dL$	Current intensity $I = - \int_L \sigma \frac{\partial U}{\partial S} dL$

σ = specific conductivity,
U = voltage.

The continuity equation for groundwater flow is equivalent to $\frac{\partial i_x}{\partial x} + \frac{\partial i_y}{\partial y} = 0$ for electric current. The function U satisfies the Laplace equation, which provides the analogy. The following counterparts can be identified.

Since H and U satisfy the Laplace equation, solutions of these equations coincide, and for H = const one has U = const, while both partial derivatives (of H and U) with respect to n are zero.

For geometric similarity and identical boundary conditions for the head function (prototype) and electric potential (model) one will arrive at analogy of the groundwater flow in a natural porous medium and electric current in a conducting medium. Hence a model made of electroconducting material provides equipotential lines which correspond to isolines of heads. The boundary conditions in the field H = const are simulated by U=const, so that a certain required voltage is provided in the model. The condition $\frac{\partial H}{\partial n} = 0$ is complied with through provision of cuts in the model.

The advantage of EHDA consists in potential solution of complex geometric schemes with various specific conductivities, which represent different permeabilities in the field. Figure 2.12 shows the principle of EHDA. The region of earth dam and foundation is made of electroconducting paper. The device includes a power pack and measuring equipment. The model potential is measured with a Wheatstone bridge. If electric current is branched into abc and adc, there will be no current between the points b and d if the resistance ratio is $r_1/r_2 = r_3/r_4$

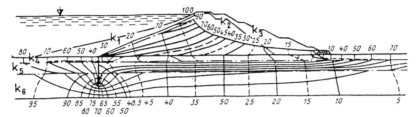

Figure 2.13. Flow net in inhomogeneous dam with foundation piling.

or if the voltage drop is proportional. The resistances r_3 and r_4 are shown on the potentiometers, while the resistances r_1 and r_2 are those in the model before and after the moving contact. By moving the selector 1 one finds the potential of point 2. In this case the voltage meter shows zero. If one takes full voltage for unity and divides the resistance into n equal parts, then one may find points with equal potentials U/n, corresponding to each step of the potentiometer; accordingly equipotential lines are generated. Streamlines are normal to equipotential lines. They can be drawn or found experimentally if conductors are identified with the boundary streamlines and paper is cut at boundary equipotentials.

The method can be used with success for plane seepage problems; if necessary, spatial problems may also be reduced to a series of plane problems. Shown in Figure 2.13 is the hydrodymamic grid (flow net) in an inhomogeneous dam on three soil layers (Filchakov 1960). The dam includes drainage and an impermeable pile which does not reach the water confining stratum.

EHDA is applicable to nonlinear problems upon the use of Khristianovich's method. In such cases it is convenient to use a facility which has a grid integrator instead of electroconducting paper; the former may consists of a discrete set of resistances. The flow net is constructed by selection on a number of models.

The first model (first approximation) is undertaken for geometric similarity with variable thickness $h_1 = ck$ in which k=permeability of soil for each zone anticipated in approximation; c=constant. Upon division of the dam body (recall it is inhomogeneous as to seepage) into subzones having permeabilities differing by a factor of 2, 3, 4, 5 etc (for instance 2, 3, 4, 5 sheets of paper are glued onto the model), the facility yields the flow net from which velocities v are computed by Darcy's law, so that they correspond to a fictitious flow (the real one does not satisfy Darcy's law). Assuming that the nonlinear law is in the form $I = \frac{1}{\kappa}v^\gamma$, the real velocities can be found by the formula $v_1 = v\frac{1}{\sqrt{\gamma}}$, and the function $L = \frac{1}{\sqrt{\gamma}}\frac{v^{\gamma-1}}{\kappa}$. may be computed. The obtained values of L provide a new model with variable thickness (variable electric conductivity) $h_2 = c/L$. Equipotential lines are again constructed by EHDA, and new v_2 are computed. The correction factors read

$$\frac{1}{l_{fic}} = \left(\frac{v}{v_2}\right)_{av}\frac{k_{av}}{k_{fic.av}}; \quad \frac{l'}{l'_{fic}} = \left(\frac{v}{v_2}\right)_{av} \tag{2.23}$$

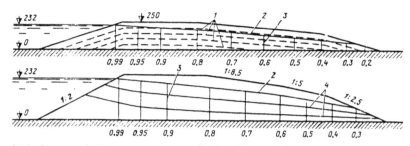

Figure 2.14. Seepage in inhomogeneous explosion-fill earth dam found by EHDA. (1) dam zones with different coefficients of permeability; (2) depression curve; (3) head isoline; (4) streamlines.

in which

l, l' = lengths of streamline and equipotential line for real flow, respectively,
l_{fic}, l'_{fic} = as above, for fictitious flow,
$k_{fic} = 1/L$ = permeability for fictitious flow,
$k_{fic.av}$ = mean value,
$k = \kappa v_2^{1-\gamma}$ = coefficient of permeability in real flow,
k_{av} = mean value.

Equation 2.23 shows that the distortion of streamlines and equipotential lines is different, so that the flow net is not orthogonal. If k_{av}/k_{fic} differs a lot from unity, the real flow is dissimilar to the EHDA one; a first approximation is reached. A new model with the distortion k_{av}/k_{fic} is to be constructed, and subsequent iterations follow. Two approximations are usually sufficient. In view of the iteration a net integrator is more practical. Figure 2.14 illustrates EHDA applied to a rock-fill dam 250 m high, with a head of 232 m.

Hele-Shaw method. This method is based on the flow of viscous liquid in a model soil structure contained in a narrow slot between two glass walls.

Twodimensional flow of viscous liquid between parallel walls is described by the following equations

$$\frac{d\,v_x}{d\,t} = -\frac{1}{\rho}\frac{\partial p}{\partial x} + \nu_\Delta v_x$$

$$\frac{d\,v_y}{d\,t} = -\frac{1}{\rho}\frac{\partial p}{\partial y} + \nu_\Delta v_y \qquad (2.24)$$

in which

ν = kinematic viscosity.

For slow motion one may neglect the inertia terms and determine the velocity components as follows

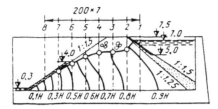

Figure 2.15. Large-scale model of rockfill dam with sand in voids.

$$v_x^* = -\frac{a^2 g}{3\nu}\frac{\partial H}{\partial x}; \quad v_y^* = -\frac{a^2 g}{3\nu}\frac{\partial H}{\partial y}$$

in which

a = slot thickness,
H = head.

If $\frac{a^2 g}{3\nu}$ is denoted by α, the so-called slot permeability factor, then Equation 2.24 reads

$$v_x^* = -\alpha\frac{\partial H}{\partial x}; \quad v_y^* = -\alpha\frac{\partial H}{\partial y} \tag{2.25}$$

The continuity equation for incompressible liquid is

$$\frac{\partial v_x^*}{\partial x} + \frac{\partial v_y^*}{\partial y} = 0 \tag{2.26}$$

Equations 2.25 and 2.26 are equivalent to Equations 2.11 and 2.12, hence there is an analogy between the motion of viscous liquid in a narrow slot and the groundwater flow in a porous medium. Laminar motion of viscous liquid is assumed, which is proved by Aravin to hold true below $Re=500$, and which is possible in a narrow slot up to 1 mm thick. Glycerine is used as the viscous liquid.

The Hele-Shaw method, employing a narrow slot, is practised in laboratory. In modelling flow in inhomogeneous soil (as to its seepage properties) one must vary the thickness of the slot to reproduce variable permeability of the model in various zones. Depression curve can be fixed on the model, along with the streamlines which may also be visualised by dyeing the liquid flowing in the model. Head isolines can be found as perpendicular to the streamlines. The origin of the potential isolines system is identified at the depression curve with $0.9H$; $0.8H$, etc.

In rare cases the groundwater flow is investigated on physical models in laboratory and on special prototype installations. High Assuan Dam was tested on a physical model; the dam consists of rock with voids filled with sand. The depression curve was identified visually through a transparent wall, and the hydraulic gradient was found with piezometers in the sections 1–8 (Fig.2.15). Physical model studies are tedious and time-consuming.

2.3 SEEPAGE STRENGTH OF SOIL

Types of seepage deformation. The water that penetrates the body and foundation of a dam brings about a mechanical action on the soil skeleton, which is characterized by seepage forces. The latter are comparable with the weight of soil. In earth dams, the seepage forces in macroscopic volumes affect the stress-strain condition of dam (which can be analysed in terms of the mechanics of continuous media), while the seepage forces in small-scale volumes act on particles of non-cohesive soil and aggregates of cohesive soil, which in turn gives rise to loss of stability and the movement of particles with the moving water. As a result, various types of seepage deformation of soil appear, depending on kind of soil (cohesive or cohesionless).

Soil suffosion is the movement of individual soil grains induced by seepage. Internal suffosion occurs when soil particles move inside osion is a dam body while external suffosion is referred to the motion of particles at the boundaries of the dam body. Suffosion is only possible in loose soil.

Contact seepage removal is the process of destruction of fine (clayey) soil at its contact with coarse material. It is caused by seepage perpendicular to the line of contact. A particular case is removal of sand at the exit of seepage water on the tailwater side or at dam slope in the exit zone.

Contact erosion is the destruction of fine soil (sandy or clayey) at its contact with coarse soil, caused by water moving parallel to the contact line.

Soil stripping is separation of cohesive soil aggregates at their contact with coarse soil, including the contact at an inverse filter.

Colmatation is the process of deposition of grains moved by percolating water; this deposition occurs in soil voids (internal colmatation) or on the surface of soil body (surface colmatation).

Chemical suffosion is the process of dissolution of salts contained therein (see Section 7.3).

Seepage deformations occur in micro- and macro-scale volumes, which brings about settlement of dam, along with sliding of slopes, piping, colmatation of inverse filters, etc. Reliable operation of an earth dam can be secured by proper selection (as to its seepage strength) of dam materials, the design and composition of filters, transition zones and drainage, all these measures aimed at preventing seepage deformation of soil.

Contact removal and contact erosion, along with soil stripping are different forms of soil suffosion which can be given a generic name of contact suffosion.

The following conditions should be observed in suffosion prevention:

1. Grain composition of soil should display the maximum number of interlocking particles (when particles cannot be removed by seepage forces).

2. The ratio of the diameter of fine particles to the clearance in voids between coarse particles should not allow fine particles to move out of the skeleton generated by coarse grains. It is obvious that coarse grains provide the supporting structure and fine particles move freely in voids; the evacuation is quite easy if fine

particles are smaller than the voids. Depending on the grain composition, suffosion can arise with the same probability as that of no failure. The ideal suffosionless soil consists of grains having the same diameter.

3. Seepage forces determined by hydraulic gradients should be limited. For high gradients, for instance if they appear at the exit of seepage water into drainage, considerable seepage forces are capable of separating and moving soil grains in the dam body and at the contact with drainage facilities; aggregates can also be detached from cracks in clayey soil if cracks appear.

The capacity of soil to resist seepage changes is referred to as the seepage strength of soil. One may distinguish the seepage strength of dam proper and the strength of its foundation. The design of a dam includes the task of selecting its structural type and material (transition zones, filters etc) that will secure the seepage strength. Transitions are a necessary element of an earth dam with water-confining constituents such as cores and screens, and make possible a reliable connection with dam body, without negative seepage changes. Inverse filters are required for connection of the dam with draining facilities and serve the same purposes, preventing at the same time the contact uplift and erosion.

In assessment of seepage strength of soil one should determine the suffosion strength, the so-called suffosivity, of soil designated as material for dam and its foundation. One should also estimate the critical local hydraulic gradients which can bring about suffosion, contact uplift and erosion. Important are also permissible hydraulic gradients and permissible percentage of possible removal of fine grains, acceptable without essential infringement on strength and stability of the structure. Such assessments are necessary for the subsequent selection of structural measures to provide seepage strength of soil.

The assessment of seepage strength in the Soviet Union is determined by the standards put forth at VNIIG and VNII VODGEO (*Guidelines* 1976 and *Guidelines* 1982).

Suffosion. The method has been proposed by VNIIG (*Guidelines* 1976) for assessment of suffosion by given soil parameters (Chapter 1); it provides the diameter of maximum seepage voids in soil by Pavchich's formula

$$d_{0max} = 0.455 \kappa \sqrt[6]{\eta} \frac{n}{1-n} d_{17} \tag{2.27}$$

in which

η = coefficient of uniformity ($\eta = \frac{d_{60}}{d_{10}}$),
d_{17} = diameter no exceeded by 17% of grains,
κ = coefficient of non-uniform distribution of particles in soil, $\kappa = 1 + 0.05\eta$.

The parameter d_{0max} determines the maximum dimension of particles which can be removed by groundwater flow as a result of suffosion:

$$d_{cimax} = 0.77 d_{0max} \tag{2.28}$$

If d_{cimax} is smaller than the minimum diameter of soil grains then the soil is not endangered by suffosion. If $d_{cimax} > d_{rmin}$ the soil is suffosive and all particles with diameters smaller than d_{cimax} can be removed from that soil. If it is accepted that no detriment is caused to the structure if the finest grains are removed, in the quantity not greater than 3–5% by weight, then such soil is practically non-suffosive. By this token the criterion of suffosivity reads

$$d_{cimax} < d_3 \div d_5 \qquad (2.29)$$

For removal of grains it is necessary that seepage rate or hydraulic gradient is higher than its respective critical value v_{cr} and I_{cr}. The critical hydraulic gradient for which particles of d_{ci}, smaller than d_{cimax}, can be removed is given by Patrashev's formula

$$I_{cr} = \varphi_0 d_{ci} \sqrt{\frac{ng}{\nu k}} \qquad (2.30)$$

in which

$\varphi_0 = 0.6(\frac{\gamma_p}{\gamma_w}-1)[0.82 - 1.8n + 0.0062(\eta - 5)] \sin(30° + \frac{\theta}{8})$,
ν = kinematic viscosity,
θ = angle between vector of flow velocity and force of gravity.

By analogy one has

$$v_{cr} = \varphi_0 d_{ci} \sqrt{\frac{ngk}{\nu}} \qquad (2.31)$$

The critical values of gradient and discharge are compared with the values observed at the structure determined by the above procedures.

By the method of VODGEO (*Guidelines* 1982), the following procedure is used to determine the value of suffosion failure gradient $I_{s.fail}$ for sandy and gravelly soils with the coefficient of uniformity $\eta > 10$, and the 10–30-% content of grains smaller than 1 mm. The grain composition of soil is taken conceptually as the skeleton, with diameters above 1 mm, filled by an aggregate of particles smaller than 1 mm, and for these fractions the grain size curves are constructed (Fig.2.16a), from which the diameters d_{10}^{sk} and d_{10}^{agg} are found. Depending on these diameters and on the angle of internal friction of the filler fraction φ_{agg} one finds the quantity $I_{s.fail}$ from the curves given in Figure 2.16b. Suffosivity of loose soil can be found from the graph provided in Figure 2.16c.

It should be noted that the VNIIG procedure gives quantities with a higher safety margin.

Contact uplift (contact suffosion). This property is determined separately for cohesionless and cohesive soils.

By VNIIG method (*Guidelines* 1976), the contact suffosion in fine loose soil at its contact with coarse soil will not take place for the grain size distributions

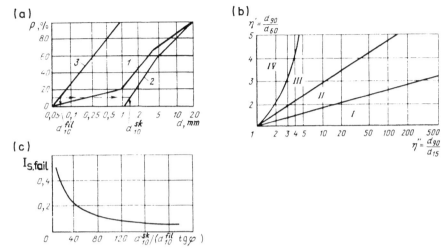

Figure 2.16. Soil suffosion assessment by VODGEO method. (a) grain size distribution of soil 1 split up into skeleton 2 and aggregate 3; (b) $I_{s.fail}$ vs. $\frac{d_{10}^{sk}}{d_{10}^{agg}} \tan \varphi_{agg}$ by Istomina's data; (c) graph for assessment of loose soil suffosion; (I) (III) zones of suffosive soil; (II) zone of non-suffosive soil; (IV) nonexistent soil.

$$\frac{D_0}{d_3} \leq 5.4 \tag{2.32}$$

in which

D_0 = mean diameter of seepage voids in the coarse soil

$$D_0 = 0.455 \sqrt[6]{\eta} \frac{n}{1-n} D_{17} \tag{2.33}$$

in which

d_3 = diameter not exceeded by 3% of soil.

The VODGEO method determines the contact suffosion of loose soil if fine soil satisfies the conditions of $\eta \leq 10$ and

$$d_{50} \geq \alpha_n D_{10} \tag{2.34}$$

in which

D_{10} and d_{50} = mean diameters of coarse soil (first filter layer) and fine soil (protected), respectively;

α_n = coefficient of transition from grain diameters to corresponding voids, usually taken as 0.155 (see below).

For $\eta > 10$ one should consider merely the conditional aggregate taken as above.

The bouyancy of unloaded soil, for instance in tailwater subject to seepage uplift by ascending flow, for $\eta \leq 10$, is determined by the exit gradient, the critical value of which is determined by Zamarin's formula

$$I_{crex} = \frac{(\rho_p - \rho_w)(1 - n)}{\rho_w} + 0.5n \qquad (2.35)$$

in which

ρ_p and ρ_w = density of soil and water, respectively.

For exit gradients $I_{ex} > I_{cr.ex}$ one should foresee the structural measures counteracting the buoyancy, for instance in the form of additional fill. If the groundwater flow exits on the tailwater slope, the stability of the latter is secured if the slope angle is taken as $m_H = 2/\tan \varphi$ (where φ = angle of internal friction).

The contact uplift of clayey soil is specific compared with loose soil, and depends much on its own condition. Two basic cases are considered:

1. Clayey soil as a continuous monolithic body;
2. Clayey soil with cracks; colmatation of filter with clayey particles is necessary for reliable seepage prevention.

Case (1) by VNIIG method. The seepage strength for cohesive soil with plasticity index $I_p \geq 0.05$ is secured if the maximum gradient of contact uplift does not exceed the acceptable one computed by Equation 2.36, for the assumed dimensions of counterseepage measures (upstream puddle blanket, screen, core)

$$I_{c.up.perm.} = \frac{1}{\varphi}[\frac{0.34}{D_{0max}^2} - 1] \qquad (2.36)$$

in which

D_{0max} = maximum seepage voids of soil underlying the counterseepage constituent (Eq.2.27),
φ = coefficient dependent on D_{0max}:

D_{0max}	0.1	0.2	0.3	0.4	0.5	0.55	0.583
φ	0.5	0.46	0.42	0.32	0.18	0.08	0.0

The VODGEO method determines the following acceptable gradient of contact uplift for clayey soil:

$$I_{c.up.perm.} = \frac{2c_t}{\gamma_w D_r k_s} \qquad (2.37)$$

in which

c_t = tensile cohesion of clayey soil,
k_s = safety factor taken because of the accuracy of determination of c_t, its value (3–5) depending on class of structure,
D_r = design diameter of dome created in clay under water due to stripping at the contact with filter.

Tensile cohesion c_t is determined experimentally, by Equation 1.51, or from Figure 1.19. The dome diameter is $D_r = D_{90}(1 + \alpha_n)$, in which α_n = coefficient depending on porosity n and the coefficient of uniformity η, determined from the

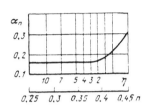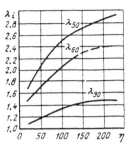

Figure 2.17. Coefficient α_n vs. porosity n and coefficient of uniformity η (left).
Figure 2.18. Coefficient of stratification λ_i vs. coefficient of uniformity η for sand and gravel soil (right).

graph in Figure 2.17 (Nedriga 1983). For consolidated soil of impoundment constituents of dams, the coefficient α_n can be taken as 0.155, and then $D_r = 1.155D_{90}$.

For sandy loam with plasticity index $I_p = 0.3$–0.5, the permissible gradient of contact head is determined as follows:

$$I_{c.up.per.} = \frac{1}{1.4(D_{90}^o)^2} \qquad (2.38)$$

in which
$\quad D_{90}^o = \alpha_n D_{90}$ and α_n is determined from the graph in Figure 2.17.

By introducing the tensile cohesion, the VODGEO method differentiates the properties of various clayey soils.

Case (2) by the VNIIG method. It provides the following condition for colmatation of filter with clayey soil of counterseepage constituent:

$$\frac{D_{17}}{d_{90}} \le \frac{26.5(1-n)}{n\sqrt[6]{\eta}} \qquad (2.39)$$

in which
$\quad D$ and d = diameters of coarse soil in the coarse grained soil of filter and clayey material, respectively,
$\quad n$ and η = parameters of filter soil.

The shortcomings of the method include the fact that dimensions of clayey soil particles are considered, and not the cohesive aggregates, although colmatation is determined by the latter, which is actually taken into account in the VODGEO method.

The VODGEO method incorporates the following condition for the colmatation of filter:

$$D_{60} < \frac{2d_{70}^a}{\alpha_n \lambda_{60}} \qquad (2.40)$$

in which

> λ_{60} = coefficient of stratification, depending on the coefficient of uniformity η and found from Figure 2.18,
> $d_{70}^a = \beta d_a'$ = design diameter of aggregate composition of eroded clayey soil,
> d_a' being taken as 0.028 mm for clay loam and 0.035 mm for clay.

The quantity β is being found as a factor depending on the moisture content at liquid limit w_L:

$w_L(\%)$	20	25	30	35	40	45	50
β	7.5	8.3	9.2	10.0	10.4	10.7	10.8

The seepage strength of clayey soil at its contact with loose soil is secured if $k \le 0.002$ cm/s.

Contact erosion. Contact erosion, much as contact suffosion, is considered separately for cohesive and cohesionless soils. The VNIIG method gives the following critical erosion gradient for fine cohesionless soil having diameters d_i greater than d_3 at its contact with coarse soil:

$$I_{cr.c.e.} = \frac{1}{\sqrt{\varphi_1}}(2.3 + 15\frac{d_i}{D_0})\frac{d_i}{D_0}\sin(30° + \frac{\theta}{8}) \tag{2.41}$$

in which

> D_0 is being found from Equation 2.33,
> θ = angle between flow velocity and the force of gravity,
> φ_1 = coefficient taken as 1 for sand-gravel-pebble soil and 0.35–0.4 for boulders.

Equation 2.41 holds true for $d_i/D_0 < 0.7$ and $Re \le 20$. The critical erosion velocity is $v_{cr.c.e} = kI_{cr.c.e}$.

The **VODGEO** method for sand and sandy gravel with d_{10}=0.1–0.57 mm and $\eta \le 10$, for gradients $I \le 1.3$ provides the mean diameter of filter grains which are not susceptible to erosion by the following formula

$$D_{50} = d_{50}(2.2 + 7.29\frac{H}{\eta}) \tag{2.42}$$

in which

> H and η = coefficients of uniformity for coarse soil (D) and fine soil (d); the ratio H/η being taken from 0.25 to 5.

The VNIIG method for cohesive soil with $I_p \ge 0.05$ stipulates that the acceptable gradient of contact erosion be as follows:

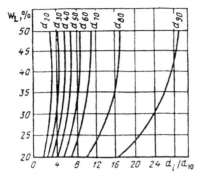

Figure 2.19. Dependence of d_i/d_{10} on liquid limit w_L upon determination of aggregate composition (Burenkova's data).

$$I_{c.e.perm.} = \frac{1}{\sqrt{D_{0max}}} \quad (D_{0max} \text{ in cm}) \tag{2.43}$$

The **VODGEO** method provides the following seepage strength at the contact of cohesive and cohesionless soils

$$d_{70}^a \geq (D_{60}^0)_k \tag{2.44}$$

$$\frac{\rho_0 d_{70}^a}{g}[\bar{v}\frac{D_{60}^0}{(D_{60}^0)_k}]^2 \leq 12 f_0 \frac{D_{50}}{D_{50} + d_{70a}} \tan \varphi \tag{2.45}$$

in which

d_{70}^a = design diameter of eroded clayey aggregates,
$(D_{60}^0)_k = 0.29 D_{60}^0 - 0.16$; $D_{60}^0 = \alpha_n D_{60}$;
$\bar{v}$ = real flow (seepage) velocity in filter voids; $\bar{v} = v/n$,
f_0 = specific energy of cohesion for clayey particles in aggregate condition,
$\varphi = 3$–$5°$.

For sandy loam soils with $I_p = 0.03$–0.05 the permissible gradient of contact erosion is found as follows:

$$I_{c.e.perm.} = \frac{5.7}{[(D_{60}^0)_k/d_a'']^{1.36}} \tag{2.46}$$

in which

d_a'' = design diameter of sandy soil aggregates, which is 0.03 mm for soil with $I_p = 0.03$, 0.13 mm for $I_p = 0.04$ and 0.22 mm for $I_p = 0.05$.

Colmatation is the process of deposition of fine grains removed by flow, in the voids of soil-protecting filter.

The VNIIG method gives the following condition of non-colmatation for the first filter layer if grain diameters d_{ci} are to be removed

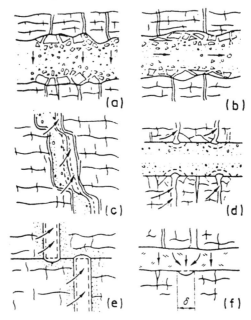

Figure 2.20. Major types of suffosion in rock.

$$D_{17}^I \geq \frac{1.1(1-n)a_*}{nC}d_{ci} \qquad (2.47)$$

in which

$C = 0.455\sqrt[6]{H}$; H = inhomogeneity factor of filter soil,
a_* = coefficient equal to 4.0 for silty soil with grains from 0.01 to 0.05,
a_* = 3.0 for fine sand with grains from 0.05 to 0.25 mm,
a_* = 2.5 for medium sand with grains from 0.25 to 0.5 mm.

The VODGEO method gives the following condition for non-colmatation of filter if the latter protects cohesive soil:

$$D_{60}^I \geq \frac{2d_{70}^a}{\alpha_n\lambda_{60}}$$

in which

d_{70}^a = diameter not exceeded by 70% of aggregates, depending on the moisture contents of clay at liquid limit w_L and diameter d_{10} (Fig.2.19).

The above formulae can be used for design of both earth dams and earth foundations.

If an earth dam is situated on a bedrock, then after analysis of seepage in the dam and foundation it is necessary to assess the seepage strength of rock, that is the possibility of removal of the material depesited in cracks, disintegration products, etc. In accordance with the Soviet standards SNiP II.16-76, the seepage strength of bedrock is determined as follows:

$$v \leq v_{cr} \qquad (2.48)$$

in which

v = discharge velocity in bedrock cracks,

v_{cr} = critical discharge velocity taken as 50 cm/s for clay, 30 cm/s for clay loam and 15 cm/s for sandy loam.

Mechanical suffosion of rock takes place in a variety of modes (Fig.2.20):

a) Removal of fine grains of the filler in cracks (filling material in cracks) by seepage flow along crack;

b) Removal of filler in the direction of flow;

c) Surface erosion of filler at the open joint between filler layer and the wall of protruding crack;

d) Erosion of filler at mouths of coinciding cracks;

e) Erosion of walls of hollow cracks in weakly cemented rock;

f) Contact uplift of clayey filler in cracks (*Design* 1984).

In assessment of seepage strength, the characteristics of crack filler must also be taken along with characteristics of groundwater flow. The design seepage-suffosion model for the foundation is derived similarly as for soil foundation, and is based on analysis of engineering and geologic findings, in particular on data on cracking in rocks (orientation of cracks, block structure, the degree of opening, etc), presence of tectonic disturbances, zones of disintegration, zones of leaching in rock massif etc. Given below are general principles for the estimation of seepage strength of rock, in accordance with the aforementioned schemes.

1. Stability of filler in large cracks or of the material of tectonic disintegration zones (schemes (a) and (b) in Figure 2.20) is assessed as for a usual non-rock soil, i.e. by comparison of the diameter of fine grains of filler d_f with the design diameter of seepage voids in soil d_o; $(d_M > d_0)$. The critical hydraulic gradient for suffosive soil with laminar seepage is found from Equation 2.30, while for turbulent flow it reads

$$I_{cr} = \varphi d_{ci} \sqrt{\frac{ng}{\nu k}} \qquad (2.49)$$

2. Suffosion stability of fine filler in clayey laminas, with respect to surface erosion on the side of an open crack (Fig.2.20c) is assessed by the hydraulic gradient which must not be above the critical value

$$I_{cr} = 1.44 \cdot 10^4 \frac{\nu^2(\delta + A)^2}{g\delta^2(\delta + B)} \qquad (2.50)$$

in which

δ = width of opening crack

A and B = hydraulic roughness parameters of crack walls; in first approximation one assumes $A=0.5$ cm and $B=1.7$ cm.

3. The lateral erosion as illustrated in Figure 2.20d is possible only for turbulent flow in these cracks. For the estimation of suffosion stability one uses the critical velocity of flow v_{cr}, below which the stability is secured

$$v_{cr} = 1.2 \cdot 10^3 \frac{\nu(\delta + A)}{\delta(\delta + B)} \tag{2.51}$$

4. The erosion of crack walls in a massif of semi-rock materials (Fig.2.20e) does not exist at velocities smaller than 1 m/s, and for the compressive strength of rock in saturated condition — not below 1 MPa.

5. Seepage uplift of clayey filler in mouths of protruding cracks (Fig.2.20f) takes place if hydraulic gradients are higher than the following critical value

$$I_{cr} = \frac{2.5 R_t}{\gamma_0 \delta} \xi \tag{2.52}$$

in which
  R_t = long-term tensile strength of filler soil,
  $\xi = \delta_*^{0.66 I_L}$,
  δ_* = dimensionless parameter of crack opening,
  I_L = liquidity index of crack filler.

It should be noted that seepage strength of bedrock is very important to construction of high dams in canyons. Reliability of water-confining constituents — such as screens, largely depends on seepage properties at the interface of structure and its foundation.

2.4 DRAINAGE AND FILTERS

Drainage. Inverse filters, including transition zones, are primary structural constituents of earth dams; they secure seepage strength and control the depression curve. We discuss herein the design and overall dimensions of inverse filters and the selection of drainage components, the primary constituent of which consists of inverse filters.

Drainage of soil is provided by a material which has permeability much higher (e.g. by two orders of magnitude) than the considered soil. Head losses in drainage are small compared with those in the entire earth dam. In many cases, the task of drainage is to reduce pore pressure in impervious components. Such means are also used to dissipate pressure in permeable layers of foundation overlain by impermeable strata. In earth dams with cores etc, the role of drainage is played by dam body made of coarse material.

A drainage structure usually consists of two parts — take-off unit and filter. The former conveys away the filtering water and is made of coarse stone, pebbles, gravel, and special drains. The filter connects the dam soil with the drainage; its

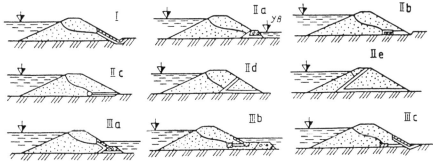

Figure 2.21. Types of drainage designs. (I) outer drainage; (IIa) drainage prism; (IIb) flat horizontal drainage and drainage band; (IId) (IIe) flat horizontal drainage in combination with sloping drainage bands; (III) combined drainage; (IIIa) sloping outer drainage and drainage prism; (IIIb) (IIIc) outer drainage and flat drainage.

chief task is to secure the seepage strength of the dam soil. A transition zone plays the role of drainage in earth dams with water-confining elements. Three types of drainage measures are distinguished, viz. outer, inner and combined (I, II and III in Figure 2.21).

Outer drainage has no peculiar varieties. It is seldom used alone; more often in combination with inner drainage. It protrudes beyond dam and does not provide the shortest seepage path, so it can affect the depression curve. Outer drainage is usually applied on the segments of earth dams being periodically flooded. The thickness of the outer drainage (together with filter) in clayey soil is taken slightly greater than the freezing stratum so as to protect the tailwater slope. The upper datum is found from the condition of freezing under highest depression curve, with inclusion of capillary rise; it also must be above the highest tailwater level under wind and wave set-up. The outer drainage is made of sufficiently uniform stones, which provide a high permeability. The size of stone D_s is found from the condition of stability of tailwater under waves. Pebbles and gravel can be arranged above the maximum level of water, in a sloping drainage. If antifreeze protection is not required, a drainage without filter should not be thinner than $3D_s$.

Inner and combined drainage provide the shortest seepage path; this is their basic property — they control the depression curve thereby. The closer the drainage to the upstream side the lower the depression curve. Since in most cases the seepage flow rate is not limited due to its low value, drainage could be located very close to the thrust slope, but this would increase its cost. The optimum location of drainage is found through analysis of technological and economic factors upon consideration of dam cost, tailwater slope, configuration of depression curve, and seepage rate.

The drainage stone prism (type IIa) is usually constructed on the river stretches of a dam if the latter is designed without cofferdams or if the river is impounded by stone dumping. If cheap stone is available, the method can also be applied on other stretches of rivers or elsewhere in similar circumstances so as to utilize the stone.

Flat horizontal drainage (type IIb), tubular drainage (type IIc), and flat horizontal drainage or band drainage in combination with inclined band drainage (type IId) are used at dry altitudes of flood plains.

Tubular drainage is widely accepted. If made of porous concrete, filter is not required. In case of drain pipes, filter is necessary. Band drainage is often employed to drain the water-confining units of earth-rockfill dams or rockfill dams. This necessity arises if semi-rock, clayey or slightly weathered rock material is subject to crushing under rolling; and also if weathering continues in dam so that the permeability falls dramatically. Sometimes drainage is arranged in the body of a dam in a seismic region.

Type IIe drainage is applied in homogeneous dams of clayey soil to intercept completely the groundwater flow. The drainage units of this type are expensive and may be recommended if sand is abundant.

Combined drainage, types IIIa, IIIb and IIIc, is in use if tailwater must be protected from waves. Type IIIb is common in hydraulic-fill dams when it is constructed after the dam.

Tubular horizontal drainage combined with vertical tubular drainage is usually applied to reduce pressure at the foundation, when the more permeable layer of bed is overlain by a less permeable soil or if it is necessary to reduce the gradients if flow exits close to dam foot.

Vertical drainage at dam foundation is made of wells aimed at intercepting groundwater flow, thus preventing suffosion of crack filler in bedrock, or uplift and suffosion in an alluvial bed. A tubular drainage can incorporate filters, especially if in alluvia. The filters can be concrete, ceramic, soil-type etc. Filters from porous concrete and plastics are most recommended.

Inverse filters (transition zones) are designed under the following conditions: filter permeability greater than the permeability of the protected soil, protection from all types of mechanical suffosion, noncolmatation by the particles removed by groundwater flow (still the removal being acceptable if it does not interfere with normal operation of dam), and prevention of the penetration of the protected soil into filter, and the filter material itself into drainage.

The first stage of inverse filter design consists of hydromechanical considerations for mean and local hydraulic gradients, in particular for flow exit into drainage. Soil is selected for the inverse filter, and the seepage strength is assessed by physicomechanical and granulometric properties. The grain size distribution and layers are designed next, followed by suffosion analysis.

Cohesionless natural (sand, gravel, pebbles) or artificial (rubble, sorted slag) materials are used for the inverse filters. The materials must contain solid and dense grains not susceptible to chemical suffosion and frost-resistant.

Two methods of inverse filter design are widely used in the Soviet Union, viz.

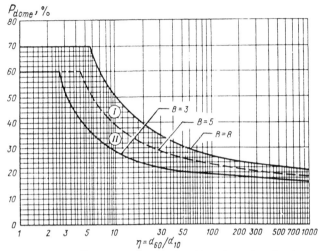

Figure 2.22. Dependence of P_r on coefficient of uniformity η in selection of design dimensions of arching particles of soil protected by inverse filter. (I) rubble region; (II) sand + gravel + pebbles.

VNIIG (*Guidelines* 1981) and VNII VODGEO (*Guidelines* 1982).

Inverse filters to protect cohesionless soil.

VNIIG method. This method concentrates on suffosivity of soil, grain size distribution for stable doming etc.

1. Suffosivity and grain size distribution. Use Equations 2.27–2.29. Artificial or quarry material is deemed suffosive under the following condition

$$\frac{d_3}{d_{17}} \geq 0.32 \sqrt[6]{\eta}(1 + 0.05\eta)\frac{n}{1 - n} \tag{2.53}$$

2. Determine size of grains of the soil protected by the inverse filter. For a non-suffosive soil the quantity d_r is found from Figure 2.22. Referring to the coefficient of uniformity η in zones I and II one determines the percentage of arching grains, P_r, and finds the design diameter of these grains by the use of the grain distribution curve. For a suffosive soil one determines the maximum hydraulic gradient in the contact layer $I_{p.max}$ and afterwards the quantity d_{ci} is computed:

$$d_{ci} = \frac{k_r I_{p.max}}{\varphi_0 \sqrt{\frac{ng}{\nu k}}} \tag{2.54}$$

in which
k_r = reliability factor = 1.1–1.5,
φ_0 factor given at Equation 2.30.

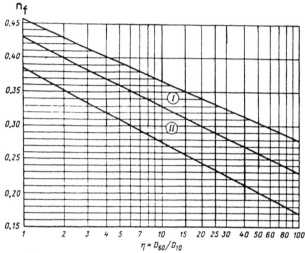

Figure 2.23. Graph for permissible porosity of inverse filter soil. (I) rubble; (II) sand + gravel + pebbles.

If $d_{ci} < d_{3-5}$ then $d_{ci.des}$ is found through P_r from the graph in Figure 2.22 on the curve $B=3$; otherwise one takes $d_{c.i.des} = Bd_{3-5}$, with $B=3-8$.

3. Ranges of applicability of nonuniform soil for inverse filters are found as follows. The permissible coefficients of uniformity D_{60}/D_{10} for inverse filter soil are below 25 for protected nonsuffosive soil and 15 for protected suffosive soil.

4. Determination of permissible interlayer coefficients. These factors for the first (I) and second (II) layers of an inverse filter read

$$\eta_M^I = \frac{D_{17}^I}{d_{ps}}; \quad \eta_M^{II} = \frac{D_{17}^{II}}{D_{ps}^I}$$

The notation d_{ps} refers to the protected soil. The real interlayer coefficient should satisfy the condition

$$\eta_M \leq \eta_{M.perm} = \frac{1}{0.25\sqrt[6]{\eta_f}}\frac{1 - n_f}{n_f} \tag{2.55}$$

in which

η_f and n_f refer to the filter.

5. The minimum coefficient of permeability of filter is

$$k_{min} \geq (2 + \sqrt[6]{\eta_f})k_{ps}$$

in which

k_{ps} = coefficient of permeability of protected soil.

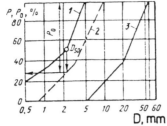

Figure 2.24. Percentage of voids in granular material with $n=0.36$ and $\alpha_n=0.155$ (Istomina & Burenkova's data). (1) protected soil; (2) filter pores; (3) filter grain size distribution.

6. Thickness and number of layers in filter. The thickness should comply with the condition

$$h_{min} \geq (5 \div 7)D_{85} \quad \text{or} \quad h_{min} \geq 5D_{90} \tag{2.56}$$

The quantity h_{min} is corrected due to penetration of fine grains in coarse skeleton.

7. Permissible porosity and density of soil in filter. The permissible porosity of protected soil and inverse filter is found in Figure 2.23 as a function of soil uniformity:

$$n_f = n_0 - 0.1 \log \eta \tag{2.57}$$

VODGEO procedure. The thickness of the filter layer protecting a granular material at a sloping interface stems from the condition of unacceptable penetration

$$h = (AD_{50} + a) \cot \omega \tag{2.58}$$

in which

$A = \frac{390}{97-P_o}$,
P_o = percentage of filter pores greater than the mean diameter of the protected soil, d_{50};
P_o being found from Figure 2.24,
a = allowance for filter thickness taken as 0.3 m,
ω = slope angle of interface.

Inverse filters for protection of cohesive soil.

VNIIG procedure. Upon selection of soil composition for the first layer of inverse filter protecting cohesive soil in a class I structure one must not permit detachment or scaling of cohesive aggregates. For structure classes II to IV, a slight defoliation is accepted to the depth $\Delta S = D_o/2$; for the first filter layer one should have $D_{o.max} \leq 15$ mm for $I_P \geq 5$.

1. Design hydraulic gradient. For a known real hydraulic gradient of the water entering the first layer of filter one finds the design gradient

$$I_{des} = 1.25 I_{exit} \tag{2.59}$$

2. Design diameter of pores in the first filter layer, in cm, reads

$$D_{0.des} = \sqrt{\frac{0.34}{\varphi I_{des} + \cos \Theta}} \tag{2.60}$$

in which

$\varphi = 0.5-1.0$;

θ = angle between the force of gravity and the vector of flow velocity, whereupon one has

$D_{0.des} \leq 0.583$ cm.

If a certain separation of aggregates in pores is acceptable (for structure classes II to IV) one gets

$$D_{0.des} = \sqrt{\frac{2.25}{\varphi I_{des} + \cos \Theta}} \tag{2.61}$$

3. Comparison of real gradient with the permissible one, I_{des} vs. I_{perm} by Equation 2.36. If the former is greater than the latter one must modify the grain size distribution so as to reduce the size.

4. Selection of the coefficient of uniformity for the first layer of inverse filter

$$\eta_{f.perm} = \frac{D_{60}}{D_{10}} \leq 50 \tag{2.62}$$

5. Prevention of separation of clayey soil into the first layer of inverse filter requires

$$D_{0.max} \leq D_{0.des}$$

6. The grain size distribution of the first layer is selected as follows

$$D_{17} \leq \frac{D_{0.des}}{\kappa c} \frac{1 - n_f}{n_f} \tag{2.63}$$

in which

$c = 0.455 \sqrt[6]{\eta}$; $\kappa = 1 + 0.05\eta$;

$$D_{10} = i D_{17} \tag{2.64}$$

in which

i is determined graphically from Figure 2.25;

$$D_{60} = \eta_f D_{10}; \quad D_{100} \leq D_{10} + 10^x D_{60} \frac{\eta_f - 1}{5\eta_f^2} \tag{2.65}$$

in which

$x = 1 + 1.28 \log \eta_f$.

The design grain size distribution curve is constructed for the first layer upon knowledge of the 10, 17, 60 and 100-percent grain diameters (Fig.2.26), and next the

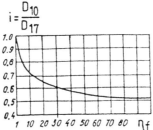

Figure 2.25. Graph for the coefficient of uniformity η_f of filter soil vs. $i = D_{10}/D_{17}$.

region of permissible grain size distribution is established for the first layer of filter. For this region, curve II prescribes the lower limit of grain size distribution. For the upper limit it is required that $D_{60.des}$ of curve II be equivalent to D'_{85} of curve I, and $D_{10.des}$ of curve II to D'_{35} of curve I. This identifies curve I. Examples and details have been provided by Pravednyy (1966), *Guidelines* (1971) and *Guidelines* (1981).

VODGEO procedure. In selection of inverse filters one considers two principal cases, viz. (1) monolithic counterseepage structure, and through cracks impossible to appear; (2) through cracks can arise in the core or screen, with their prevailing orientation from upstream side to tailwater.

In the first case the computation is based on the condition of contact uplift, with the use of hydraulic gradient of Equation 2.37 and the quantities c_t and D_r. It is recommended that the selection of the first layer of the inverse filter for protection of cohesive soil of screen or core should allow for a certain penetration, so that the design grain size distribution must be the coarse one generated as a result of the stratification.

The coefficient of stratification (separation) is taken as

$$\lambda_i = \frac{D_{i.st}}{D_{i.in}} \tag{2.66}$$

in which

$D_{i.st}$ = grain diameter of stratifying soil,
$D_{i.in}$ = grain diameter of original (initial) soil.

For gravelly soil with $\eta = 10–250$, the quantities $D_{60.st}$, $D_{50.st}$ and $D_{30.st}$ can be found by Equation 2.65 upon use of the coefficient λ_i from Figure 2.18; in approximation one has $D_{10.st} = 0.1 D_{60.st}$. The grain size distribution of the second filter layer results from the assumption that the first layer is the original design layer of protected soil.

For the second design case one employs Equations 2.43 and 2.46.

The required width of the first layer of filter is

$$b_f = R + a \tag{2.67}$$

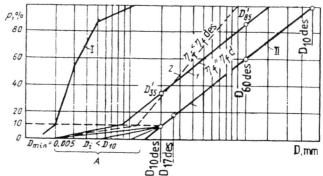

Figure 2.26. Design soil grain distribution and designation of permissible region for first filter layer. (I) cohesive (clayey) soil protected with filter; (II) design grain size distribution of soil with $\eta_{f.des}$ (lower boundary); (1) upper limit of zone with $\eta_{f.des}$ employing $D'_{35} = D_{10.des}$ and $D'_{35} = D_{60.des}$; (2) boundary of possible deviation of the upper limit; (A) area of possible dimensions of fine fractions $D_i < D_{10}$ within filter (from $D_{min} = 0.005$ mm to D_{10}).

in which

R = maximum radius of flow spreading upon exit from cracks,

$R = 1.33 H_{cr} \sqrt{\frac{1}{\ln(8 H_{cr}/\delta)}}$

H_{cr} = depth of open crack,

δ = width of open crack,

$a = 0.3–0.5$ m = safety margin of filter width.

In dams with screens, the thickness of the subsequent filter layer is chosen on the condition of nonpenetration and is taken as h computed from Equation 2.58; the second layer protects the first one so that the first layer is regarded as the protected soil. Details and examples are given by Istomina et al. (1965).

In selection of grain size distribution under the condition of colmatation in cracks, which is likely in thin counterseepage structures of high earth-rockfill dams, one must remember that cracks do not necessarily appear at the entire screen (core)-filter interface, and therefore it is only on parts of these interfaces that a finer filter is required.

In any case one should thoroughly investigate the possible use of local soil as material for filters, as this dramatically decreases the cost of dam. Effectiveness of drainage is the primary condition of reliable dam operation; hence adequate selection of grain size distribution and high quality of dumping operations are prerequisites of successive performance of dams.

CHAPTER 3

Soil consolidation in dam body and foundation

3.1 THEORETICAL BACKGROUND

In the seepage theory outlined in Chapter 2 it was assumed that the motion of liquid was due to the difference of heads in a rigid non-deforming porous medium; thus the soil compressibility was neglected. In reality soil is compressible, which must be considered in its consolidation, in particular if considerable quantities of water are contained in voids.

In general, soil is a multiphase porous medium consisting of mineral particles making up the soil skeleton, together with water and air in soil voids. Because of these three components it is customary to refer to soil as a three-phase medium; if pores are fully saturated with water soil is a two-phase medium, and in the absence of water it is simply a single-phase medium.

If a certain load is applied to soil, its consolidation is accompanied by changes in porosity; they depend not only on compressibility of soil skeleton but also on the behaviour of water in pores — the so-called pore water. Obviously, compression of soil and reduction of its porosity require elimination of water from the pores. The external load brings about stresses in the soil skeleton and pressures in the pore water and pore air, referred to as pore pressure.

The soil condition with excess pressure in the pore water due to external load is referred to as unstabilized. Generation of pressures or additional piezometric heads gives rise to unsteady forced seepage. The process of unsteady seepage is completed upon dissipation of pore pressure and the transition of soil into stabilized condition. Consolidation of soil, that is its densification and elimination of water, depends on compressibility and permeability of soil and its water content. In highly permeable soil, pore pressure does not arise in practice as its dissipation is very rapid. On the other hand, the dissipation in almost impermeable soils is a very long-term process.

If the air content in soil voids is high, the consolidation can proceed without pore pressure and with a constant water content. Therefore consolidation computations are conducted only for foundations and structures (hydraulic, civil or industrial) consisting of cohesive soil with high water contents.

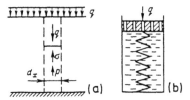

Figure 3.1. Schematic diagram of one-dimensional consolidation (a) and rheological model for one-dimensional consolidation of two-phase soil (b).

Consolidation of soil is controlled not only by the physico-mechanical properties of soil but also by the size of a consolidating area, as the dissipation of pore pressure depends on the length of seepage paths.

The following factors must be considered in consolidation analysis: temporal variation of the settlement of structure and its foundation, linked to gradual dissipation of pore pressure due to external load; stability of cohesive slopes; bearing capacity of foundation in unstabilized condition. Hence the construction of cohesive dams and foundations requires a proper consolidation forecast.

An external load applied to an unconsolidated soil is borne by both skeleton and pore water. In the simplest one-dimensional case (Fig.3.1b), the unit pressure on an elementary area can be decomposed into skeleton stress and water pressure: $q = \sigma + p$. Sometimes, the skeleton stress is referred to as effective pressure σ_{ef}, while the water pressure is called neutral pressure σ_{neut}; so that one has $q = \sigma_{ef} + \sigma_{neut}$ and accordingly q is total pressure.

The consolidation process can be illustrated by the simplest rheological model (Fig.3.1b): a cylinder with a hollow piston, elastic spring and liquid.

The exerted external pressure q is taken over by the spring and the liquid. At first, when the liquid can be regarded incompressible, the force is transferred entirely to the liquid, and one has $q = \sigma_{neut} = p$. The pressurized liquid is forced to flow through openings in the piston, and the spring (which simulates the soil skeleton) becomes compressed. The degree of this compression varies in the process of consolidation (the compression grows and so do the stresses in soil skeleton), while the pore pressure falls. The process is stabilized for $q = \sigma_{ef}$ and $p=0$.

In an unconsolidated soil, the shear resistance, determined by the normal stresses in the skeleton shear plane, as described by Coulomb, is lower : $\tau = (q - p) \tan \varphi + c = \tan \sigma + c$. Hence it becomes evident that the consolidation problem is very important in stability analysis.

In mathematical description of the consolidation process one must consider the rheological properties of each soil phase, the interaction of the phases, and the variation of phase relationships in the process of consolidation. Two basic theories of consolidation are presently available in soil mechanics: seepage theory and the theory of volumetric forces. They differ as to the stress-strain relationships used and the treatment of the interaction of soil phases.

The seepage theory of consolidation is a simplified version of the general theory of volumetric forces, in which the primary rheological equation of soil state is expressed through a compression curve, and the interaction of soil phases is reduced to the simplest equation of equilibrium, by which the specific load applied to soil consists of effective and neutral pressures. The third of the soil phases — air obeys the Boyle-Mariotte law for the transformations of volume and pressure. Dissolution of air in water is described by the Henry law. Such an approach simplifies the formulation and solution of the problem. The seepage theory of consolidation was formulated by Terzaghi in 1925 and was later improved by Soviet scientists (Florin 1939, Gersevanov & Polshin 1948, Gorelik 1975, Ter-Martirosyan 1973, Tsytovich 1973, Verigin 1965) and alien researchers (cf. Florin 1939, Gersevanov & Polshin 1948, Gorelik 1975, Tan Tuong Kie 1957, Ter-Martirosyan 1973, Tsytovich 1973, Verigin 1965, etc).

The theory is sometimes named after Terzaghi, Gersevanov and Florin. By virtue of its relative simplicity, the seepage theory of consolidation will henceforth be devoted most attention.

The theory of volumetric forces is based on the assumption that the interactions between soil phases, skeleton and water, in the process of consolidation, are manifested as volumetric forces due to pressures in the pore water. The stress-strain relationships in soil skeleton are taken from the linear theory of elasticity, while the interaction of the phases is displayed as the volumetric forces due to the action of water on soil skeleton. In this theory, too, the phase transformations in soil are given by the seepage law and the continuity of mass for water and skeleton in unit volume. The theory of volumetric forces was formulated by Biot and Florin in the years 1935–1941 (Biot 1941, Biot 1956, Florin 1939, Yokoo et al. 1971, etc). Therefore it is sometimes referred to as the Biot-Florin theory, while the associated consolidation problem, based on this theory has been given the name of conjugated problem of elasticity and consolidation. Mathematical modelling in this case is much more sophisticated, which makes it difficult to solve practical problems.

Clayey soils possess the property of creep, which may be included in the equations of state for soil skeleton. Because of creep, the consolidation of soil skeleton under loading, even in the absence of pore pressure, is a temporal — and not instantaneous — process. Therefore primary and secondary consolidation is distinguished. Primary consolidation is the one caused by compression of pore water, while secondary consolidation is linked to viscous deformations of soil skeleton. The terms are not entirely adequate as creep of soil skeleton is manifested immediately upon application of a force.

The phenomenon of soil consolidation can be illustrated by consolidation of clay in a compression apparatus (Fig.3.2a). Measurements of pore pressure are usually conducted with a thin hollow needle introduced into a sample or with a pore pressure gauge. The load is deemed instantaneous. If a sample has the disturbed structure, with small initial density and full saturation (all voids filled up

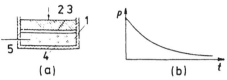

Figure 3.2. Schematic compression apparatus (a) and temporal variation of pore pressure in the apparatus (b). (1) oedometer; (2) press; (3) drainage; (4) soil; (5) pore pressure gauge.

with water), the 'instantaneous' load gives rise to 'instantaneous' increase in pore pressure, up to the maximum possible, i.e. specific pressure exerted on the sample ($p_{max} = q$). Gradual dissipation of pore pressure and growing settlement of press take place afterwards. If the soil in sample is dense and unsaturated, then the pore pressure grows gradually up to $p_{max} < q$, with subsequent dissipation. Instantaneous settlement of press takes place at the beginning, growing until pore pressure becomes subject to dissipation. It is assumed that the instantaneous settlement of press results from instantaneous compression of the gaseous phase in soil voids, while the growth of pore pressure is linked to creep of soil skeleton. These processes of consolidation are illustrated in Figure 3.2b.

3.2 BASIC RELATIONSHIPS OF THE TEORY OF CONSOLIDATION

One-dimensional problem of seepage theory for soil consolidation (Florin 1948, Florin 1961, Gersevanov & Polshin 1948). In the simplest consolidation problem one assumes that soil is fully saturated with water, the compressibility of water and soil skeleton is negligible, soil permeability may be deemed constant, i.e. independent of soil porosity or creep of soil skeleton, and the original hydraulic gradient can also be neglected.

The theory of consolidation, like the seepage theory (see Chapter 2), deals with a certain fictitious, continuous seepage-type flow of pore water, characterized by discharge and seepage velocities, $\bar{u}$ and u, respectively, related by $\bar{u} = (1 - n)u$, where n = soil porosity.

A basic equation of consolidation may be derived from continuity equations for liquid and solid phases. These equations read that, for a fully saturated soil, the volume of water leaving an elementary control volume of soil is equal to the volume of solid particles entering the control volume. From the motion of liquid in an elementary control volume, along the z-axis (Fig.3.3) it folows that the water contents varies in the process of consolidation as follows

$$u_z dx dy dt - (u_z + \frac{\partial u_z}{\partial z} dz) dx dy dt = -\frac{\partial u_z}{\partial z} dx dy dz dt \tag{3.1}$$

in which
$\quad u$ = velocity of liquid phase.

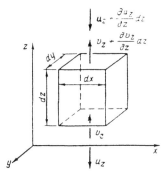

Figure 3.3. Variation of water content and volume of soil skeleton upon consolidation of elementary volume.

Taking into account that soil porosity in an elementary volume changes in time dt from n to $n + \frac{\partial n}{\partial t} dt$, and the volume of pores is related to the volume of inflowing (outflowing) water, the continuity equation for the liquid phase reads

$$\frac{\partial u_z}{\partial z} + \frac{\partial n}{\partial t} = 0 \tag{3.2}$$

From an analogous assumption for the motion of soil skeleton one has

$$\frac{\partial v_z}{\partial z} + \frac{\partial m}{\partial t} = 0 \tag{3.3}$$

in which
   v = velocity of soil skeleton.

Summing up Equations 3.2 and 3.3, with $n+m=1$, one has

$$\frac{\partial u_z}{\partial z} + \frac{\partial v_z}{\partial z} = 0 \tag{3.4}$$

Since both solid and liquid phase move in the process of soil consolidation, Darcy's equation of the type given in Equations 2.3 and 2.8 can be written down as follows

$$u_z - \epsilon v_z = -k \frac{\partial H}{\partial z} \tag{3.5}$$

in which
   ϵ = void ratio (Section 1.1),
   H = head in consolidation problems.

Equation 3.5 accounts for motion of soil skeleton during consolidation.
   Equilibrium equation assumes the form already discussed for the case of constant load q applied to a layer of soil (Fig.3.1a)

$$q = \sigma + p \tag{3.6}$$

which means that the total stresses in water and skeleton are constant (in time). Equation 3.6 sometimes reads (* for stable conditions):

$$\sigma + p = \sigma^* + p^*$$

Equation of soil state is taken as a compression relationship:

$$\epsilon = \epsilon_0 - \alpha\sigma \qquad (3.7)$$

in which
  ϵ_0 = original soil void ratio,
  α = coefficient of compressibility (Eq.1.24).

Head and pore pressure are interrelated as indicated in Chapter 2:

$$H = \frac{p}{\gamma_w} + z \qquad (3.8)$$

Porosity and void ratio are interrelated as follows:

$$n = \frac{\epsilon}{1+\epsilon}; \quad m = \frac{1}{1+\epsilon} \qquad (3.9)$$

The above equations and relationships may be reduced to the following system of partial differential equations, if the variable $1+\epsilon$ is replaced by the constant $1+\epsilon_m$, in which the latter denotes mean void ratio over a given interval of consolidation:

$$\frac{\partial\epsilon}{\partial t} = (1+\epsilon_m)\frac{\partial}{\partial z}k\frac{\partial H}{\partial z} \qquad (3.10)$$

or, with inclusion of Equations 3.6-3.8

$$\frac{\partial H}{\partial t} = \frac{1+\epsilon_m}{\gamma_0\alpha}\frac{\partial}{\partial z}k\frac{\partial H}{\partial z} \qquad (3.10a)$$

If permeability is regarded as constant during consolidation, then for a certain consolidation factor $c_v = \frac{k(1+\epsilon_m)}{\alpha\gamma_w}$ the consolidation equation reads

$$\frac{\partial H}{\partial t} = c_v\frac{\partial^2 H}{\partial z^2} \qquad (3.11)$$

Solution requires boundary and initial conditions. The boundary conditions for the head function are identical with those for the seepage theory, viz. the head function is given at permeable boundaries while for impermeable ones one has $\partial H/\partial z = 0$.

  As a rule, solution of consolidation problems involves determination of excess piezometric head in a consolidating medium, $H = p/\gamma_0$. On the strength of Equation 3.6 and the assumption on complete transfer of load to water at the time when load is applied, the initial condition for excess head reads

$$H_0 = \frac{q}{\gamma_0}$$

The problem can be more complex if various physical factors intervene, so that Equation 3.10 assumes a more sophisticated form. Consider a number of the most important cases.

1. If strong cementation bonds exist in soil, one considers compressibility of water and mineral particles, the equation of consolidation assumes form of Equation 3.11, and the consolidation factor reads

$$c_v = \frac{k(1 + \epsilon_m)}{\alpha \gamma_0 [1 + \frac{\epsilon_m}{\alpha}(\frac{1}{K_w} - \frac{1}{K_s})]} \tag{3.12}$$

in which

K_w = bulk modulus of compressibility of water, about 10^5 MPa,
K_s = bulk modulus of compressibility of mineral grains, about 10^6 MPa.

One may show that for $\alpha > 0.01$ MPa^{-1} the compressibility of water and soil skeleton can be neglected.

The original pore pressure at $t=0$ reads

$$p_0 = \frac{q}{1 + \frac{1}{\alpha}\left(\frac{\epsilon-0}{K_w} + \frac{1}{K_s}\right)} \tag{3.13}$$

in which

ϵ_o = soil void ratio immediately before load is applied.

2. Considering the variation of permeability during consolidation one gets the following equation of consolidation

$$\frac{\partial H}{\partial t} = (1 + \epsilon_m)\frac{dk}{d\epsilon}\left(\frac{\partial H}{\partial z}\right)^2 + \frac{1 + \epsilon_m}{\gamma_0}\frac{dk}{d\epsilon}\gamma_{sub}\frac{\partial H}{\partial z} - \frac{k(1 + \epsilon_m)}{\gamma_0\frac{d\epsilon}{d\sigma}}\frac{\partial^2 H}{\partial z^2} \tag{3.14}$$

If the relationship between permeability and void ratio is assumed linear one has

$$k = k'_\Phi - \frac{k'_\Phi - k''_\Phi}{\epsilon' - \epsilon''}(\epsilon' - \epsilon) \tag{3.15}$$

in which the quantities with primes and double primes are attributed to the loads σ' and σ'', then Equation 3.14 is reduced to an equation with constant coefficients

$$\frac{\partial H}{\partial t} + \alpha\left(\frac{\partial H}{\partial z}\right)^2 + \beta\frac{\partial H}{\partial z} + \delta\frac{\partial^2 H}{\partial z^2} = 0 \tag{3.16}$$

in which

$$\alpha = -(1 + \epsilon_m)\frac{k'_\Phi - k''_\Phi}{\epsilon' - \epsilon''}; \quad \beta = \frac{\gamma_{sub}}{\gamma_w}\alpha; \quad \delta = -\frac{1 + \epsilon_m}{\gamma_w}\frac{k'_\Phi - k''_\Phi}{\epsilon' - \epsilon''}\frac{\sigma'' - \sigma'}{\ln k'_\Phi/k''_\Phi}$$

For the selected interval from σ' to σ'' the respective permeabilities and void ratios are determined experimentally.

3. In consideration of soil skeleton creep, the compression relationship in Equation 3.7 for one-dimensional creep may be replaced by a formula representing the change in porosity till time t due to unit load applied at time τ_1:

$$\epsilon(t) = \epsilon(\tau_1) - \sigma(\tau_1)\delta(t, \tau_1) - \int_{\tau_1}^{t} \frac{d\sigma}{d\tau}\delta(t, \tau)d\tau \tag{3.17}$$

in which

$$\delta(t, \tau) = \alpha_0 + \sum_{1}^{m} \alpha_n \left[1 - \exp^{-\kappa_n(t-\tau)}\right] \tag{3.18}$$

The notation α_0 stands for the part of strain arising immediately upon application of load, while the other component, referred to as creep measure, grows gradually with time. In approximate computations Equation 3.18 is taken as

$$\delta(t, \tau) = \alpha_0 + \alpha_1 \left[1 - \exp^{-\kappa_1(t-\tau)}\right] \tag{3.18a}$$

The parameters α_0, α_1 and κ_1 are determined experimentally in long-term compression tests (Meschyan 1967, 1985).

With inclusion of Equations 3.17 and 3.18 the consolidation equation 3.10 becomes

$$\alpha_0\gamma_w \frac{\partial^2 H}{\partial t^2} + \gamma_w\kappa_1(\alpha_0 + \alpha_1)\frac{\partial H}{\partial t} = (1 + \epsilon_f)\frac{\partial}{\partial z}k\left(\kappa_1\frac{\partial H}{\partial z} + \frac{\partial^2 H}{\partial z^2}\right) \tag{3.19}$$

If creep strains grow rapidly ($\kappa_1 \to \infty$), Equation 3.19 is reduced to Equation 3.10a, with $\alpha_0 + \alpha_1 = \alpha$. Equation 3.19 is a second-order equation and therefore requires, along with the usual initial condition $H(0, z) = q/\gamma_w$, yet another initial condition

$$\frac{\alpha_0}{\alpha_1\kappa_1}\frac{\partial H}{\partial t} - \frac{q}{\gamma_w} + H = \frac{1 + \epsilon_m}{\gamma_w\kappa_1\alpha_1}\frac{\partial}{\partial z}k\frac{1 + \epsilon_m}{\gamma_w\kappa_1\alpha_1}\frac{\partial}{\partial z}k\frac{\partial H}{\partial z} \tag{3.20}$$

4. The consolidation problem for a three-component soil medium is very important because the presence of air entrapped in soil affects considerably the mode of pore pressure dissipation and magnitude of pore pressure.

In derivation of consolidation equation for a three-phase soil one makes use of the Boyle-Mariotte law for volume versus pressure and the Henry law for dissolution of air in water under pressure. Upon many simplifications the consolidation equation reads

$$\frac{\partial H}{\partial t} = \frac{k(1 + \epsilon_{mean})}{a\gamma_w\omega}\frac{\partial^2 H}{\partial z^2} \tag{3.21}$$

in which

$\omega = 1 + \frac{\beta(1+\epsilon_{mean})}{\alpha}$,

β = coefficient of volumetric compressibility equal to $\frac{s+\alpha n}{p_o+p}$,

s = fraction of air in unit volume,

n = fraction of water in unit volume,

α = Bunsen coefficient of absorption,

p_o = original pore pressure.

An adequate value of β, with variable s and p, is taken as average for a certain pressure range, from minimum to maximum, by analogy to the earlier case of void ratio. Hence the effect of entrapped air on the soil consolidation is considered in a certain quasi-two-phase model, consisting of skeleton and water-air mixture, the average compressibility of which is determined by β.

The initial condition for three-phase soil reads

$$H_0 = \frac{q}{\gamma_w \omega_0}, \quad \omega_0 = 1 + \frac{\beta(1 + \epsilon_0)}{\alpha}$$

5. Consolidation equations can be obtained in terms of the original hydraulic gradient in soil I_o, structural strength of soil, nonlinear stress-strain relationships, etc (Didukh 1970, Florin 1953a, b, Shirinkulov & Dasibebov 1966).

Plane and spatial problems of seepage consolidation. The differential equations derived above for consolidation in one-dimensional formulation are readily expandable to two- and three-dimensional cases if the compression line is taken as $\epsilon = \epsilon_0 - \frac{\alpha}{1+n\xi}\Theta$, with $n=1$ for plane problem and $n=2$ for three-dimensional problems, along with the following notation: ξ = coefficient of side pressure, Θ = total normal stress.

The consolidation equation in the simplest case, for example three-dimensional consolidation of a two-phase medium reads

$$\frac{\partial H}{\partial t} = \frac{k(1 + \epsilon_{mean})(1 + 2\xi)}{3\gamma_w a} \left(\frac{\partial^2 H}{\partial x^2} + \frac{\partial^2 H}{\partial y^2} + \frac{\partial^2 H}{\partial z^2} \right) \tag{3.22}$$

In the plane case $\frac{1+2\xi}{3}$ is replaced by $\frac{1+\xi}{2}$, and the second derivative of H is eliminated from the right-hand side.

In two- and three-dimensional problems, the original distribution of excess pore pressure is determined with Florin's hypothesis, by which the total stress in soil corresponds to stable conditions and does not vary in time. The hypothesis in the general case yields the following equation of equilibrium

$$\sigma_x + \sigma_y + \sigma_z + 3p = \sigma_x^* + \sigma_y^* + \sigma_z^*$$

For $t=0$ one has

$$H(x, y, z, 0) = \frac{\sigma_x^* + \sigma_y^* + \sigma_z^*}{3\gamma_w} \tag{3.23}$$

The above hypothesis makes it possible to determine the original distribution of head for a known stress distribution in stabilized soil.

Equation 3.22 can be made more complex through inclusion of the gaseous phase in soil, creep of soil skeleton, etc as done above for the one-dimensional case.

More general consolidation equations in terms of the seepage theory of consolidation are derived upon cosideration of soil creep and ageing (Florin 1953).

Detailed effects of creep and compression of pore water were dealt with by Ter-Martirosyan (1973). Gorelik (1975) put forth a system of equations for consolidation with s variable during consolidation. Two differential equations have been derived for two unknown functions p and s. In the simplest case, one equation may be considered for p, with subsequent inclusion of air volume related to head, which is possible if the variation of moisture content and air content during consolidation is neglected. In general, the relationships become more sophisticated in view of the nonlinear relationships between porosity and stresses in soil skeleton and between permeability and soil porosity.

Model of volumetric forces is based on the assumption that the process of consolidation embodies interactions of two media, soil skeleton (in general, with air bubbles) and pore water. Soil skeleton is regarded as a linearly deformable medium, for which the differential equations of the theory hold. The interaction forces between the two media arise in the course of consolidation — mass or volume forces due to buoyancy of soil grains in water and groundwater flow. The motion of pore liquid is described by the Darcy-Gersevanov equation. Hence the one-dimensional consolidation problem of ponderable soil is described by the system of equations:

$$\frac{\partial \sigma_z}{\partial z} + \gamma_w \frac{\partial H}{\partial z} + \gamma_{sub} = 0; \quad \frac{\partial \sigma_z}{\partial t} = \frac{1 + \epsilon_m}{\frac{\partial \epsilon}{\partial \sigma_z}} \frac{\partial}{\partial z} k \frac{\partial H}{\partial z} \tag{3.24}$$

For uniform loading on the surface of imponderable soil, with $\gamma_{sub} = 0$ and $\frac{\partial \epsilon}{\partial \sigma_z} = -a$, the system in Equation 3.24 transforms into Equation 3.10.

The design model of volumetric forces for static and dynamic problems of the theory of consolidation has been generalized by Zaretskiy who included plastic and rheological properties of soil. Details are discussed in Chapter 4 in terms of stress-strain relationships for earth dams. Conjugated problems of consolidation and elasticity have been dealt with by Frenkel (1944), Gersevanov & Polshin (1948), Gorelik (1975), Kerchman (1974) and Nikolayevskiy (1962).

3.3 SOLUTIONS OF THE THEORY OF CONSOLIDATION

Solutions to consolidation problems aimed at finding excess pressures and settlement of foundations and earth structures are found in terms of the differential equations derived in the theory of consolidation, for given boundary and initial conditions. Moreover, a fairly simple method exists which yields pore pressure in foundation or earth structure; it does not consider the dynamics of pore pressure in soil but instead the maximum pore pressure in closed-system soil is taken. Computations by the method of 'closed system' are sometimes referred to as the 'method of compression curve'. Details for three-phase soil are presented by Gorelik (1975).

The following criterion is critical in computations by the method of closed system:

$$c_1 = \frac{k(1 + \epsilon_0)t}{\gamma_w \alpha h_0)^2} \leq 0.0156$$

in which

> t = time of loading growth up to its maximum for foundations, or time of dam
>
> construction for consolidation computations of dams,
>
> h_o = thickness of soil layer in foundation with one-side drainage, or its half for two-side drainage; for homogeneous dams one has $h_0 = m_{up}h_d$,
>
> h_d = dam height, m_{up} = upstream slope factor, while for dams with cores h_o reads $b/2$,
>
> b = core width.

Unlike consolidation factor c_v, the parameter c_1 includes t and h_o, thus dissipation of pore pressure during dam construction or loading of a soil layer is possible to encompass.

Order of excess pressure computations.

1. Determine the so-called 'ultimate' excess pore pressure (when the gaseous phase is entirely dissolved in soil):

$$p_n = \frac{p_0}{\alpha}\left(\frac{1}{D_{S0}} - 1\right) \tag{3.25}$$

in which

> p_o = initial pore pressure prior to loading,
>
> D_{S0} = initial degree of saturation of soil,
>
> α = Bunsen's coefficient of absorption, depending on temperature T:

$T\,°C$	0	10	20	30	40	50
$\alpha \cdot 100$	2.86	2.24	1.83	1.54	1.32	1.14

2. Determine void ratio:

$$\epsilon_n = w_0\frac{\gamma_p}{\gamma_w}; \quad \epsilon_n = D_{S0}\epsilon_0 \tag{3.26}$$

in which

> γ_p = unit weight of soil particles,
>
> w_0 = initial water content of soil (fractions of unity),
>
> D_{S0} = initial degree of saturation.

3. Employ the compression equation to determine the effective consolidation stress $\sigma_{sk} \equiv \sigma_{ef}$.

4. Determine load 'limit':

$$q_n = p_n + \sigma_{sk}$$

5. Compare the load q applied to soil with the computed one. If the former is greater, the excess pore pressure is given as $p = q - \sigma_{sk}$, while in the opposite case the values of p and σ_{sk} are found graphically as a solution to the transcendental equation

$$\varphi_1(\sigma_{sk}) = \varphi_2(\sigma_{sk}) \tag{3.27}$$

in which

$$\varphi_1 = 1 - \frac{p_0}{q + u_0 - \sigma_{sk}}; \quad \varphi_2 = \frac{\epsilon_0 - \epsilon(\sigma_{sk})}{\epsilon_0 - K_0}; \quad K_0 = (1 - \alpha)\frac{w_0\gamma_p}{\gamma_w}$$

The graph on the left of Equation 3.27 is called 'inventory graph' (Fig.3.4), while the RHS graph is 'relative compression graph', which should be constructed to the same scale, up to the value of σ_{sk} corresponding to $\varphi_2 \leq 1$.

Values of p and σ_{sk} are found by putting the inventory graph (on stencil) on the graph of relative compression so as to place the end of the abscissae axis of the former graph on the latter graph at $\sigma_{sk} = q$. The abscissa of the intersection of both curves on the lower axis of relative compression gives σ_{sk}, while the upper axis yields p.

An example of p and σ_{sk} computations by Gorelik (1975) is illustrated in Figure 3.5. The point A marks the intersection of both curves, φ_1 and φ_2.

Validity of the 'closed system' scheme is proved satisfactory in view of good agreement of experimental and computed pore pressures. Laboratory measurements of pore pressure in compression tests conducted on samples 11.3 cm in diameter and 2 cm high (Nichiporovich & Tsybulnik 1963) on the so-called consolidation stands, 30 cm high and 30 cm in diameter (Bushkanets et al 1973) have confirmed good coincidence of both sets of data. The tests were carried out on samples of loessial clay loam with liquid limit of 28% and plastic limit of 20%. The initial water content in various series of tests varied from 17% (initial unit weight of dry soil being 17 to 17.5 kN/m³) to 27% (15 to 15.5 kN/m³, respectively). Instantaneous pressure of 1 and 2 MPa, or stepwise up to 1 MPa, was exerted on the sample. Compression tests were conducted in parallel, and compression curves were constructed for the twin samples. Experimental pore pressure factors, α_{op} and α_p are compared in Figure 3.6 with their computed counterparts (Balykov et al. 1973).

The field investigations performed by Gidroproyekt on Charvak Dam have resulted in graphs of load q and pore pressure p measured by a gauge installed in the core of the dam during construction versus the values computed by the method of closed system, for various water contents w_o. It may be seen that the agreement is fairly good (Fig.3.7).

Analytical methods are usually used in simpler one-dimensional cases. For two- and three-dimensional cases, analytical solutions may be obtained only in some particular configurations of loading and consolidated areas. However, the use of the analytical solutions is very helpful for analysis of various effects on the process of consolidation.

Consider as an example a solution of one-dimensional problem of the seepage theory of consolidation as given in Equation 3.11. The equation is of parabolic type (Tikhonov & Samarskiy 1966) such as those describing the processes of heat transfer, diffusion, etc.

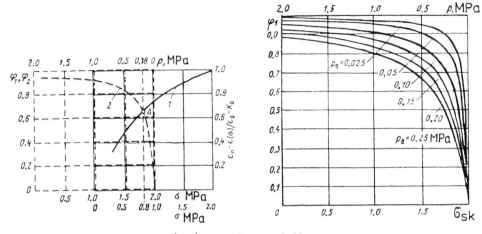

Figure 3.4. Inventory graph $\varphi_1(\sigma_{sk})$ for various p_o (left).

Figure 3.5. Example of graphical definition of p_0 and σ_{sk} (right). (1) relative compression graph; (2) inventory graph. Upper scale of σ_{sk} for curve 2, lower scale for curve 1.

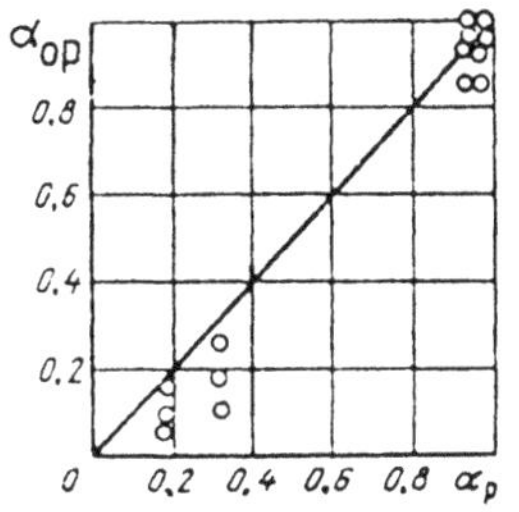

Figure 3.6. Pore pressure factors α_{0p} and α_p computed and measured in 'closed system'.

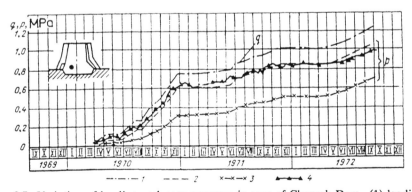

Figure 3.7. Variation of loading and pore pressure in core of Charvak Dam. (1) loading due to overlying soil layers; (2) pore pressure graph computed for water content of 19.7% (50-% confidence); (3)pore pressure computed for water content of 18.3% (90-% confidence); (4) pore pressure by records of probe deployed in dam core (•).

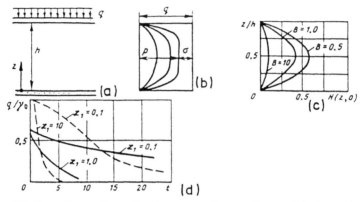

Figure 3.8. Solution of one-dimensional seepage theory of consolidation. (a) schematic layout; (b) distribution of excess pressure in soil layer; (c) distribution of initial excess head with inclusion of soil creep and the parameter $B = \frac{k(1+\epsilon_m)\pi^2}{\gamma_w \kappa_1 \alpha_1 h^2}$; (d) head change at midlayer with (solid line) and without (dashed line) creep.

An analogy takes place between steady flow of water in a porous medium, described by the Laplace equation, and steady electric current in a conductor (Section 2.2); similar analogy occurs between steady seepage of pore water in the consolidation process and temperature variation. The latter analogy can be useful in exploration of the consolidation process, formulation of initial and boundary conditions, etc. The equation of heat conduction is expressed through Equation 3.11, with conductivity characteristics replacing c_v.

Equation 3.11 is solved by the well-known method of separation of variables (the Fourier method), whereupon the function $H(z,t)$ is presented as two functions of single variable: $H(z,t) = Z(z) \cdot T(t)$. In one-dimensional problems (Fig.3.8a), for the initial condition $H(z,0) = q/\gamma_w$ and boundary conditions $H(0,t) = H(z,t)=0$ i.e. drawdown, a solution to Equation 3.11 reads

$$H = \frac{4q}{\pi\gamma_w} \sum_{i=1,3,5,\ldots}^{\infty} \frac{1}{i} \exp - \left(\frac{c_v i \pi^2 t}{h^2}\right) \sin \frac{i\pi z}{h} \tag{3.28}$$

in which

h = depth (thickness) of soil stratum.

Obviously, for t tending to infinity one has $H=0$, which corresponds to complete dissipation of the excess pore pressure.

Settlement of the stratum surface as a result of consolidation is also a function of time, and can be found by summing up the strains e_z in the vertical direction:

$$s(t) = \int_0^h e_z(t)dz = \int_0^h \frac{a\sigma}{1 + \epsilon}dz \approx \frac{1}{1 + \epsilon_m} \int_0^h \sigma \, dz$$

Variation of σ over the stratum can be determined from the equlibrium:

$$\sigma(z,t) = q - \gamma_w H(z,t)$$

Then for s(t) one obtains

$$s(t) = \frac{aqh}{1 + \epsilon_m} \left[1 - \frac{8}{\pi^2} \sum_{i=1,3,5,\ldots}^{\infty} \frac{1}{i^2} \exp\left(-\frac{c_v i^2 \pi^2 t}{h^2}\right) \right] \qquad (3.29)$$

The quantity in the brackets is called consolidation degree, which characterizes the part of total settlement of soil layer $s = \frac{aqh}{1+\epsilon_m}$ taking place till time t.

Shown in Figure 3.8b is the temporal variation of pressure and stresses in soil skeleton during consolidation. In practical computations by Equation 3.28 and Equation 3.29 it is sufficient to keep a few terms of the series as their convergence is very strong.

For negligible instantaneous strain and importance of creep, i.e. for $\alpha_o = 0$ one should consider the following equation:

$$\frac{\partial H}{\partial t} = c\left(\kappa_1 \frac{\partial^2 H}{\partial z^2} + \frac{\partial^2 H}{\partial z^2 \partial t}\right) \qquad (3.30)$$

in which

$c = \frac{k(1+\epsilon_m)}{\gamma_w \kappa_1 \alpha_1}$ for initial condition,

$$H = c\frac{\partial^2 H}{\partial z^2} + \frac{q}{\gamma_w} \qquad (3.31)$$

which produces the solution of the one-dimensional problem for the earlier boundary conditions:

$$H(z,t) = \frac{4q}{\pi\gamma_w} \sum_{i=1,3,5,\ldots}^{\infty} \frac{1}{(1+c\alpha_i^2)} \times \exp\left(-\frac{\gamma_1 c\alpha_i^2}{1+c\alpha_i^2}t\right) \sin\frac{i\pi z}{h} \qquad (3.32)$$

in which

$\alpha_i = i\pi/h$.

Florin (1948) analysed this case and showed that inclusion of creep produces corrections of pore pressure distributions both at the moment of incipient loading and during the entire period of consolidation. With inclusion of Equation 3.32 the initial condition in Equation 3.31 yields

$$H(z,0)\frac{4q}{\pi\gamma_w} \sum_{1,3,5\ldots}^{\infty} \frac{\sin\frac{i\pi z}{h}}{\left[1 + \frac{i^2\pi^2}{h^2}\frac{k(1+\epsilon_m)}{\gamma_w\kappa_1\alpha_1}\right]i} \qquad (3.33)$$

From Equation 3.33 it follows that the initial pressure distribution depends on permeability and compressibility of soil and the process of creep strain growth. For $\frac{k}{\kappa_1\alpha_1} \rightarrow \infty$ and $H(z,0) \rightarrow \infty$ the consolidation takes place without pore pressure growth. For $\frac{k}{\kappa_1 \alpha_1} \to 0$ and $H(z,0) \to q/\gamma_w$ one encounters the common consolidation problem. For intermediate values of $\frac{k}{q_1 \kappa_1}$ there occurs partial transfer of pressure to liquid, and the graphs of initial head are not rectilinear (Fig.3.8c). From Equation 3.32 it follows that for $\kappa_1 \to \infty$, i.e. no creep, solutions to Equations 3.32 and 3.28 are identical. In this case the coefficient α_1 plays the role of the consolidation factor. From Figure 3.8d it may be seen that for various κ_1 the dissipation of pore pressure is slower than in the case when the creep of soil skeleton is not considered, although the initial values are lower.

Analysis of solutions to consolidation problems with creep of soil skeleton should be supported by experimental data. The latter proves that, in tests on soil samples of different height in compression apparatus, the initial coefficient of pore pressure is below unity, even for fully saturated soil. Moreover, the consolidation factor n introduced by Maslov (1982) who described the ratios of consolidation times for samples of different height, $t_1/t_2 = (h_1/h_2)^n$ for n in the range between 0 and 2, while in the seepage theory of consolidation one has $n = 2$.

If the arrested air is taken into consideration, the solution of one-dimensional problem of consolidation yields for excess pressure:

$$H(z,t) = \frac{4q}{\pi \gamma_w \omega_0} \sum_{i=1,3,5...}^{\infty} \frac{1}{i} \times \exp \left(-\frac{i^2 \pi^2 c_0 t}{h^2} \right) \sin \frac{i \pi z}{h}$$

in which

$$c_0 = \frac{k(1 + \epsilon_m)}{\gamma_w a \omega_0}; \quad \omega_0 = 1 + \frac{\beta(1 + \epsilon_m)}{a}$$

Analytical solutions of two-dimensional and three-dimensional problems can be obtained only for the simplest cases of loading. This is due to the fact that the initial distribution of excess head $H_0(x,y,z) = \frac{\sigma_x^* + \sigma_y^* + \sigma_z^*}{3\gamma_w}$ is provided by the solution of the elasticity problem corresponding to the stable conditions. The relationships for the stresses σ_x^*, σ_y^*, σ_z^* in sophisticated load configurations become very complex so that their integration is quite troublesome. The problems are even more complex if skeleton creep is to be included. Methods of operational calculus (Sneddon 1955) are harnessed to tackle consolidation equations such as Equation 3.19.

In consolidation computations for water-confining structures, the plane problem of consolidation can be reduced to one-dimensional problem upon assumption that the outflow of pore water takes place only in the horizontal direction. This assumption is justified for narrow cores when the seepage in the vertical direction (towards moving ceiling of the place layer) is negligible, or if there is seepage anisotropy (horizontal permeability much higher than the vertical one). Such assumptions have made it possible to solve the problem of consolidation in closed form (Nichiporovich & Tsybulnik 1963). In addition, soil properties by Nichiporovich & Tsybulnik (1963) are constant, and so are mean stresses, in

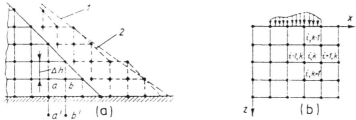

Figure 3.9. FDM grid. (a) rectangular grid for dam core; (b) FDM grid for consolidation computations; (1) core slope; (2) fictitious slope contour.

the horizontal cross-section of a counterseepage unit; it is assumed that the form changes do not affect pore pressures or volume change.

Numerical methods have paved their way in solution of the consolidation problems. Finite-difference method (FDM) is utilized both in seepage approach to the consolidation problem and in solution of conjugated problems (Florin 1948, Zaretskiy 1983). Soil consolidation, particularly for earth dams and foundations, is widely computed within FDM formulation of the basic equation. Figure 3.9a illustrates how a dam core is represented by a rectangular grid with Δx and Δz increments. In the particular case shown in the drawing the grid is quadratic ($\Delta x = \Delta z = \Delta h$). In numerous cases the real slope is replaced by a fictitious contour.

The finite-difference equation for consolidation of a two-phase soil in a two-dimensional situation reads

$$H_{t+\Delta t,i,k} = \left(1 - 2\frac{\alpha}{m^2} - 2\alpha\right) H_{t,i,k} + \frac{\Theta^*_{t+\Delta t,i,k} - \Theta^*_{t,i,k}}{2\gamma_w} +$$
$$+ \frac{\alpha}{m^2}(H_{t,i-1,k} + H_{t,i+1,k}) + \alpha(H_{t,i,k-1} + H_{t,i,k+1}) \qquad (3.34)$$

in which

$H_{t+\Delta t,i,k}$ = head at the node i, k at time $t + \Delta t$,

$H_{t,i,k}$ = head at node i, k at time t,

$H_{t...}$ = heads at nodes adjacent to node i, k at time t (Fig.3.9b),

$\Theta^*_{t+\Delta t,i,k} - \Theta^*_{t,i,k}$ = increment of total stresses at node i, k during the time step Δt, with inclusion of the effect of overlying soil strata ($\Theta^* = \sigma^*_x + \sigma^*_z$),

$$\alpha = \frac{c_v \Delta t}{(\Delta z)^2} \leq \frac{1}{2(1 + 1/m^2)} \qquad (3.35)$$

where

$m = \Delta x/\Delta z$.

The condition in Equation 3.35 makes possible the selection of the time step necessary for stability of solution. In the simplest case the condition reads 1-

$2a/m^2 - 2\alpha = 0$. It may also be a multiple of the time between placement of the design soil layers having thickness $\Delta z (\Delta h)$.

For a three-phase medium, the denominator of the second term on RHS of Equation 3.34 is modified by the factor $\omega_o = 1 + \frac{\beta(1 + \epsilon_m)}{\alpha}$ which accounts for the arrested air. Then Equation 3.35 reads

$$\alpha = \frac{c_v \Delta t}{\omega_0 (\Delta z)^2}$$

For the sake of simplicity it may be assumed that the sum of stresses in soil skeleton in stable state can be approximated by $\Theta^* = (1 + \xi)\gamma h$, in which h stands for elevation of soil layer above a given point. In more accurate computations, the stress state of core in stable condition (with the factor of stepwise construction included) can be found independently by solving the static problem for the entire dam profile.

The boundary conditions of the problem are:

1. Excess head in pore water is zero at dam crest, tailwater edge of core and at the upstream part of core edge above water in reservoir;

2. Head is equal to the vertical distance from water level in reservoir to the given point (h_1) all over the upstream part of core edge below the datum of tailwater (upon filling of reservoir during construction);

3. At core foot, if it is a water-confining stratum, the condition $\frac{\partial H}{\partial z} = 0$ is assumed, that is equal heads at a or b and a' or b', respectively. For permeable foundation, one takes $H = 0$ in the absence of water in reservoir during construction; linear distribution of heads, from maximum at the upstream edge of core (depth of water in reservoir) to zero on the tailwater edge of core, is assumed in the case of stagewise construction.

Settlement of core during construction and operation of dam is determined by summing up the settlements of all elementary soil layers Δz located on the vertical line passing through a given point, and below it:

$$S = \sum_{n=1}^{N} \Delta S_n; \quad \Delta S_{t,i,k} = \frac{\epsilon_{t,i,k} - \epsilon_{0,i,k}}{1 + \epsilon_{0,i,k}} \Delta z \tag{3.36}$$

in which

N = number of soil layers below a given point.

The void ratio $\epsilon_{t,i,k}$ is found from the depression curve, as a function of the effective stress acting in soil skeleton:

$$\sigma_{z,t,i,k} = \sigma^*_{t,i,k} - \gamma_w H_{t,i,k}$$

FDM is also applicable for consolidation computations with inclusion of soil creep (Goldin 1966). FDM for consolidation computations of couterseepage measures, in terms of the model of volumetric forces (conjugated problem of the theories of elasticity and consolidation) was applied by Zaretskiy & Lombardo (1983). The

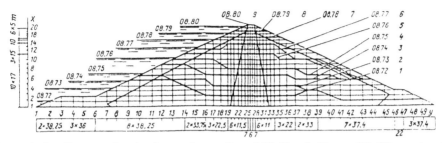

Figure 3.10. Computational scheme for earth-rockfill Nurek Dam. (1)–(9) construction stages.

soil model was three-phase, with air effect pronounced through a modified modulus of elasticity, and with different permeabilities in vertical and horizontal directions, depending on soil density (volume change); the stress-strain relationship for soil skeleton was based on the theory of plastic strenghtening.

The procedure was used for computations of pore pressure and settlement of Nurek Dam, and the results have been compared with the field data measured. Real strength and deformation figures for soils of all zones of the dam were used, as obtained from triaxial tests, upon consideration of all techniques and schedules employed for the dam and the reservoir. The tests were carried out on saphedobian sandy loam, filled into the dam core, with gravel and pebbles of fills, and rock. The apparatus was PTS-300 of Gidroproyekt design. Dependence of permeability on soil density was also investigated.

The design scheme of Nurek power plant (Zaretskiy & Lombardo 1983), which incorporates time schedule for filling of dam zones and impoundment of the reservoir, is shown in Figure 3.10. Design values of piezometric heads have been compared with pore pressures measured with strain gauges during construction of the dam. Figure 3.11 shows that the overall picture of heads in the dam core, by computations and measurements, is similar. Maximum heads (shown as circles) for the points of the lower core are close to each other (they differ by 7 to 8%).

The computations have shown that dissipation of excess pore pressure and stabilization of seepage take place roughly after 20 years of dam operation. Graphs of temporal variation of piezometric head in the dam core at datums of 42 m and 120 m are depicted in Figure 3.12. Good agreement of the computed and measured data has also been reached for settlements of dam core and tailwater fill.

The finite-element method (FEM) finds more and more applications in computations of consolidation in terms of both theories, seepage and volumetric forces. This is due to the fact that FEM easily embodies peculiar properties of geometry, strength, strain, seepage inhomogeneity, etc. FEM is as convenient as FDM, and may be implemented with the aid of modern computers of large memory and high speed. The background of the method, and its variational formulation may be

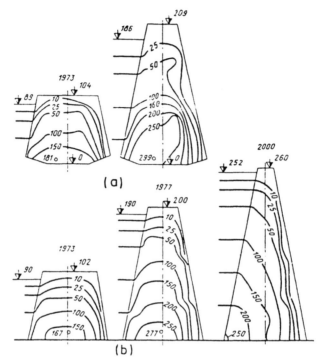

Figure 3.11. Piezometric heads in dam core of Nurek Dam during construction in the years 1973–1980 and at steady seepage (the year 2000) by prototype data (a) and from computations (b).

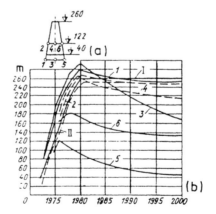

Figure 3.12. Temporal variation of piezometric heads at typical locations of dam core, datums 40 m and 122 m. (a) configuration of points (1–6); (b) results of computations; (I) dam construction graph; (II) reservoir impoundment graph.

found in Zienkiewicz (1975), where all pertinent formulae are clearly outlined.

In terms of the seepage theory of consolidation, with inclusion of the arrested gas and the creep of soil skeleton, the basic relationships for FEM are given by Bugrov (1975). Matrix equations for plane FEM problems can be found by minimizing, with respect to the vector of displacements of FE nodes in plane $\{q\}$ and the vector of excess pore pressure $\{p\}$, of the following functional given at t (Rozin 1978):

$$
\begin{aligned}
E = {} & \frac{1}{2} \int\limits_S \int \{e\}^T \{\sigma\} dS + \frac{1}{2} \int\limits_S \int e_{gas} p \, dS + \\
& + \sum_0^t \left(\frac{1}{2} \int\limits_S \int \{I\}^T \{v_0\} dS \right) \Delta t - \int\limits_S \int \{U\}^T \{G\} dS - \\
& - \int\limits_{l_1} \{U\}^T \{f\} dl - \sum_0^t \left(\int\limits_{l_2} \{p\}^T \{V\} dl \right) \Delta t
\end{aligned}
\tag{3.37}
$$

The first two terms correspond to the strain energy ($e_{gas} - \beta p$), the third term gives the energy loss in flow within S, $\{I\}$ is the vector of pore pressure gradients, and $\{v_0\}$ stands for the vector of seepage velocities of water relative to soil skeleton, while $\{G\}$ and $\{f\}$ incorporate the work of external volumetric and surface (at the boundary l_1) forces, respectively. The last term describes the concentrated flux $\{V\}$ at the boundary l_2 (the sign 'T' denoting transposition). The conditions of minimum E with respect to q and p (stationary energy for possible variations of q and p) make it possible to obtain the system of resolving equations for FEM (details given by Bugrov 1975).

FEM was used for consolidation computations of Mica Dam in Canada (Einsenstein & Zaw 1977). The stagewise construction of the dam, the interaction of core and thrust prisms, and the nonlinear stress-strain relationship in the soil was taken in consideration.

In conclusion, it may be said that consolidation of slightly permeable compressible soil as a dam foundation must be analysed because settlement of dam and stability of slope depend on consolidation of foundation. In the case of weak, fully saturated soil of foundation one often faces landslide hazards at dam slope and foundation, the latter in unstable condition, for coefficients of pore pressure close to unity.

Analysis of the consolidation of foundations is conducted by the same methods used for earth dams. The growth of the loading on foundation due to the weight of soil in dam, and the variation of boundary conditions at the outer interface (surface of foundation) is included in the analysis. The variation of pore pressure in the foundation can be determined not only by computations but also by measurements in soil, yielding excess pore pressure. Analysis of piezometric data measured provides corrections for dam construction and such a rate of dam growth for which there is no hazard of slope instability due to increase in pore pressure in the foundation. This procedure was used in the construction of a cooling pond for the Uglegorsk scheme on the foundation the upper layers of which consisted of silty clay loam with high water contents (Yevdokimov et al. 1977).

3.4 DESIGN GUIDELINES

Consolidation computations of water-confining constituents of earth-rockfill dams, homogeneous dams of slightly permeable soil, and foundations made of clayey soil constitute a compulsory stage of design. These computations provide excess pore pressure and settlements. It is impossible to assess slope stability and the bearing capacity of foundations without the knowledge of pore pressure (Florin 1948, 1961).

The present state of the theory of consolidation permits pore pressure computations with inclusion of numerous factors — elastic, plastic and rheologic properties of soil, seepage anisotropy as a function of compaction, stagewise construction, and variation of head at boundaries of consolidated areas.

Complete consideration of all above factors involves complicated mathematical problems and respective methods and algorithms for solution. Such an approach would be appropriate in design of unique earth dams. In many practical cases it is reasonable to resort to the simplest design schemes and utilize simple consolidation models yielding sufficiently reliable results.

Simple design schemes are also rational in view of many inaccurate estimates of soil properties intervening in the models, particularly permeability. In terms of the seepage theory of consolidation, with an order-of-magnitude assessment of the coefficient of permeability and computation of the consolidation factor by a smoothed depression curve, it becomes advisable to use an experimental value of consolidation factor instead. The 'closed system' model provides a reasonably overestimated pore pressure.

Abundant field data on the growth of pore pressure available for the structures constructed so far points to the occurrence of considerable pore pressures in many instances (particularly if unconsolidated soil of high water content is filled in a dam), which indicates that the problem of consolidation must be explored in depth. Underestimation of pore pressure may have catastrophic consequences.

CHAPTER 4

Stress-strain state and strength of dam soil

The stability of dam slopes in primary and secondary combinations of loading, the growth of pore pressure, the settlement of core (screen) due to consolidation, the settlement of fills (thrust prisms) of coarse grained soil etc may be predicted by various design methods.

Analysis of the stress-strain state of earth dams combines utilization of the afore-mentioned local problems and special hypotheses. The transition from the stress-strain relationship to slope stability or evaluation of cracking is a very complex task and involves a number of additional assumptions. However, by undertaking this analysis one may manage to include the effect of stagewise construction of dam on the stress-strain relationship, which is impossible by the less general methods. In the wake of progress in the computer-aided design and construction of the large dams of Nurek, Rogun and others, it has become feasible to include the stress-strain state of earth dams as a crucial component of modern design of large (and smaller) dams.

Analysis of the stress-strain relationships in earth dams is so complex because strain properties of soil depend on many factors (cf. Section 1.3) — acting mean stress $\sigma = 1/3(\sigma_x + \sigma_y + \sigma_z)$, components of stress deviator, loading history of a given stress field, time elapsing from the instant of incipient loading etc. All these factors are incorporated in the models discussed in Chapter 1. The models may be implemented by numerical methods — FEM, FDM, variational-difference etc.

FEM (Rasskazov & Vitenberg 1972), FDM (Gun 1971, Zaretskiy 1983), and FEM with minimization of the energy functional by the method of local variations (MLV) (Belakov 1983, Rasskazov 1977, Rasskazov & Belakov 1982) have been used most widely. Any exact explicit solution to the stress-strain problem of an earth dam is practically impossible to obtain in view of the complexity of soil models and the growth of the area of solution (indirect inclusion of the loading path). FDM in variational formulation is shown in detail by Zaretskiy (1983), while FEM is presented by Rasskazov & Vitenberg (1972) and elsewhere. Discussed below is FEM with minimization of the energy functional, by the method of local variations, and explicit FDM in time domain, which are the most prospective methods for problems with loading paths in terms of nonlinear soil models.

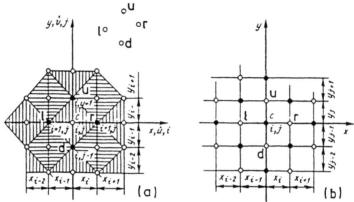

Figure 4.1. Element of finite-difference approximation. (a) determination of stresses and strains at point C; (b) determination of displacements at point C.

4.1 VARIATIONAL-DIFFERENCE CALCULUS IN STRESS-STRAIN PROBLEMS OF EARTH DAMS

The variational-difference method elaborated at Gidroproyekt (Zaretskiy 1983) makes possible the solution of stress-strain problems for earth dams under static and dynamic loading. The method is used in combination with equations derived in the theory of plastic consolidation of soil (Section 1.4).

The design method is so constructed that solution of static problems is obtained from an algorithm for the dynamic problem. The procedure has been formulated for plane strain problems. Since the problem is dynamic, and the static solution is a particular case, the strains e_x, e_y, e_z are replaced by the strain rates $\dot{e}_x, \dot{e}_y, \dot{e}_z$, the displacements $\dot{u}, \dot{v}$ are substituted for the displacements u, v etc.

The stress-strain relationship is derived on the assumption of small strains, which usually holds true for earth dams. If the assumption is not complied with, then each stage of construction must be computed separately, with respective corrections for coordinates. One has:

$$\dot{e}_x = \frac{\partial \dot{u}}{\partial x}; \quad \dot{e}_y = \frac{\partial \dot{v}}{\partial y}; \quad \dot{e}_{xy} = \frac{1}{2}\left(\frac{\partial \dot{u}}{\partial y} + \frac{\partial \dot{v}}{\partial x}\right) \qquad (4.1)$$

The problem is time-explicit; the time increment Δt is a very small quantity, so that the following assumption may be added: stress tensor components vary little, hence the increments of plastic strains $\Delta e_x^p, \Delta e_y^p, \Delta e_{xy}^p$ are also small in comparison with the plastic strains due to the earlier strain stage; this in turn justifies the conclusion that the layout of the load surface does not change within Δt. Total strain consists of elastic and plastic components. As noted in Section 1.3, this decomposition enables simple and fairly accurate computational schemes. This approach has been quite widespread.

The aforementioned equations are supplemented by the variational equation of Lagrange:

$$\int_D \int [(\rho\ddot{u} + k\dot{u} - F_x)\delta u + (\rho\ddot{v} + k\dot{v} - F_y)\delta v]\,dx\,dy +$$

$$+ \int\int_D (\sigma_x\delta\dot{\epsilon}_x + \delta_y\dot{\epsilon}_y + 2\sigma_{xy}\delta\dot{\epsilon}_{xy})dx\,dy - \int_l (P_{xn}\delta u + P_{yn}\delta v)dl = 0 \qquad (4.2)$$

in which

ρ = density,

k = friction factor, proportional to velocity (it is needed only in ideal elastic problems),

F_x, F_y = components of volumetric forces,

P_{xn}, P_{yn} = components of contour stresses.

The problem is solved for homogeneous initial conditions:

$$\begin{aligned}
\sigma_x(x,y,0) &= \sigma_y(x,y,0) = \sigma_{xy}(x,y,0) = 0 \\
\dot{u}(x,\,y,\,0) &= \dot{v}(x,y,0) = 0
\end{aligned} \qquad (4.3)$$

The structure of finite-difference grid is depicted in Figure 4.1. The set of all grid nodes can be split up in two classes: (1) set of nodes S for which the sum $i + j$ is odd (open circles in Figure 4.1), and (2) set of nodes U, for which $i + j$ is even (full circles).

Components of stress and strain rate tensors are determined at S, and components of the vector of displacement rate are found for U nodes. It must be emphasized that the dynamic problem is solved in the grid, and the static problem is a partial case only. Therefore, respective rates, and not stresses, strains or displacements are found at the nodes. In order to avoid double indices, the following notation is proposed:

$i + 1, j$ = point [r] (to the right of C);

$i - 1, j$ = point [l] (to the left of C);

$i, j + 1$ = point [a] (above C);

$i, j - 1$ = point [b] (below C).

Equation 4.3 in finite differences reads:

$$\begin{aligned}
\dot{e}_x^C &= \frac{\dot{u}_p - \dot{u}_l}{x_i + x_{i-1}}; \quad \dot{e}_y^C = \frac{\dot{v}_a - \dot{v}_b}{y_j + y_{j-1}} \\
e_{xy}^C &= \frac{\dot{u}_r - \dot{u}_l}{y_j + y_{j-1}} + \frac{\dot{v}_a - \dot{v}_b}{x_i + x_{i-1}}
\end{aligned} \qquad (4.4)$$

After rates of total strains, with inclusion of the stress σ_{ij} for the preceding instant, $t - \Delta\iota$, are found, together with viscoelastic strains and other parameters of soil model, one determines the stress-strain condition at time t in the following sequence:

1. Determine components of stress tensor (in invariant form) upon assumption that there is no increase in plastic strain over a given time step;

2. Modify the configuration of the limit surface at given time (cf. Section 1.5);

3. Correct configuration of various zones making up the plane (σ, σ_i), i.e. the stress-strain condition due to the earlier step, and fix the loading surface;

4. Compute stresses at t and rates of plastic and elastoviscous strains.

The method is supplemented by pore pressure and consolidation computations, which are very important in analysis of earth-rockfill dams under seismic forcing, as the seismic pore pressure growing in the rubble fill (thrust prism) does reduce the stability.

The pore pressure is found as follows:

$$\frac{\partial}{\partial x}\left(\bar{k}_x \frac{\partial p_w}{\partial x}\right) + \frac{\partial}{\partial y}\left(\bar{k}_y \frac{\partial p_w}{\partial y}\right) = \frac{n}{E_{0w}}\dot{p}_w - \dot{e} \qquad (4.5)$$

in which

$\bar{k}_x, \bar{k}_y$ = relative coefficient of permeability in horizontal and vertical direction, respectively:

$$\bar{k}_x = \frac{k_x}{\gamma_w g}; \quad \bar{k}_y = \frac{k_y}{\gamma_w g}$$

g = acceleration due to gravity,

E_{ow} = modulus of elasticity of aerated liquid,

$\dot{e}$ = rate of total volume changes:

$$\dot{e} = \dot{e}_x^e + \dot{e}_y^e + \dot{e}_x^p + \dot{e}_y^p$$

$e_x{}^e, e_x{}^p$ = components of elastic strains; and

E_{ow} depending on water contents:

$$E_{0w} = \frac{p_a - p_w}{1 - D_S + \frac{D_S(p_a - p_w)}{E_0}} \qquad (4.6)$$

in which

D_S = degree of saturation,

E_o = 2000 MPa = modulus of elasticity of deaerated liquid.

For full saturation one has $D_S = 1$ and $E_{ow} = E_o$. In general, the degree of saturation depends on pressure:

$$D_S = \frac{D_{s0}(p_a - p_w)}{P_a - (1 - \beta)D_{S0}p_w} \qquad (4.7)$$

in which

p_a = atmospheric pressure,

p_w = pore pressure,

$\beta \approx 0.021$ = Henry's constant of air dissolution in water,

D_{s0} = initial coefficient of saturation.

It must be remembered that in reality one has $E_{ow} \ll E_o$ because pore water contains fairly high quantities of dissolved gas and entrapped air.

The pressure p_w in Equation 4.5 is determined over the set of nodal points of S-type, i.e where stresses are determined.

The stability of the solution sought is secured if the following condition is satisfied (Zaretskiy 1983):

$$\Delta \tau \leq \frac{2\Delta x}{\sqrt{2}} \sqrt{\frac{E_{0y}}{\rho} \left(1 + \frac{1}{\frac{E_{0y}k\Delta t}{\Delta x^2}} + \frac{E_{0y}n}{E_{0w}}\right)} \qquad (4.8)$$

in which

E_{oy} = modulus of bulk elasticity over unloading area if soil is taken as a linearly deformable medium,

n = soil porosity.

Obviously, for the x and y increments varying with x and y , and for variable coefficients of permeability (minimum of k_x and k_y) and E_{ow}, the criterion in Equation 4.8 should be complied with for the most unfavourable conditions.

The above procedure permits inclusion of anisotropic properties of soil by the use of empirical formulae. For instance, to find the coefficient of permeability k_x one usually uses Equation 1.17 for clayey soil or Equation 1.15 for loose soil. If one has data showing $k_x = k_y$ one introduces the relation $k_y = ck_x$, with c much below 1, to be found experimentally (the x-axis coinciding with the orientation of stratum boundary, or horizontal in case of soil filling). For c below 0.05 one may assume that E_y goes to zero, and seepage becomes one-dimensional. Such an assumption makes the consolidation problem much simpler (cf. Chapter 3).

The seismic design problems of earth structures by the methods presented by Zaretskiy (1983) are dealt with in Chapter 5.

4.2 FINITE ELEMENT METHOD COMBINED WITH THE METHOD OF LOCAL VARIATIONS (PLANE STRAIN PROBLEM)

Combination of FEM and MLV was first used by Kudryavtsev et al. (1970) and was inspired by the advantages of FEM. The latter is regarded as a decomposition of an area into elements and a respective system of stress, strain and displacement functions. The method of minimizing the energy functional of the type in LHS of Equation 4.2, without inertia terms, can be different in various applications. For instance, an explicit stepwise scheme is illustrated in Section 4.1.

The stress-strain relationship in a dam body can be assumed as linear; in this case the minimization of the energy functional yields a system of linear equations (Fradkin 1973, Rasskazov & Vitenberg 1972, Zienkiewicz 1975). Extensive literature is devoted to this subject but we shall not dwell on this topic as the stress-strain relationship in soil is usually nonlinear.

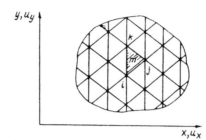

Figure 4.2. Decomposition into triangular elements and numbering of nodes in element m (plane problem).

In the case of nonlinear stress-strain relationship there is no way to reduce the minimization problem to a system of linear equations. By using the method of elastic solutions one may solve the problem by iterations for a series of linear equations (Dolezhalova 1983, Kulhawy & Dunkan 1972, Rasskazov & Vitenberg 1972) but the convergence of the iteration process remains an open problem. Aside from the method used by Lombardo (1973), a lot of attention is paid to the method of local variations (Chernous'ko 1965, Chernous'ko & Bonichuk 1973).

Energy model of soil has been presented in Chapter 1. Hereinafter this model will be deemed a system of physical equations describing the behaviour of soil under loading in the zone of active loading, in Equation 1.31, and unloading in Equation 1.34, in combination with the strength condition in Equation 1.33. The method of local variations is another method of minimization, constructed by analogy to Equation 4.2 and upon approximation by finite elements, along with nonlinear relationships.

The use of MLV in combination with FEM decomposition produced the energy model of soil, which in turn yielded a variety of stress, strain, and stability problems for earth-rockfill dams.

FEM can be used with different elements; triangular elements are however preferred in nonlinear problems since strain energy increments due to changes in the strain tensor e_{ij} in a triangular element are found in the simplest manner (Fig.4.2):

$$\delta E_m = \sigma_{ij}\delta e_{ij}\omega \qquad (4.9)$$

in which

E_m = strain energy of m-th element,
ω = area of element.

Equation 4.9 follows from the linearity of displacement function in element:

$$\begin{aligned} u_x &= \alpha_1 + \alpha_2 x + \alpha_3 y \\ u_y &= \alpha_4 + \alpha_5 x + \alpha_6 y \end{aligned} \qquad (4.10)$$

The linear function of coordinates ensures continuity of displacement not only at nodes of the triangular elements but also at the boundary. The function of displacement is as linear along the interface of two neighbouring elements as in both elements. Since the end values of the functions (at nodes on the boundary) coincide, so do the displacement functions themselves.

Strains in an arbitrary triangular element (*m* in Figure 4.2) read:

$$e_x = \frac{\partial u_x}{\partial x} = \alpha_2; \quad e_y = \frac{\partial u_y}{\partial y} = \alpha_6$$

$$e_{xy} = \frac{\partial u_x}{\partial y} + \frac{\partial u_y}{\partial x} = \alpha_3 + \alpha_5 \tag{4.11}$$

i.e. strains are constant and do not depend on coordinates; hence stresses do not depend on coordinates, either. Therefore the strain energy is simple to determine by Equation 4.9. Strains are easily found by nodal displacements.

If horizontal displacement u_i of node i in Figure 4.2 is known, and the remaining displacements are zero, the following system of linear equations may be written:

$$
\begin{aligned}
u_i &= \alpha_1 + \alpha_2 x_i + \alpha_3 y_i \\
0 &= \alpha_1 + \alpha_2 x_j + \alpha_3 y_j \\
0 &= \alpha_1 + \alpha_2 x_k + \alpha_3 y_k \\
0 &= \alpha_4 + \alpha_5 x_i + \alpha_6 y_i \\
0 &= \alpha_4 + \alpha_5 x_j + \alpha_6 y_j \\
0 &= \alpha_4 + \alpha_5 x_k + \alpha_6 y_k
\end{aligned}
\tag{4.12}
$$

By solving the system in Equation 4.12 with respect to α_2 and α_3 one finds in agreement with Equation 4.11:

$$\alpha_2 = \alpha_x = \frac{y_k - y_j}{\Delta} u_i; \quad e_y = \alpha_6 = 0$$

as the second system of equations has the trivial solution:

$$\alpha_4 = \alpha_5 = \alpha_6 = 0; \quad \alpha_3 = e_{xy} = \frac{x_j - x_k}{\Delta} u_i$$

in which
Δ = absolute determinant equal to double area of the triangular element:

$$\Delta = \begin{vmatrix} 1 & x_i & y_i \\ 1 & x_j & y_j \\ 1 & x_k & y_k \end{vmatrix} = (x_j - x_i)(y_k - x_i) - (x_k - x_i)(y_j - y_i) \tag{4.13}$$

For known strains and stress-strain relationships one may find components of the stress tensor, independent of coordinates inside the given element, and subsequently the strain energy by Equation 4.9. If strain tensor components depend on coordinates, the stress tensor components will also depend on coordinates, and the procedure of strain energy computations in an element will become more complex (see Section 4.3) than in the considered case. Therefore triangular elements in plane problems are preferred in nonlinear soil models.

Effectiveness of the solutions has given rise to a new formulation of the problem of the stress-strain relationships in earth-rockfill dams; a system of iterative design has been worked out for various structures, and technologically as well as economically the optimum design may be chosen for a dam (see Chapter 9).

Solution of the stress-strain-displacement problem with finite triangular elements is sought among linear functions inside each element, continuous over the entire grid (Fig.4.2).

By analogy to FEM, represent the external load as concentrated forces at nodes (Q_{qk}). Let one of the variating displacement components u_{qk} undergo displacement $\bar{u}_{qk}$ along one of the coordinate axes chosen, and the remaining components be zero. This corresponds to the situation when only one of the nodes is displaced in one of the two directions (two-dimensional problem), by a quantity $\bar{u}_{qk}$, and the remaining nodes are fixed in both directions.

By differentiating the displacement at one node in one direction one induces the variation of some terms in the following expression

$$E = \int_G \sigma_{ij} e_{ij}(\bar{u}_{qk}) dG - Q_{qk}\bar{u}_{qk} \qquad (4.14)$$

These terms are just the ones linked to the elements surrounding a given node.

The inertia terms are skipped in Equation 4.14, compared with Equation 4.2, and the relationship reads for displacements, and not displacement rates. Surface forces are concentrated at nodes.

Integration in Equation 4.14 can be replaced by summation:

$$E = \frac{1}{2} \sum_1^M \sigma_{ij} e_{ij}(\bar{u}_{qk}) \omega_m - Q_{qk}\bar{u}_{qk} \qquad (4.15)$$

in which

ω_m = area of m-th triangular element,
M = number of elements surrounding the node qk.

Compute E for the following displacements: $\bar{u}_{qk} = u_x; \; u_x \pm h; u_y; \; u_y \pm h$, in which h = step of variations. From the given combination take the values in the directions x and y which bring about reduction of the functional in Equation 4.15. Then take the variations of displacements at the second node. Pass over all nodes in both directions. If at least at one node the displacement $\bar{u}_{qk}$ causes decrease in E for $u_x \pm h$ or $u_y \pm h$, then all nodes are passed at the step $\pm h$ as long as consecutive passage over the area (n-th) at the step $\pm h$ yields the minimum functional at all nodes for $\bar{u}_{qk} = u_x$ or u_y. At this point, for the n-th iteration, with respect to displacements, the problem is deemed solved with an accuracy of $\pm h$. The step of variations is then reduced by the factor $\alpha < 1$, usually 0.5, and one takes $h_{n+1} = \alpha h_n$. The procedure is repeated with the smaller step. It is stopped if h_n becomes smaller than a certain predetermined quantity $| h |$.

Upon completion of the variational procedure one ends up with an approximate solution of accuracy h_o. Higher accuracy may be attained with a finer grid, for repeated iterative computations.

As shown by Chernous'ko (1965), the convergence of the method is ensured for $h/\tau \to 0$ along with $\tau \to 0$, in which τ = size of regular grid. The convergence problems have been elaborated at length by Chernous'ko & Bonichuk (1973).

Each passage at each node involves at least three operations of variations ($u_x, u_x +h, u_x - h, u_y - h, u_y +h$), and at most five operations ($u_x, u_x +h, u_x - h, u_y - h, u_y +h$), as the procedure may be interrupted when the minimized functional decreases for the first time (for instance, $E^+ < E$, in which E^+ = value of E at $u_x + h$, and E corresponds to u_x).

Transition from displacements to strains is implemented by usual formulae for triangular elements (Zienkiewicz 1975):

$$
\begin{aligned}
e_x &= [(y_k + y_j)u_{xi} + (y_i - y_k)u_{xj} + (y_j - y_i)u_{xk}]/\Delta \\
e_y &= [(x_j - x_k)u_{yi} + (x_k - x_i)u_{yj} + (x_i - x_j)u_{yk}]/\Delta \\
e_{xy} &= [(x_j - x_k)u_{xi} + (x_k - x_i)u_{xj} + (x_i - x_j)u_{yk} + \\
&\quad +(y_k - y_j)u_{yi} + (y_i - y_k)u_{yj} + (y_j - y_i)u_{yk}]/\Delta
\end{aligned}
\tag{4.16}
$$

in which

$$
\Delta = \begin{vmatrix} 1 & x_i & y_i \\ 1 & x_j & y_j \\ 1 & x_k & y_k \end{vmatrix} = (x_k - x_i)(y_j - y_i) - (x_j - x_i)(y_k - y_i)
\tag{4.17}
$$

$x_i, x_j, x_k, y_i, y_j, y_k$ = respective coordinates of given triangular element;
$u_{xi}, u_{xj}, u_{xk}, u_{yi}, u_{yj}, u_{yk}$ = respective horizontal and vertical displacements of the i, j, k element.

Transition from strains to stresses depends on a model chosen.

If the stagewise construction is incorporated in computations, Equation 4.16 reads

$$
\delta e_x = [(y_k - y_j)\delta u_{xi} + (y - i - y_k)\delta u_{xj} + (y_j - y_i)\delta u_{xx}]/\Delta
\tag{4.18}
$$

in which
δ = increase in strains and displacement due to stages of construction.

Stagewise construction of dam is reflected in FEM in increasing mass (volumetric) forces due to the growth of dam. This is of particular importance if gravity structures are built when the self-weight is comparable with surface forces.

Construction of high and very high structures for large reservoirs stipulates inclusion of construction stages in computation of surface forces (pressure of water). The acting load differs from those computed on the assumption of instantaneous actions. The latter assumption is even less acceptable for materials which do not obey Hooke's stress-strain relationship, i.e. if the principle of superposition does not hold — soil is such a material.

If simultaneous action of all forces on a structure is assumed (which for a heavy structure is equivalent to the assumption on instantaneous action) then the displacement of any point is defined as a difference of coordinates before and after load is applied.

For structures built stage by stage (growing area), displacements are conveniently defined as differences of coordinates in given stage (obviously, if the point has already been generated), and at the instant when the point appears in the constructed area. Hence the displacement of an earth dam is deemed the displacement of deep benchmarks. In particular, displacements of points at dam crest due to the self-weight will always be zero by this definition. It is merely the effect of soil creep which brings about displacement of crest during operation of dam. The definition for the growth period is convenient while comparing results of computations with field data. At the same time, it does not characterize the usual real deformability of structure since the changes of subareas are not counted before a given point is generated. Full displacements in the growth area may be found by integration of displacement graphs coordinate by coordinate (areas of the graphs are summed up).

The equations of equilibrium for the growth area, in tensor notation, read

$$\frac{\partial \dot{\sigma}_{ij}}{\partial x_j} = 0 \quad \text{or} \quad \dot{\sigma}_{ij,j} = 0 \tag{4.19}$$

(and for instantaneously built structures this equation becomes $\sigma_{ij,j} = 0$) in which $\dot{\sigma}_{ij}$ = stress rate.

Equation 4.19 in the coordinate form reads

$$\frac{\partial \dot{\sigma}_{xx}}{\partial x} + \frac{\partial \dot{\sigma}_{xy}}{\partial y} = 0; \quad \frac{\partial \dot{\sigma}_{yx}}{\partial x} + \frac{\partial \dot{\sigma}_{yy}}{\partial y} = 0$$

In the case of discretely growing stresses (volumetric forces) Equation 4.19 becomes

$$\frac{d\delta\sigma_{ij}}{dx_j} + \delta f_i = 0 \tag{4.20}$$

in which

δ = stress increment (volumetric forces) over step (stage of construction).

Hence the equilibrium conditions for the growth area are satisfied with regard to stress increments.

In stagewise design it is assumed that the bedrock is incompressible and there are no displacements at the interface. This assumption is confirmed by field findings at El-Infiernillo Dam (Marsal & Arellano 1965). Shear always occurs inside the weaker material, what is supported by experimental evidence.

If motion is possible in vicinity of an interface, this will be reflected in the elemental decomposition of the area.

Once a solution is obtained with an accuracy of h_o, the computation may be repeated for a finer grid. The procedure of making the grid finer and finer is stopped if the response of the system to the minimum step is negligible.

The above procedure necessitates minimization of the functional at each stage of construction, but it is more economic with respect to computer memory.

If one computes the functional increment δE (Eq.4.15) at each point of displacement variations (and this is very convenient as the soil model is differential in this case), then values of the functional must not be found for $u'_{\eta k} = u'_{xk} or u_{yk}$, for one has

$$\delta E = \sum_{1}^{m} \delta\sigma_{ij}\delta\delta_{ij}(h)\omega_m - h\delta Q_{qk} \tag{4.21}$$

in which

  h = step of variations,
  δQ_{qk} = increment of forces concentrated at nodes during a given stage of construction,
  $\delta\sigma_{ij}$ = increment of stress tensor in a given stage,
  $\delta\delta e_{ij}(h)$ = increment of strain tensor due to a given step of displacement variations.

The quantity $\delta\delta$ points to the conventional character of the operation — finding strain rates of second order with respect to total strains, as single δ denotes strain rate at a given stage of construction. Each value of δ δe_{ij} obtained is related to the chosen model $\delta\delta\sigma_{ij}$. Summation at each element of the stress increment $\delta\delta\sigma_{ij}$, with earlier values of $\delta\sigma_{ij}$, due to earlier variations, yields a new $\delta\sigma_{ij}$ for subsequent variations. Similar procedure is followed for $\delta\delta e_{ij}$ and δu_{ik}. For sufficiently small h the deformation process is described well, as in this case the function $\sigma_{ij} = f(e_{ij})$ is given as a large set of piecewise-linear segments. This approach is suitable for incremental soil models, to which the energetics soil model indeed belongs.

In terms of Equation 4.21, the number of variations at a point is at least two and at most four, i.e. time savings are at the cost of memory, since one has to keep the bulk of $\delta\sigma_{ij}$, and not only σ_{ij}.

The algorithm of stress-strain computations for earth-rockfill dams with inclusion of stagewise construction consists in solving a series of variational problems. Let us discuss the algorithm for MLV applied to these problems.

Decompose the area considered (a dam or a dam and compressible foundation) into triangular elements, as done in FEM. Complex design of dam, and often the layered structure of foundation exclude a regular grid, although it would be desirable for a homogeneous dam. Components of stresses and strains at element e are denoted by the upper index r.

The functional δE is presented in the form:

$$\delta E = \sum_{1}^{m} J_{ij} - k_{ij} \tag{4.22}$$

$$J_{ij} = (\delta\sigma_x^r + \frac{1}{2}\delta\delta\sigma_x^r)\delta\delta e_x^r(h) + (\delta\sigma_y^r + \frac{1}{2}\delta\delta\sigma_y^r(h) + 2(\delta\sigma_{xy}^r + \frac{1}{2}\delta\delta\sigma_{xy}^r)\delta\delta e_{xy}^r \quad (4.23)$$

in which

$k_{ij} = Q_{qk}h$,

q = number of given node,

k being i or j, as might result from the direction of variations.

The forces Q_{qk} at a node consist of volumetric forces (weight and/or buoyancy) and pressure of water on the upstream side exerted on the thrust edge of a counterseepage element. The vertical component of the volumetric force due to the weight of soil is

$$Q_{qj} = \sum_1^m \frac{1}{3}\gamma_r\omega_r \quad (4.24)$$

in which

m = number of elements belonging to node q in given stage of construction,

j = direction index for force along x-axis,

γ_r = unit weight of soil in element r,

ω_r = area of element r taken as $0.5 \mid \Delta \mid$, see Equation 4.12.

Inclusion of the consecutive zone gives rise to Q_{qj} only at the nodes belonging to this zone. Rising water level on the upstream side (by zones) generates buoyancy at the nodes corresponding to the zones of water level variation. In this case one has

$$Q_{qj} = \frac{1}{3}(\gamma_{sub.r} - \gamma_r)\omega_r$$

The forces Q_{qi} arise if water level in reservoir alters at the edge of the core of a counterseepage unit. The forces Q_{qi} are merely due to rising head. Moreover, the vertical component of water pressure caused by the slope of the upstream edge of the core, Q_{qj}, is added.

Sequence of the nodes at which variations are produced is given by the order of elements.

The problem is particular because the soil model is incremental and has limitations (flow condition). Hence the result depends on loading path, which is determined by stages of construction and the geometric forms. In addition, the order of passage over the nodes and variations in directions produces a certain loading 'history'.

The difficulties may be avoided by linearization of stress-strain relationship within a stage, but this generates considerable inaccuracies. Another way is to reduce the initial step of variations but this requires longer times of computations. The method presented is attractive, despite the hypotheses made, because the direction of variations may be suited to the real construction sequence for earth dams, which even softens the assumption on instantaneous construction of a consecutive zone.

The experience also shows that the solution does not fall in 'resonance' but instead follows the desired setup of displacement with changing step of variations.

For variations with step $\pm h$, the strain components $\delta\delta e_x$, $\delta\delta e_y$, $\delta\delta e_{xy}$ are determined in each case for all elements. The strain $\delta\delta e_z$ (normal to drawing) is always zero by definition for plane strain problem. Computation of strains permits determination of increments of volumetric strains

$$\delta\delta e = \delta\delta e_x + \delta\delta e_y$$

and the diagonal components of strain deviator

$$\delta\delta\epsilon_x = \delta\delta e_x - \frac{1}{3}\delta\delta e; \quad \delta\delta\epsilon_y = \delta\delta e_y - \frac{1}{3}\delta\delta e; \quad \delta\delta\epsilon_z = -\frac{1}{3}\delta\delta e$$

By initial data for active loading and $k_s > 1$ one resorts to Equation 1.31 and finds components of stress deviator

$$\delta\delta S_x = 2|\sigma|^{1-n}\left\{f(v)\frac{E_0}{n}e^{-b(1-\bar{k})} + D_{S0.dam}\bar{k}\left[1 - e^{-b(1-\bar{k})}\right]\right\}\delta\delta\epsilon_x \qquad (4.25)$$

....

For soil one may assume $f(v)$ about 1.5.

For collective Γ, and for the condition $k_s > 1$ or $\bar{k} > 0$ one finds the increment of volumetric strain due to shear and then uses Equation 1.31 to compute mean stress increment

$$\delta\delta\sigma = \frac{E_0\delta_{ij}}{n|\sigma|^{n-1}}\left[\delta\delta e - \text{sign}\,(\Gamma - \Gamma_0)\frac{M\delta\delta\Gamma}{|\sigma|}\right] \qquad (4.26)$$

$$\delta\delta\Gamma = \sqrt{\frac{2}{3}}\sqrt{2(\delta\delta\epsilon_x)^2 + 2(\delta\delta\epsilon_y)^2 - 2\delta\delta\epsilon_x\delta\delta\epsilon_y + 1.5\delta\delta\epsilon_{xy}} \qquad (4.27)$$

Find components of stress tensor increments

$$\delta\delta\sigma_x = \delta\delta\sigma + \delta\delta S_x \qquad (4.28)$$

....
....

Using Equation 4.22 one finds the sign of δE. For a positive value one changes the sign of h and the process of variations is repeated. Upon completion the following set is formulated for each element belonging to the node in a given stage of construction

$$\delta\sigma_x^n = \delta\sigma_x^{n-1} + \delta\delta\sigma_x$$

....

....

$$\sigma_x^n = \sigma_x^{n-1} + \delta\delta\sigma_x$$

....

....

$$e_x^n = e_x^{n-1} + \delta\delta e_x \qquad\qquad (4.29)$$

....

....

$$E_{vd}^n = E_{vd}^{n-1} + \sigma^n\delta\delta e$$
$$E_d^n = E_d^{n-1} + \sigma_x^n\delta\delta e_x + \sigma_y^n\delta\delta e_y + 2\sigma_{xy}^n\delta\delta e_{xy} - \sigma^n\delta\delta e$$
$$k_s = \frac{U_0 - E_m}{E_d^n}$$

The upper index n denotes the current variation, and $n - 1$ denotes the respective quantity in the element after the preceding variation. E_{vd} is the energy of volumetric strains in element, E_d = energy of form changes of element due to stress deviator, and k_s = safety factor (strength safety) of soil in element, by Equation 1.33.

Pass to variations in the perpendicular direction and to next node. For $k_s = 1$ the stress increment upon loading is possible only due to cubic compression by Equation 1.31, but without dilatancy (this is feasible for small h only). For unloading one uses Hooke's law (Eq.1.34).

Operation of an earth-rockfill dam becomes particular for the action of horizontal forces along a certain surface inside the area (upstream edge of counterseepage unit), while the particular features of the material of the upstream prism include its incapability of taking load.

In solutions with minimization of the functional by MLV the above difficulty may be overcome without a conflict with continuity of displacements. The nodes on the upstream edge of core are characteristic of horizontal forces induced by rising water level. At these nodes, the generation of δE upon variations of displacements in horizontal direction is implemented only for the core elements, while those for vertical displacements only due to elements of core or upstream prism (if buoyancy exceeds pressure on upstream edge), depending on sign of the vertical force. This approach is possible in view of the clear-cut definition of construction and loading stages and the other stages, linked to rising water on the upstream side, i.e. due to one of the primary construction factors — sequence of construction.

The upstream water pressure forces are taken as surface ones, exerted on the upstream edge of core or screen, because the elastoplastic strains in earth-rockfill dams are displayed rapidly, while the volumetric forces induced by seepage in clayey cores are by far slower. On the other hand, one often needs evaluation of pore pressure to predict cracking in core (Section 4.4).

The growth of strains in core is assessed by soil properties found in 'closed system' tests. This implies soil is a quasihomogeneous medium. Extremely slow outflow of water from a clayey core (screen) justifies this assumption for the stress-strain condition of dams. Strains and settlements due to consolidation may be found separately from consolidation problems.

One must stress that the general solution given by Zaretskiy (1983) in its treatment of consolidation holds primarily for inclusion of pore pressure in the upstream prism under seismic effects.

Limited speed of computers does not permit very fine grids for this class of problems. Solutions must be treated as approximate because of finite h and constant grid size τ.

4.3 FINITE ELEMENT METHOD COMBINED WITH METHOD OF LOCAL VARIATIONS (SPATIAL PROBLEM)

While solving spatial stress-strain problems for earth dams one must decide on a number of factors — shape of element, input system, definition of strain energy, determination of principal stresses, printing mode etc.

In the wake of the two-dimensional analysis in Section 4.2, one may think that an element with coordinate-independent strain tensor components should be preferred. In the spatial problem, tetrahedron becomes a counterpart of triangle; decomposition into tetrahedrons and processing of such data is however inconvevient, although some solutions are available (Fradkin 1973). For convevient comparison of two- and three-dimensional problems we use a pentagon (Fig.4.3b) which is generally an irregular feature for stress-strain problems in earth dams (bases are not mutually parallel).

Again for the sake of comparison, the x- and y-axis are preserved as in the two-dimensional formulation, and the z-axis is oriented from the LHS bank of canyon to RHS (Fig.4.3a, b).

In a given element, displacements depend on coordinates, by analogy to Equation 4.10:

$$\begin{aligned}
u_x &= \alpha_1 + \alpha_2 x + \alpha_3 y + \alpha_4 z + \alpha_5 xz + \alpha_6 yz \\
u_y &= \alpha_7 + \alpha_8 x + \alpha_9 y + \alpha_{10} z + \alpha_{11} xz + \alpha_{12} yz \\
u_z &= \alpha_{13} + \alpha_{14} x + \alpha_{15} y + \alpha_{16} z + \alpha_{17} xz + \alpha_{18} yz
\end{aligned} \tag{4.30}$$

The number of constants in Equation 4.30 should be equal to the number of nodes (6) multiplied by the measure of space (3); i.e. 18.

In order to find constants from displacements of nodes one must formulate three systems of linear algebraic equations, six equations in each system, by analogy to Equation 4.12:

$$\{u\} = \{\alpha\}[x] \tag{4.31}$$

in which

$\{a\}$ = column of unknown constants from Equation 4.30,

$[x]$ = matrix of node coordinates.

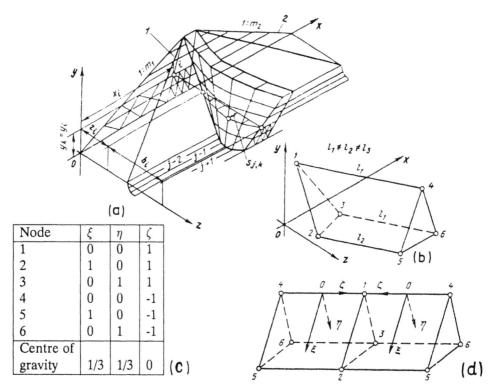

Node	ξ	η	ζ
1	0	0	1
2	1	0	1
3	0	1	1
4	0	0	-1
5	1	0	-1
6	0	1	-1
Centre of gravity	1/3	1/3	0

(c)

Figure 4.3. Design scheme of dam and its FEM approximation (spatial problem). (a) FEM decomposition; (b) finite element; (c) values of local coordinates at characteristic points of element; (d) adjacent elements and their local coordinate systems; (1) dam trace; (2) projection of dam profile on the plane Oxy; (i) number of node in profile; (j) number of vertical section; (k) number of horizontal section.

Since the matrix $[x]$ in some cases does not have an inverse (Zienkiewicz 1975), the solution becomes complicated; therefore normalized coordinates are preferred in determination of the strain functions. The transformations may be found in Bezukhov & Luzhin (1974), Kolar et al. (1975) and Zienkiewicz (1975). In standardized coordinates, the characteristic points of an element have strictly defined local coordinates, the axes of which are rigidly connected with each element. The configuration of the axes of the standardized system can be chosen arbitrarily. In the element considered, the axes of local coordinates ξ, η, ζ are oriented as shown in Figure 4.3b, but the local numbers of nodes themselves change with the node for which the variations of displacements are taken. The latter is assigned the number 1 (Rasskazov & Belakov 1982). Accordingly, the opposite node on the same side may be attributed No.4 etc, as in Figure 4.3c.

Characteristic points of element have their standardized coordinates as indicated in Figure 4.3d. Geometrically, the standardization may be regarded as a transformation of an element in the coordinates ξ, η, ζ into a regular right-angled triangular prism with unit sides of the base and double-unit height.

The transition from local to global coordinates, and vice versa, is given as follows:

$$
\begin{aligned}
x &= p_1 x_1 + p_2 x_2 + p_3 x_3 + p_4 x_4 + p_5 x_5 + p_6 x_6 \\
y &= p_1 y_1 + p_2 y_2 + p_3 y_3 + p_4 y_4 + p_5 y_5 + p_6 y_6 \\
z &= p_1 z_1 + p_2 z_2 + p_3 z_3 + p_4 z_4 + p_5 z_5 + p_6 z_6
\end{aligned}
\tag{4.32}
$$

The coefficients $p_1, p_2, \ldots, p_6$ are functions of standardized coordinates:

$$
\begin{aligned}
p_1 &= \frac{1}{2}(1 - \xi - \eta)(1 + \zeta); \quad p_4 = \frac{1}{2}(1 - \xi - \eta)(1 - \zeta) \\
p_2 &= \frac{1}{2}\xi(1 + \zeta); \quad p_5 = \frac{1}{2}\xi(1 - \zeta) \\
p_3 &= \frac{1}{2}\eta(1 + \zeta); \quad p_6 = \frac{1}{2}\eta(1 - \zeta)
\end{aligned}
\tag{4.33}
$$

By analogy to Equation 4.32 the displacement of any point is given in terms of angular displacements:

$$
u_x = p_1 u_{x1} + p_2 u_{x2} + p_3 u_{x3} + p_4 u_{x4} + p_5 u_{x5} + p_6 u_{x6}
\tag{4.34}
$$

....

Any function φ determined in local coordinates (displacements, strains, stresses etc) can be presented in standardized coordinates through the following differential equations

$$
\begin{aligned}
\frac{\partial \varphi}{\partial \xi} &= \frac{\partial x}{\partial \xi}\frac{\partial \varphi}{\partial x} + \frac{\partial y}{\partial \xi}\frac{\partial \varphi}{\partial y} + \frac{\partial z}{\partial \xi}\frac{\partial \varphi}{\partial z} \\
\frac{\partial \varphi}{\partial \eta} &= \frac{\partial x}{\partial \eta}\frac{\partial \varphi}{\partial x} + \frac{\partial y}{\partial \eta}\frac{\partial \varphi}{\partial y} + \frac{\partial z}{\partial \eta}\frac{\partial \varphi}{\partial z} \\
\frac{\partial \varphi}{\partial \zeta} &= \frac{\partial x}{\partial \zeta}\frac{\partial \varphi}{\partial x} + \frac{\partial y}{\partial \zeta}\frac{\partial \varphi}{\partial y} + \frac{\partial z}{\partial \zeta}\frac{\partial \varphi}{\partial z}
\end{aligned}
\tag{4.35}
$$

Denote

$$
\begin{vmatrix}
\dfrac{\partial x}{\partial \xi} & \dfrac{\partial y}{\partial \xi} & \dfrac{\partial z}{\partial \xi} \\
\dfrac{\partial x}{\partial \eta} & \dfrac{\partial y}{\partial \eta} & \dfrac{\partial z}{\partial \eta} \\
\dfrac{\partial x}{\partial \zeta} & \dfrac{\partial y}{\partial \zeta} & \dfrac{\partial z}{\partial \zeta}
\end{vmatrix} = [I]
\tag{4.36}
$$

Then Equation 4.35 will read

$$
\left\| \begin{matrix} \frac{\partial \varphi}{\partial \xi} \\ \frac{\partial \varphi}{\partial \eta} \\ \frac{\partial \varphi}{\partial \zeta} \end{matrix} \right\| = [I] \left\| \begin{matrix} \frac{\partial \varphi}{\partial x} \\ \frac{\partial \varphi}{\partial y} \\ \frac{\partial \varphi}{\partial z} \end{matrix} \right\|
\tag{4.37}
$$

Components of the strain tensor for a spatial problem can be written as

$$
e_{xx} = \frac{\partial u_x}{\partial x}; \quad e_{yy} = \frac{\partial u_y}{\partial y}; \quad e_{zz} = \frac{\partial u_z}{\partial z}
$$

$$
e_{xy} = \frac{\partial u_x}{\partial y} + \frac{\partial u_y}{\partial x}; \quad e_{xz} = \frac{\partial u_x}{\partial z} + \frac{\partial u_z}{\partial x}; \quad e_{yz} = \frac{\partial u_y}{\partial z} + \frac{\partial u_z}{\partial y} \tag{4.38}
$$

If the functional φ in Equation 4.37 is deemed a function of displacements, then Equation 4.37 may be presented as follows to find strains:

$$
\begin{Vmatrix} \frac{\partial \varphi}{\partial x} \\ \frac{\partial \varphi}{\partial y} \\ \frac{\partial \varphi}{\partial z} \end{Vmatrix} = [I]^{-1} \begin{Vmatrix} \frac{\partial \varphi}{\partial \xi} \\ \frac{\partial \varphi}{\partial \eta} \\ \frac{\partial \varphi}{\partial \zeta} \end{Vmatrix} \tag{4.39}
$$

It is crucial to determine the elements of the inverse matrix $[I]^{-1}$. They are given through the global coordinates as provided in Equations 4.32 and 4.36:

$$
[I] = [p'] \begin{vmatrix} x_1 & y_1 & z_1 \\ x_2 & y_2 & z_2 \\ x_3 & y_3 & z_3 \\ x_4 & y_4 & z_4 \\ x_5 & y_5 & z_5 \\ x_6 & y_6 & z_6 \end{vmatrix} \tag{4.40}
$$

$$
[p'] = \begin{vmatrix} \frac{\partial p_1}{\partial \xi} & \frac{\partial p_2}{\partial \xi} & \frac{\partial p_3}{\partial \xi} & \frac{\partial p_4}{\partial \xi} & \frac{\partial p_5}{\partial \xi} & \frac{\partial p_6}{\partial \xi} \\ \frac{\partial p_1}{\partial \eta} & \frac{\partial p_2}{\partial \eta} & \frac{\partial p_3}{\partial \eta} & \frac{\partial p_4}{\partial \eta} & \frac{\partial p_5}{\partial \eta} & \frac{\partial p_6}{\partial \eta} \\ \frac{\partial p_1}{\partial \zeta} & \frac{\partial p_2}{\partial \zeta} & \frac{\partial p_3}{\partial \zeta} & \frac{\partial p_4}{\partial \zeta} & \frac{\partial p_5}{\partial \zeta} & \frac{\partial p_6}{\partial \zeta} \end{vmatrix} \tag{4.41}
$$

Differentiate Equation 4.33 to get

$$
[p'] = \frac{1}{2} \begin{vmatrix} -(1+\xi) & (1+\xi) & 0 & -(1-\zeta) & (1-\zeta) & 0 \\ -(1-\xi) & 0 & (1+\xi) & -(1-\zeta) & 0 & (1-\zeta) \\ (1-\xi-\eta)\xi & \xi & \eta & -(1-\xi\eta) & -\xi & -\eta \end{vmatrix} \tag{4.42}
$$

If the elements of the matrix $[I]$ are denoted by d_{ij}, then combination with Equations 4.40 and 4.42 yields

$$
\begin{aligned}
d_{11} &= [-(1+\zeta)x_1 + (1+\zeta)x_2 - (1-\zeta)x_4 - (1-\zeta)x_5]/2 \\
d_{12} &= [-(1+\zeta)y_1 + (1+\zeta)y_2 - (1-\zeta)y_4 - (1-\zeta)y_5]/2 \\
d_{13} &= [-(1+\zeta)z_1 + (1+\zeta)z_2 - (1-\zeta)z_4 - (1-\zeta)z_5]/2 \\
d_{21} &= [-(1+\zeta)x_1 + (1+\zeta)x_3 - (1-\zeta)x_4 - (1-\zeta)x_6]/2 \\
d_{22} &= [-(1+\zeta)y_1 + (1+\zeta)y_3 - (1-\zeta)y_4 - (1-\zeta)y_6]/2 \\
d_{23} &= [-(1+\zeta)z_1 + (1+\zeta)z_3 - (1-\zeta)z_4 - (1-\zeta)z_6]/2 \\
d_{31} &= [(1-\xi-\eta)x_1 + \xi x_2 + \eta x_3 - (1-\xi-\eta)x_4 - \xi x_5 - \eta x_6]/2 \\
d_{32} &= [(1-\xi-\eta)y_1 + \xi y_2 + \eta y_3 - (1-\xi-\eta)y_4 - \xi y_5 - \eta y_6]/2 \\
d_{33} &= [(1-\xi-\eta)z_1 + \xi z_2 + \eta z_3 - (1-\xi-\eta)z_4 - \xi z_5 - \eta z_6]/2
\end{aligned} \tag{4.43}
$$

It is comfortable to examine the stresses and strains at the centre of gravity of an element. By putting in Equation 4.43 the standardized coordinates of the centre of gravity (Fig.4.3d) one obtains elements of the matrix $[I]$ for the centre of gravity:

$$
\begin{aligned}
d_{11} &= (-x_1 + x_2 - x_4 + x_5)/2 \\
d_{12} &= (-y_1 + y_2 - y_4 + y_5)/2 \\
d_{13} &= (-z_1 + z_2 - z_4 + z_5)/2 \\
d_{21} &= (-x_1 + x_3 - x_4 + x_6)/2 \\
d_{22} &= (-y_1 + y_3 - y_4 + y_6)/2 \\
d_{23} &= (-z_1 + z_3 - z_4 + z_6)/2 \\
d_{31} &= (x_1 + x_2 + x_3 - x_4 - x_5 - x_6)/6 \\
d_{32} &= (y_1 + y_2 + y_3 - y_4 - y_5 - y_6)/6 \\
d_{33} &= (z_1 + z_2 + z_3 - z_4 - z_5 - z_6)/6
\end{aligned}
\tag{4.44}
$$

Denoting elements of $[I]^{-1}$ by r_{ij} and taking $y_1 = y_4, y_2 = y_5, y_3 = y_6$ (Fig.4.3a) one obtains d_{32} from Equation 4.44. The matrix $[I]$ is of third rank so that the inverse operation is not difficult:

$$
r_{ij} = (-1)^{i+j} \frac{[A_{ij}]}{[A]}
\tag{4.45}
$$

in which

$[A_{ij}]$ = algebraic complement to the element, $[A_{ij}] = [I_{ij}]$,
r_{ij} = matrix generated from $[I_{ij}]$ by subtraction of the row and column with the indices i and j, respectively,
$[I]$ = determinant of $[I_{ij}]$.

Taking into account $d_{32} = 0$ one obtains

$$
r_{11} = \frac{d_{22}d_{23}}{[I]}; \quad r_{12} = -\frac{d_{21}d_{33}}{[I]}; \quad r_{13} = \frac{d_{21}d_{23} - d_{22}d_{13}}{[I]}
$$

$$
r_{21} = \frac{d_{21}d_{32} - d_{23}d_{31}}{[I]}; \quad r_{22} = \frac{d_{11}d_{33} - d_{31}d_{13}}{[I]}
$$

$$
\tag{4.46}
$$

$$
r_{23} = \frac{d_{11}d_{23} - d_{13}d_{21}}{[I]}
$$

$$
r_{31} = \frac{d_{21}d_{32} - d_{22}d_{31}}{[I]}; \quad r_{32} = \frac{d_{12}d_{31}}{[I]}; \quad r_{33} = \frac{d_{11}d_{22} - d_{12}d_{21}}{[I]}
$$

The determinant

$$
[I] = d_{11}d_{22}d_{33} + d_{12}d_{23}d_{31} - d_{31}d_{22}d_{13} - d_{12}d_{21}d_{33}
\tag{4.47}
$$

also includes $d_{32} = 0$.

For determination of strains at the centre of gravity of the element one must differentiate by local coordinates (Eq.4.34) with inclusion of Equation 4.33, as it has been done in determination of $[I]$ (Eq.4.43):

$$\frac{\partial u_x}{\partial \xi} = \frac{\partial p_1}{\partial \xi} u_{x1} + \frac{\partial p_2}{\partial \xi} u_{x2} + \frac{\partial p_3}{\partial \xi} u_{x3} + \frac{\partial p_4}{\partial \xi} u_{x4} + \frac{\partial p_5}{\partial \xi} u_{x5} + \frac{\partial p_6}{\partial \xi} u_{x6} \qquad (4.48)$$

or upon use of Equation 4.42

$$\frac{\partial u_x}{\partial \xi} = \frac{1}{2} [-(1+\zeta) u_{x1} + (1+\zeta) u_{x2} - (1-\zeta) u_{x4} + (1-\zeta) u_{x5}] \qquad (4.49)$$

In the method of local variations, the operations are simultaneous only at one node (always No.1) and in one direction. Thus

$$\frac{\partial u_x}{\partial \xi} = \frac{1}{2} (1+\zeta) u_{x1} \qquad (4.50)$$

and substitution of ξ for the centre of gravity of the element (O) yields

$$\frac{\partial u_x}{\partial \xi} = -\frac{1}{2} u_{x1} \qquad (4.51)$$

By analogy one obtains similar relations for the other elements of the RHS of the column in Equation 4.39:

$$\frac{\partial u_x}{\partial \eta} = -\frac{1}{2} u_{x1}; \quad \frac{\partial u_x}{\partial \zeta} = \frac{1}{6} u_{x1}$$

in which

u_{x1} = value of varied displacement.

The relationships between strains and displacements are derived in similar manner.

In accordance with Equation 4.39 for components of $[I]^{-1}$ in the form of Equation 4.46 one determines the strain components at the centre of gravity:

$$\begin{aligned}
\frac{\partial u_x}{\partial x} &= \frac{1}{2} \left(-r_{11} - r_{12} + \frac{r_{13}}{3} \right) u_{x1} \\
\frac{\partial u_x}{\partial y} &= \frac{1}{2} \left(-r_{21} - r_{22} + \frac{r_{23}}{3} \right) u_{x1} \qquad (4.52) \\
\frac{\partial u_x}{\partial z} &= \frac{1}{2} \left(-r_{31} - r_{32} + \frac{r_{33}}{3} \right) u_{x1}
\end{aligned}$$

Variations of the displacements u_{y1}, u_{z1}, with Equation 4.52, yield respectively u_y, u_z and u_y, u_z.

Hence one obtains all components as stipulated in Equation 4.38 for the outlined calculus of variations, in one of the directions, by global coordinate axes.

The algorithm for strains of the centre of gravity is as follows: select a node; find displacement (as in plane problem); select elements adjacent to the node; compute d_{ij} by Equation 4.44 and r_{ij} by Equation 4.46 in each element; find derivatives of displacements by global coordinates, by Equation 4.52; and finally determine components of the strain tensor at the centre of gravity of the element.

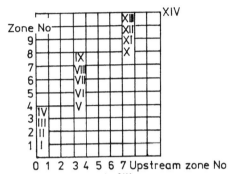

Figure 4.4. Graph of dam zoning and reservoir filling.

The remaining part of the algorithm is as for the plane problem but the calculus of variations proceeds in three directions.

Some difficulties arise in the search for the strain energy. It must be noted that the projection of the element on the plane xOy is a triangle, so that the components of the strain tensor vary only along $z(\zeta)$. Integration of the strain energy by Equation 4.21 proceeds along z by the Simpson method (Zienkiewicz 1975), with inclusion of the stresses and strains of the adjacent element (along the z-axis) which contains the varied node. In other respects, the FEM using minimum energy functional does not differ from the method of local variations.

One may note a certain difficulty arising in the determination of principal strains and stresses, compared with the plane problem; nonetheless the Schwarz-Pickard method (Section 5.4) can be used to find the eigenvalues and eigenvectors; any other method involving diagonalization of matrices may also prove suitable.

4.4 RESULTS OF SOLUTIONS TO STRESS-STRAIN PROBLEMS FOR EARTH DAMS

Stress-strain conditions of earth dams in stagewise construction. Consider the growth of stresses and strains by an example of a 300-m dam with central core and thrust prisms of pebbles, the external zones of which are made of rock. The plane strain problem was solved by the method described in Section 4.2 (FEM using MLV). Consider two cases of dam construction by horizontal and sloping layers. In construction by horizontal layers (Rasskazov 1977) one builds a dam by layers reaching full width (Fig.4.8a). Construction by sloping layers (zones) involves horizontal layers for the upstream prism and core, and the construction of the tailwater slope in proportion to the growth of the dam, in layers parallel to the tailwater slope (Fig.4.8b).

The construction of a dam by inclined layers is usually applied to very high dams, while a hydraulic scheme is put in operation at low heads and the first tier is operated at minimum scope of works. Filling of soil in both cases is implemented by fine horizontal layers, with subsequent rolling. Hence in both cases of construction the physico-chemical properties of soil are identical, and the differences consist in loading path for different sequences of construction (one must distinguish the stage of operation and the zone of operation; the former includes either the zone of operation or usual stepwise impoundment of water on the upstream side).

Figure 4.4 illustrates the graph of dam construction by zones. The coordinate axis shows zones of operation, for which the numbers up to 4 inclusive coincide with numbers of operation stages. The abscissae axis shows the zones to the level of which water was impounded on the upstream side. For instance, stage V was associated with impoundment from zone 0 to 3 inclusive. Further, up to stage IX the dam was again constructed up to number 8 inclusive. In stage X the upstream level was raised from third to seventh zone etc. The final stage (XIV) was linked to rise from the seventh zone to the tenth zone.

Figure 4.5a–d shows results of the solution of the stress-strain problem for dam construction by horizontal layers, while Figure 4.5e illustrates the solution for horizontal and inclined layers. The results are presented in this manner because the character of graphs and isolines does not expose considerable qualitative differences; the discrepancies are mostly quantitative. More details for the results are discussed below for various points of the structure.

Vertical displacements in dam core (Fig.4.5a) are much higher than those in the thrust prisms, which is primarily due to high deformability of core material and also because of the transfer of pressure (vertical) exerted by water on the upstream side and displayed on the thrust edge of the core. The transfer of pressure to the core brings about prevailing displacements, notably of the upstream edge of the core. Shown in Figure 4.5a in vertical cross-section are the graphs of vertical displacements during construction of the dam. The graphs were constructed without displacements due to the own weight of a zone during its construction, that is the consolidation in the course of construction was not fixed. Such graphs somehow provide pictures of displacements for benchmarks placed at the top of each zone upon its completion, i.e. after the changes due to the weight of soil in a given zone are complete.

It is typical that the maximum of vertical displacement is located at $H/3$, in which H = height of dam. That maximum is usually at $1/3$ to $1/2H$. The same result is got from field measurements on constructed facilities. It may be said in advance that the maximum vertical displacement with inclusion of spatial configurations in a rather narrow cross-section (having the factor $m \leq 2.5$) is displaced somewhere above $H/3$ but below $H/2$.

Horizontal displacements in plane problem (Fig.4.5b) have their maxima at the upstream edge of core and are uniformly distributed over the height of the core. Absolute values of horizontal displacement about core axis are 75–80%, or even 100% of the vertical displacement. The damping of the horizontal displacement on the tailwater prism is insignificant, which in general coincides with field findings. Solution to the spatial problem yields data on damping of horizontal displacements.

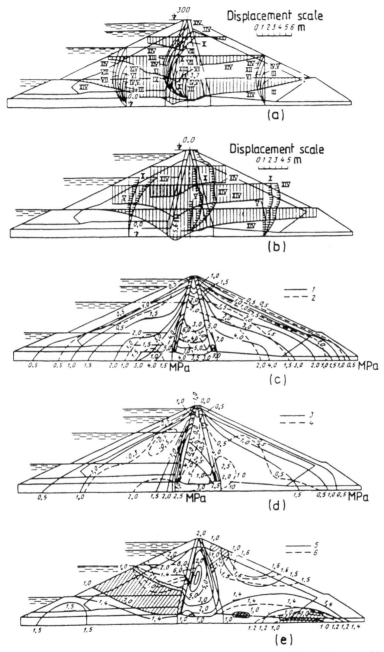

Figure 4.5. Stress-strain state of dam accounting for construction scheme. (a) vertical displacements at end of construction; (b) horizontal displacements at end of construction; (c) isolines of normal stresses σ_y (1) and σ_x (2); (d) isolines of stresses σ_z (3) and σ_{xy} (4); (e) isolines of reliability factors; (5) construction by horizontal layers; (6) construction by sloping layers.

Isobars σ_y (Fig.4.5c) are very interesting. The difference in deformation properties of dam soils brings about considerable unloading of core and concentration of stresses in the transition zone. The effect of core 'hanging' on thrust prisms (arc effect) is known since long owing to field observations on dams, and also from solutions of stress-strain problems for earth-rockfill dams by the photoelasticity method. It must be stressed that the arc effect also occurs in homogeneous fills, which is displayed in decrease in σ_y in comparison with γh over fill axis and increase in σ_y above γh in the slope zone.

The arc effect is very important because the decrease in vertical stresses with regard to γh may reach 60% and more, depending on the ratio of strain properties of the core and prisms, and core width. In the solution shown the arc effect reaches about 25%. Such considerable unloading can cause that σ_y on the upstream edge of the core is lower than the respective value of $\gamma_w h$, which is the pore pressure p_w on the thrust edge.

It must be noted that, because of the peculiar features of core construction, when edges are shaped as a 'pine' as a result of layer by layer compaction, permanent conditions of cracking occur due to different strain properties of core and transition zones; therefore this process should be analysed not at the thrust edge of core but along its axis and σ_y should be compared with the respective p_w. It is the cracking strength of the central zone of core that controls its monolithic properties. Therefore the comparison of σ_y at the thrust edge with $\gamma_w y$ provides a certain margin of safety.

In general, the reliability factor due to cracking on horizontal planes is determined as follows:

$$k_s^{cr} = \frac{\sigma_y + c_t}{p_w} \qquad\qquad (4.53)$$

in which

> k_s^{cr} = cracking reliability factor,
> c_t = tensile cohesion (Eq.1.51; Fig.1.19),
> p_w = pressure of water at given point.

At present the permissible value of k_s for dams is not limited by standards. It seems that k_s must be above 1.1 for dams of all classes since they accomodate transition zones designed for 'self-healing' of cracks if the latter arise. Hence the presence of the additional system of protection makes it possible to lower the permissible value k_s versus usual values of k_s for a given class of structure.

In consideration of possible cracking on horizontal planes one may reduce c_t because in layer-by-layer compaction it is the horizontal planes for which c_t is slightly below that for vertical planes, or even close to zero, although some recent investigations have shown that c_t is only slightly reduced on horizontal planes.

Essential unloading can bring about k_s by Eq. 1.58 below 1. In this case the continuity of core can be disturbed and cracks will generate, so that hydraulic discontinuity appears. This process seems to have occurred on Boulderhead Dam

(Fig.4.6). Hence the analysis of stress-strain conditions provides a check on possible generation of hydraulic failure.

Isobars are shown in Figure 4.5c. Concentration of stresses σ_y on the upstream edge of core is a characteristic result of the action of water on the upstream side and the pressure of earth on the upstream thrust prism.

Diagrams of τ_{xy} (Fig.4.5d) supplement the above conclusions on the operation of a structure — they show considerable concentration of τ_{xy} at the interface of core and transition zones. Isobars σ_z (stresses normal to the plane of drawing, Figure 4.5.d) make it possible to evaluate the condition of hydraulic failure of core on vertical planes. The value of σ_z is close to σ_3 almost over the entire lower part of the area, i.e. the Lode-Nadai parameter by Eq. 1.26 is about -1, which corresponds to the work of soil in triaxial apparatus; there are however zones in which this parameter is zero or even 1. More details are provided below.

Isolines of reliability factors are given in Figure 4.5e. Analysis of slope stability is the most responsible stage of dam design. The method described in Sections 1.3 and 4.2. has singled out the safety factor for soil strength at each point of structure, which lays the basis for identification of limit state zones, i.e. the so-called differential strength safety factor may be determined. Transition to stability (strength) indicators for the entire structure, inferred from the distribution of differential safety factors at individual points is in general an ambiguous task. In principle, this task remains unsolved without additional assumptions.

One of the additional assumptions can stipulate that the possible failures surface be a smooth curve. Taking a series of possible smooth curves, by differentiating values of k_s one finds a mean weighed value of k_s for each of them. A smooth curve with minimum k_s will provide the stability factor for the structure. This procedure may be used for the upstream and tailwater slopes. The most hazardous ones are shown in Figure 4.5e, along with k_s at boundaries of the zones.

As seen in Figure 4.5e, the most hazardous surface is the one passing through the zone of limit state on the upstream thrust prism, foundation of core, and zones of limit state in the tailwater thrust prism. The mean value of k_s along this line is about 1.5. If the hazardous surface embodies above 50% of the volume of dam, one may introduce an indicator of overall stability of structure. The dam considered is practically of uniform strength as the stability factors for the upstream and tailwater slopes are close to 1.5.

Evaluation of the overall and local stability in design practice is executed by engineering methods based on some assumptions on the form of the failure surface (see Chapter 6).

The basic differences arising in the distribution of k_s, as controlled by the construction sequence, are observed at the crest of the tailwater thrust prism. If a dam is constructed by sloping layers, a diverging zone of limit state appears at the interface of the core and the tailwater thrust prism. The dynamics of the growth of limit state and the causes of the generation of this zone are exemplified below.

Hence the construction of a dam by sloping layers gives rise to a hazardous

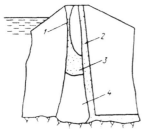 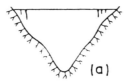 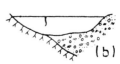

Figure 4.6. Cracking in Boulderhead Dam (Great Britain) (left). (1) failure hole; (2) transition; (3) failure zone; (4) moraine clay loam core.

Figure 4.7. Crack occurrence (right). (a) at crest of canyon — vertical planes stretching from upstream side to tailwater; (b) at dam crest, due to different deformability of foundation zones.

zone with regard to local stability in view of horizontal displacements towards tailwater; in terms of cracking in gravel and pebbles or rock, which is quite real, it is known that the so-called interlocking occurs (virtual cohesion). All these arguments point to the conclusion that traditional methods of stability assessment for slopes (Chapter 6) are principally inadequate to identify details of dam operation, for the problems are simplified at the stage of formulation.

Cracking in earth dams. Shear cracks appear if soil reaches its shear strength limit. They are hazardous in view of soil loosening upon shear (dilatancy) and due to increase in permeability in the zone of limit state of core. As a result, the zone can become a filter for the consolidated region of soil, and mean hydraulic gradients will be higher in the core, which in turn will give rise to local failures in core (screen). These zones require permanent inspection as to the composition and filling of filters for transitions. The pertinent information may be obtained from stress-strain problems solved for dams and axial cross-section of dams. The zones of limit state are shown in Figure 4.5e.

Detachment cracks usually appear at top of dam; they are dangerous because they may enhance erosion by water on the upstream side.

Cracks are commonly generated at flanks of dams if the cross-section factor m is below 2 (Fig.4.7a) — at crest above the inflexion of canyon profile; they stretch from the upstream side to the tailwater (Fig.4.7b).

Detachment cracks also appear in the longitudinal cross-section, parallel to dam axis, as a result of different strain properties of core and thrust prisms.

Information on cracking in coarse grained soil and at interfaces of coarse grained soils and dam cores has appeared recently. The opening of cracks reaches 20–40 cm, their depth being up to 10–15 m, whereupon the rubble keeps an inverted slope form. Such cracks are usually stretching along dam crest. Their occurrence testifies to the action of interlocking in coarse grained soil.

Solution of stress-strain problems for plane situations along dam crest provide distorted results as they do not include, for instance, water pressure on the upstream side. Spatial problems must be solved in evaluation of the operation of dam.

Cracks are sometimes generated after a structure is put into operation. A large dam is constructed several years, and if detachment cracks were generated due to tensile stresses during construction, along with unacceptable elongations, they would have been eliminated on site. Soil has creep properties (deformability under load in time domain), whereupon the closer the soil condition to its limit state, the more intensive the creep properties; consolidation additionally occurs in soil. As the result, once the construction is completed the process of heterogeneous deformation and redistribution of stresses takes place — tensile strength limit is reached in some zones and cracks are generated. If stress-strain design does not include creep properties, the above process cannot be singled out, but still the crack-hazard zones may be identified.

Experimental data and field observations have proved that soils with plasticity index $I_P < 15$ are more susceptible to cracking than soils with fine fractions and higher plasticity index, which are less vulnerable to tensile strains and cracking.

Wet and underconsolidated soil is less vulnerable to cracking than well consolidated soil of low moisture content. These findings make it possible to recommend the construction of the upper part of core in high dams of more plastic soil (if available), and avoidance of oversaturated and unconsolidated clayey soils.

Distribution of stresses in counterseepage units of dams (core or screen) make also possible more accurate evaluation of pore pressure (see Chapter 3).

The above data characterizes the condition of dam at the end of the construction period. It is interesting to analyse the dynamics of the growth of stress-strains and other characteristics of the structure work during its construction. To this end Figure 4.8 shows with numerals the points at which the growth of stresses and k_s were considered. The diversification in the selection of the points has been caused by the specific formulation of FEM solutions in which displacements are determined at nodes and stresses at the centres of gravity of the elements.

Growth of displacements stresses, reliability factor and other parameters during construction of dams. Figure 4.9 illustrates the results obtained for the growth of displacement components during stagewise construction. The abscissae axis identifies the stages corresponding to rise in water level in reservoir (Fig.4.8).

Note that in the absence of substantial differences between the results obtained for sloping and horizontal layers, the results for sloping layers were drawn only in Figures 4.9–4.11.

One does not identify any essential qualitative difference at points 1–8. A significant quantitative difference in vertical displacements for the last stage of construction may be seen only for points 4 and 5. These points are located in the dam zones in which the sequence of construction should not control the operation of dam. The highest discrepancy in displacements is noted at points of tailwater

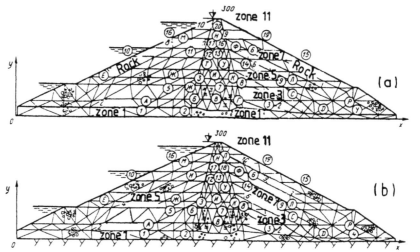

Figure 4.8. Finite-element decomposition upon dam construction. (a) horizontal layers; (b) sloping layers.

thrust prism and core (13, 14, 15, 17, 18, 19, 20); it exceeds one metre.

The most important 'interference factor' is the motion of water on the upstream side. Rise of water level in the core brings about not only the growth of horizontal displacements but also decrease in vertical displacements (points 7, 13 and 20) — 'displacement' in its common sense, whereupon the higher the point in the core, the more important this effect. For sloping layers the 'displacement' effect is more pronounced. On the other hand, for the points on the upstream thrust prism (1, 5, 6, 10, 11, 12, 16 and 17), the rise in water level on the upstream side is accompanied by increase in vertical displacements, although they are minute.

Horizontal displacements increase considerably with rising water level on the upstream side, but observed at the slope points are considerable displacements caused also by the growth of the structure, particularly if the construction proceeds by horizontal layers.

It is worthwhile to note that changes in the design graphs of operation parameters for stagewise construction are similar — by the character of the curves — to the graphs obtained from field observations; they illustrate the dynamics of the growth of these parameters under fairly complex loading conditions.

The dynamics of stresses in stagewise construction, for various schemes of construction, is illustrated in Figure 4.10.

At the points belonging to the dam core (C, J, N, P, T, U) one observes smooth growth of stress components, whereupon the effect of the sequence of construction on the stress condition is not exposed at points on the axis of the lower part of core (C, U). This effect is pronounced in the upper part of the core, and is especially strong for σ_y at point P.

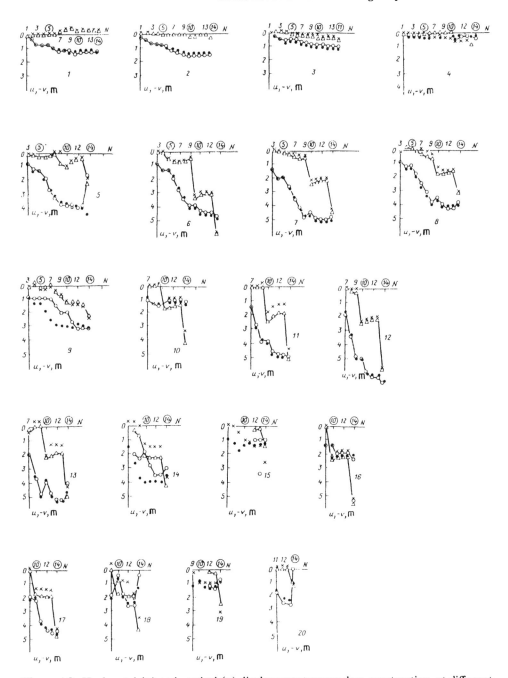

Figure 4.9. Horizontal (u) and vertical (v) displacements upon dam construction, at different points (cf. Figs 4.4, 4.8). (o) (Δ) v&u for sloping layers; ($\bullet$) ($\times$) v&u for horizontal layers.

Values of $\gamma_w h$ are also given in the graphs for points C, J, N, P, which sheds light on the dynamics of the arc effect. It is seen that the arc effect begins growing from the instant when dam reaches at least 30% of its height above the points considered.

Variation of k_s at given points is different. At lower points on core axis (C) k_s grows rapidly with the growth of dam. Impoundment of reservoir in the fifth and tenth stage produced no real change in k_s, but the last stage of impoundment brought about significant decrease. At a slightly higher point on core axis the first stage of impoundment (10th) brought about consolidation of soil, and subsequent filling (14th) caused reduction of k_s. Even higher along the axis (points N and P) the motion of water caused strong decrease in k_s.

Points T and U are situated symmetrically on the upstream and tailwater edges of the core at mid-height of dam. The sequence of construction controls these points very strongly, particularly the tailwater one (U). The construction by horizontal layers largely increases the strength of soil at this point, and increasing upstream water level brings about improved work of soil.

Points D, E, R, S, K, L, O, V belong to the tailwater thrust prism. The effect of construction sequence on stress condition and k_s is pronounced in different manner for various points. At points E, R, S, L the operation of soil worsens if the dam is constructed by horizontal layers. At point D the effect of the construction sequence is not observed. At points K, O, V the construction by horizontal layers improved the work of soil, but it must be stated that points E, R, S, L are inside the tailwater thrust prism while points O, V control the stability of the upstream part of the prism.

Point V is of particular interest because it is located in the zone of limit state upon construction by sloping layers. The growth of the limit state at point V occurs at rising water on the upstream side, but in the construction by horizontal layers the tangential stresses are positive and σ_x and σ_y are higher, which enhances the capacity of soil to strengthen during construction. Upon construction by sloping layers the tangential stresses are negative, and σ_x and σ_z are smaller by absolute value than in the first case. In the end this does not yield a necessary margin of safety, and the soil reaches its limit state upon rise of water in the final stage of construction. It must also be emphasized that σ_z at these points is the maximum normal stress, i.e. the wedging action of soil is here pronounced to a large extent.

Points A, B, G, I, M belong to the upstream thrust prism. The effect of construction sequence is of no practical importance. The rising water on the upstream side brings about submergence of soil, which in most cases produces lower normal stresses, higher tangential stresses (B) and lower k_s.

However, tangential stresses do not grow at every point (G, H, I) and k_s is reduced in this case due to more intensive decrease in σ_x upon growth of σ_y and σ_z. Point M is an exception — it is close to the upstream slope, and water level rise brings about increase in σ_x and σ_z for almost unchanged σ_y for the maximum

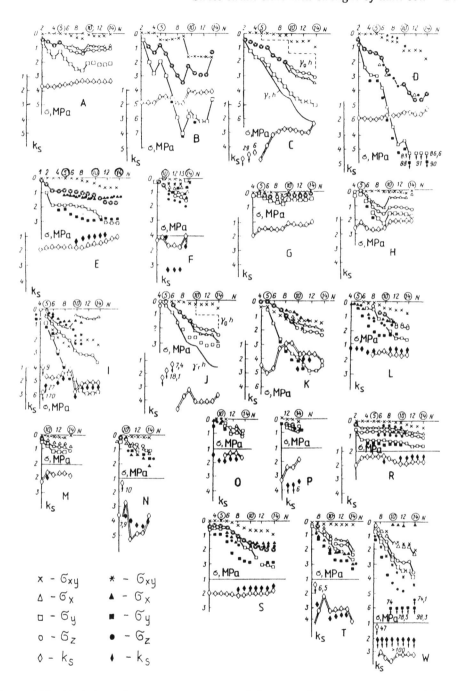

Figure 4.10. Variation of stress components and k_s upon dam construction, at different points (cf. Fig.4.8). Sloping layers: $(\times)$ σ_{xy}; $(\triangle)$ σ_x; $(\square)$ σ_y; $(\circ)$ σ_z; $(\diamond)$ k_s. Horizontal layers: $(\star)$ σ_{xy}; $(\triangle)$ σ_x; $(\square)$ σ_y;$(\bullet)$ σ_x; $\diamond$ k_s.

normal stress. Reduction in components of the stress deviator brings about growth of k_s.

Figure 4.11 makes possible evaluation of the discrepancies in the configuration of the planes of principal stresses and strains and the notation of the principal planes for stagewise construction; it also provides the Lode-Nadai parameter for stresses, λ and the parameter of loading trajectory μ for fixed λ (Eqs 1.26 and 1.27).

These parameters determine the loading path.

Figure 4.11 shows the variation of all parameters of loading path in the course of dam construction by sloping layers and upon reservoir filling.

Consider first the lower points on the core axis (C, J). For point C one has a characteristic coaxiality of the stress tensor and strain tensor, which is typical of the points close to dam foundation (D, E, R, A), when the discrepancy of plane angle (up to 10–15°) is insignificant, occurs only in initial stages of loading, and decreases to 0–5° upon subsequent loading. The parameter λ is close to -1, and μ varies from 1 to 0 (it should be remembered that $\mu = 0.33$ is constant in standard triaxial tests). The highest rotation of the axes of principal stresses and strains at point C reaches 15°, which occurs in the stages of rising water level. The parameter λ at this point varies little, form -1 to -0.6. The variation of μ is considerable, from 0.4 to 3.2, which indicates the leading increase in mean stress, that is hardening of soil (for cubic compression $\mu \to \infty$). Cases of pure cubic compression do not occur in the structure, but yet for $\mu > 3.0$ the stress condition may practically be considered close to cubic compression.

At point J located higher at the core axis, a considerable discrepancy in the configuration of the planes of principal stresses and strains (up to 60°) takes place in initial stages of loading. Because of the compressibility of the underlying soil stratum, this discrepancy is gradually smoothed out with increasing load, and becomes as small as 2° in the final stage of loading. Rotation of the principal axis of stresses reaches 60°. The parameter μ at this point is rather small (from -0.6 to -0.8), but there occurs a certain 'breakthrough' to -0.2 upon first filling of the reservoir. Variations of the parameter μ are similar to those at point C.

The higher the point at core axis (N, P), the stronger the variation of all parameters of loading path. The loading of the overlying soil is insufficient to bring closer the planes of principal stresses and strains. At point P the discrepancy of the slope angles reaches 44°, and the rotation itself is 70°. Behaviour of the parameter λ is also more complex. At the same point it varies in the range of +0.8 to -1.0, and again returns to +0.3.

The soil at edges of the core (points T and U) faces substantial variations of the parameters of loading path, particularly at the tailwater edge (U), where the misalignment of the planes reaches 74°, the parameter λ is zero, and μ varies from +1 to +2 and more. The angles of rotation of the planes reach 30°. It must be noted that the work of soil improves if the angles of principal planes do not coincide; this is caused by increasing energy of form change.

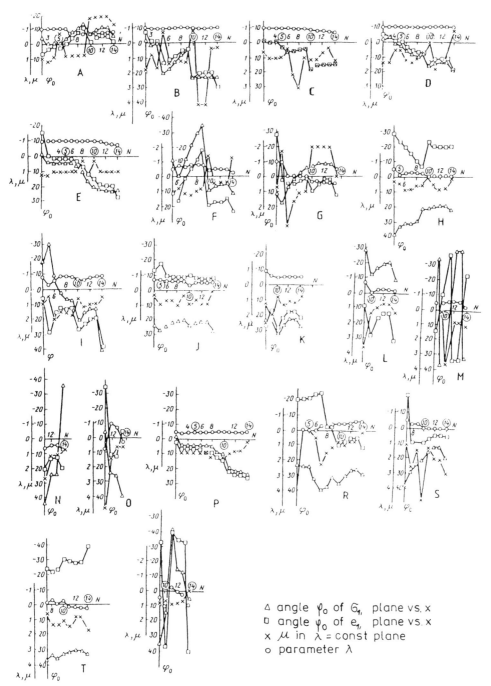

Figure 4.11. Variation of slope angle φ_0 of the planes of principal stresses and principal strains, parameter λ and coefficient k_s upon dam construction by sloping layers, at different points of dam. ($\triangle$) φ_0 for plane of σ_1 vs. x-axis; ($\square$) φ_0 for plane of e_1 vs. x-axis; (x) μ in plane λ = const; (o) parameter λ.

Points S, L, O, V are most severely subjected to the effect of the loading path. High revolutions of the planes of principal stresses (up to 70°) are a typical feature. The misalignment of the principal planes reaches 60° for high variation of the Lode-Nadai parameter. For instance, the parameter λ at point V varies from 0.4 to -0.4 and back again to +0.1. The same parameter at point O is positive (from 0.4 to 0.6), and at point P it becomes negative, from -1.0 to -0.5. The dominant value of μ is about +1.0, although there are significant deviations.

For points A, G, H, I, M, which are located in the upstream prism, negative and zero values of λ are characteristic, along with the transition from positive to negative μ for rising water on their upstream side (unloading), or in the end, dramatic decrease in μ (at point M).

The discrepancy of the angles of principal planes reaches 40°. Their sharp rotations are mostly associated with filling of reservoir and reach 50°, although at some points with low stresses the rotation may reach 30 or 40° due to the growth of dam. Hence the analysis of soil work in the structure upon its construction reveals complex spatial trajectory of loading. The complexity of the loading trajectories is particularly strong in rather weakly loaded zones and at interfaces of various materials. Considerable misalignment of the tensors of stresses and strains is exposed.

Computations of stress-strain conditions in earth dams under plane deformation make it possible to expose numerous properties of an earth dam: the effect of construction sequence on the working capacity of the dam; the effect of pore pressure on cracking in dam core; the growth mechanisms for displacements, stresses, strength reliability factor at various points of dam; discrepances between principal planes of stresses and strains etc. However, in the solution of plane problems many factors, primarily topographic, do not appear. Moreover, there is no way of assessing the arc effect in dam core on the formation of the stress condition and, first of all, on the growth of stresses σ_2, which is crucial to the strength of core material at its contact with rock bank. All these findings exemplify the importance of the analysis of stress-strain condition in earth dams in spatial formulation.

The algorithm of solution for three-dimensional problems is given in Section 4.3. The system of coordinates and decomposition into elements (Rasskazov & Belakov 1982) is illustrated in Figure 4.3a.

Analysis of the spatial work of a dam may be illustrated by the example of an earth-rockfill dam 335 m high with slightly inclined core (Figs 4.12, 4.13). The configuration of dam soil and the soil properties are similar to their counterparts in the dam for which the stress-strain conditions in plane problem were analysed above, see Figure 4.13a.

The versed arc of the core at the bed of the canyon is 10 m and that at the crest is 45 m.

The construction of dam by horizontal layers was analysed.

Graphs of vertical (u_y), horizontal (u_x) and transverse (u_z) displacements are shown in Figure 4.13a–c. The horizontal displacements in the core for two cases

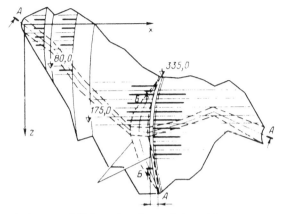

Figure 4.12. Schematic layout of 335-m dam in spatial problem.

of spatial configuration — rectilinear core in plan and an increased arc factor with the above parameters, are presented in Figure 4.13d. It must be noted that the central section $A - A$ is the vertical cross-section along the river bed. It is projected on the coordinate plane xOy.

The highest vertical displacements are reached at the central cross-section (Fig. 4.13a). On the left side (which is steeper), higher displacements u_y, than those on the right side, are encountered. Higher slopes of the bank correspond to higher vertical displacements. This tendency is also supported by field evidence.

The absolute maximum of vertical displacements, 2.39 m i.e. about 1/3 of dam height, has been found in the central cross-section on the tailwater thrust prism. Solution of the plane strain problem for the central cross-section of dam at the same place has given - 3.2 m. The same relation is observed for the displacement u_y obtained for plane and spatial problems, as computed by other methods.

The maximum vertical displacement observed in the field at Nurek Dam, which is compatible as to the scale and type of material with the computed values, was 2.5 m. Note that in the solution for this dam in terms of the plane strain problem, the maximum u_y was 3.7 m (Fig.4.5a), so that it was overestimated. In terms of spatial formulation, the vertical displacements are considerably reduced in narrow cross-sections. The character of the class of vertical displacements in the cross-section $B - B$ along core axis (Fig.4.13c) sheds light on the reduction of the displacements u_y compared with the plane problem — the structure is 'hanging' on canyon banks.

Horizontal displacements u_x are most conspicuous in the dam central cross-section. This displacement reaches 3.26 m in the upper part of the core on the thrust edge (Fig.4.13b). About the same value was obtained in solution of the plane problem, but the spatial formulation provides fast damping of the horizontal displacement inside the core — at the core axis they are below 0.4–0.5 m (Fig.4.13d).

The above findings are explained in terms of the principal difference of the plane and spatial schemes — in the former a dam works as a bedrock-supported

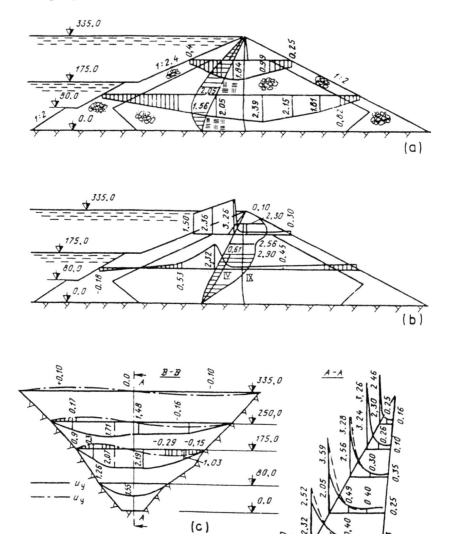

Figure 4.13. Displacements in dam body. (a) vertical u_y in central section $A - A$; (b) horizontal u_x in central section; (c) u_y and u_z in section $B - B$ of Figure 4.12 — along core axis; (d) u_x in core (cross-section $A - A$).

cantilever, and under the action of water on the upstream side there is a tendency towards rotation of the entire dam profile. In the spatial problem, the work of the structure is more sophisticated — first of all the structure is supported at the banks, and secondly there is an effect of river bends (Fig.4.14), which enhances the damping of horizontal displacements in the core and obviously comes closer to the reality.

The above remarks bring one to the conclusion that the horizontal displacements are considerable (1.5 m on the upstream slope at 250 m datum) on the upstream thrust prism, particularly in its upper part. The increase in horizontal displacement on the tailwater prism, due to rising water on the upstream side, was not practically noticeable. It must also be stated that the horizontal displacement measured at the axis of the core of Nurek Dam does not exceed 0.6–0.8 m.

Hence the damping effect for horizontal displacement in core, observed in the field on a number of dams, may be analysed theoretically in terms of three-dimensional models.

Comparison of horizontal displacements in dam core, in terms of its configuration in plan view, shows that a minor arc effect may bring about reduction of horizontal displacements at the thrust edge, by a factor of 1.5 and more (Fig.4.13d).

Transverse displacements u_z are directed from banks towards the central section (Fig.4.13c). The highest displacements, above 1 m, are found at the thrust edge of core in the upper part of dam (Fig.4.14b, c), which is caused by the transverse component of loading by water on the arched thrust edge. The displacements u_z, much as the horizontal displacements u_x are rapidly damped in the core and do not exceed 0.2 m at the axis of core (Fig.4.13c).

The displacements u_x and u_z in horizontal sections of dam depicted in Figure 4.14 are of utmost interest. In the upstream thrust prism, at the lower cross-section (Fig.4.14a), the displacements are generally towards the tailwater. At higher datum the vectors of displacements may be oriented towards both upstream and tailwater side, on the left and right bank, respectively, which is attributed to the protruding left bank.

The displacement vectors shown at random in Figure 4.14 illustrate the displacement trajectories in the dam body — they encircle the left bank and are afterwards directed along the river. Within the tailwater thrust prism one has a similar protrusion of the right bank, but it is less pronounced and does not produce reverse displacements.

The stress condition of dam body is primarily characterized by the arc effect between canyon banks. This phenomenon is clearly illustrated by the distribution of stresses σ_y in the cross-section at the core axis (Fig.4.15a). One may note the non-uniformity of stress distributions σ_y in the upper half of the dam core. A certain unloading occurs at the centre of the core (Fig.4.15b). The unloading is explained by 'hanging' of core on thrust prisms; the magnitude of σ_y is compatitible with the water pressure on the upstream side. However, as seen from Figure 4.15b, this unloading is associated with the tailwater edge of core, and the stresses σ_y at

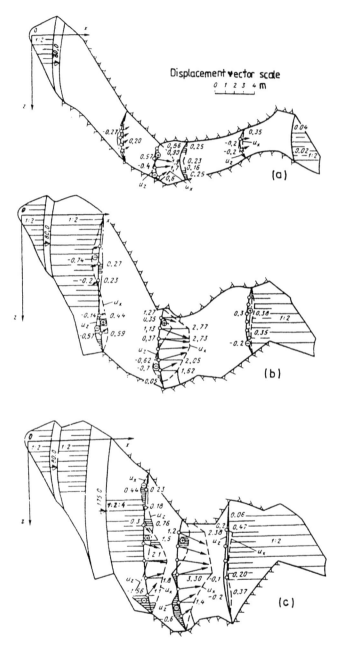

Figure 4.14. Displacements u_x, u_z in the xOy plane at datums 80, 130 and 250 m above canyon bed. Arrows show displacement vectors.

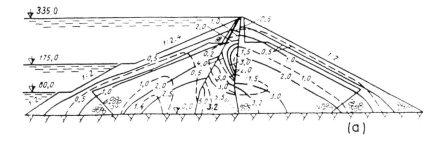

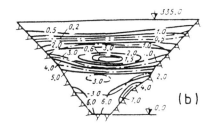

Figure 4.15. Stress isolines in dam body. (a) σ_x and σ_y in central section $A - A$; (b) σ_y and σ_z in axial core section $B - B$; dashed line for σ_x, solid line for σ_y, dash-dot line for σ_z.

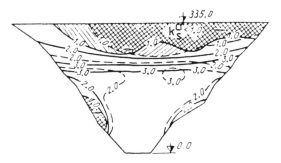

Figure 4.16. Isolines of cracking safety factors k_s^{cr} in horizontal planes at thrust edge of core. Dashed line for rectilinear axix $A = 0$; solid line for arching with $A = 45\ m$.

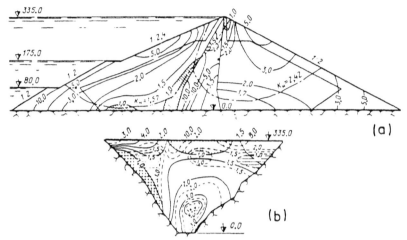

Figure 4.17. Isolines of reliability factor. (a) central section $A - A$; (b) tailwater transition $(B - B$ in Figure 4.12); solid line prior to upstream water level rise.

the thrust edge are higher because of the presence of the vertical component of load on the sloping thrust edge of core.

Further on, stresses σ_y increase rapidly with growing depth. There is concentration of σ_y at the right-hand side (7 MPa), attributed to local shallowing of the bank. Nevertheless, the arc effect in the core is preserved over the entire height of dam, and there occur zones linked to the central section (Fig.4.15a, b) where σ_y does not exceed $0.75\gamma y$; they are encountered in the lower part of the core.

Distribution of stresses σ_z in core (Fig.4.15a) is similar — low level of stresses is observed at the crest of core, and unloading is slightly above dam half-height. The stresses σ_z are greater at the thrust edge of core than at the core axis.

Shown in Figure 4.16 is the distribution of k_s by Equation 4.53 over horizontal planes at the thrust edge of core. An arc effect added to the core reduces the hazardous zone, which shows that this measures can be taken to improve the cracking stability of core.

Shown in Figure 4.17b are two families of k_s isolines in the tailwater transition $(B - B)$ — prior to ultimate filling of reservoir and after inpoundment. Although the values of k_s in this cross-section are fairly high, there are some zones of limit soil condition at the flanks and crest. It should be noted that the soil at flanks is often in limit state.

Rising upstream water level reduces the size of these limit zones, and the limit zone at the right-hand-side flank disappears due to increase in components of stresses σ_x and particularly σ_z, if water level rises. At the same time, for rising water level, a limit state zone is generated at dam crest. A similar zone of limit state at the crest, along the tailwater edge of core (Fig.4.17b) shows that a longitudinal crack may be generated at the crest, as indeed observed in some dams.

Values of k_s in core (Figs 4.17a, 4.18) are high. In the lower part of core, values of k_s increase in proximity of the thrust edge, and the reverse process takes place in the upper part.

A limit state zone, resulting from unloading due to action of water on the thrust edge of core, is observed on the upstream thrust prism about the prism-core interface (Fig.4.17a). In horizontal cross-sections (Fig.4.18a, b) one sees that the limit zone also stretches to the flanks. Values of k_s in the thrust prism sharply increase at end sections of dam; protruding flanks of canyons constitute supports for the dam body and bring about concentration of normal stresses, counteract displacements, and increase stability of the structure. Upon comparison of k_s for various models, the most dangerous surface in plane and three-dimensional problems was observed for $k_s = 1.68$ and 2.42, respectively.

Similar increase in stability also occurs for the upstream slope — if three-dimensionality is considered in narrow and bent (in plan) cross-section under static forces, a significant margin of stability appears.

Another problem resolved through analysis of the three-dimensional work of dam consisted in identification of the arc effect pronounced in working capacity of dam core (Belakov 1983). Analysis of the arc effect versus cracking in dam core shows that, for $A/L \approx 0.07$, the most favourable conditions are created (A = versed sine of arc at crest, L = curved length between flanks at crest level). High arc ratio brings about worse working conditions of core at banks, although k_s increases about the central cross-section, which is due to appearance of considerable force acting in the direction of the z-axis, which reduces the force at the core-bank interface. However, the absence of the arc effect causes reduction in the core-flank contact stresses, and results in reduction of σ_z at the interface. An optimum value of A must be found in each specific case.

4.5 UTILIZATION OF STRESS-STRAIN COMPUTATIONS IN DAM DESIGN

The analysis of stress-strain conditions has the following design implications.

1. The displacements obtained in computations at various instants of construction are used for assessment of the additional volumes of earth necessary for construction of dam at any stage of construction. The displacements may be used for quality control — if displacement components are much above the prediction, this becomes a warning indicating that the textural density was reduced or the quarry was changed. Errors in prediction are obviously possible but the warning must be taken seriously.

2. Stress components σ_y and σ_z make possible the assessment of cracking conditions in dam core. Slope stability is analysed as an outcome of stress computations, with inclusion of stagewise construction. Other simplified methods are useless in this respect.

Analysis of stress conditions (in three-dimensional problems) provides the versed

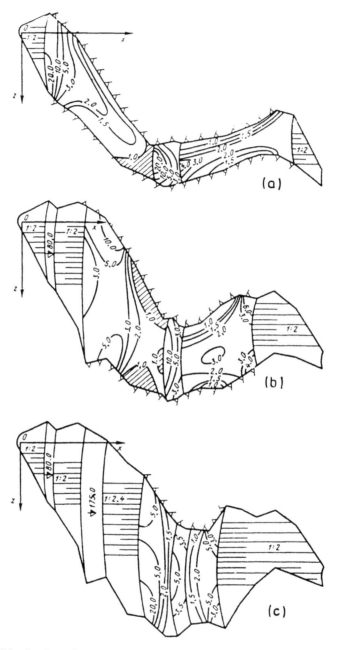

Figure 4.18. Distribution of reliability factor in horizontal dam sections. (a) (b) (c) sections at datum of 80, 130 and 250 m, respectively.

sine of core, which is also impossible by other methods. To date the arc parameters of core have been assumed by rough and arbitrary methods.

3. Dynamics of stress conditions at various points of dam permits identification of various causes of unfavourable phenomena in dams, for instance the growth of limit state zones, with possible subsequent elimination of these factors.

4. Transformation of limit states zones in the upstream prism, connected with deformability of core, may be analysed upon solution of the problem. The thrust force exerted by the upstream prism of the core occurs if a narrow limit state zone stretches from the foot of core to the slope, and plays the role of a 'lubricant' for the overlying soil of the prism.

5. Solution of stress problems for dams permits identification of minimum acceptable width of core below which 'hydraulic failure' may occur. These problems are impossible to solve by other simplified methods.

6. Solution of stress-strain problems for earth dams provides a sound basis for optimization of earth dam design (Chapter 9).

CHAPTER 5

Seismic stability of earth dams

5.1 SOME REMARKS ON EARTHQUAKES

More than 15% of the Soviet territory is exposed to strong earthquakes. Hazardous seismic zones are concentrated in mountainous systems such as the Carpathian Mountains, the Crimea, the Caucasus, Middle Asia, Peribaykal, Chukotka, the Far East, Sakhalin, and Kamchatka.

The Earth's active seismic zones are shown in Figure 5.1 (Bolt 1981). The abundance of earthquake focuses is correlated with symmetric (in plan view) chains of islands located in the Pacific Ocean and to the east of the Carribean Sea. The Aleutian Islands, stretching from Alaska to Kamchatka, and Japanese Islands are examples. On the opposite side of the Pacific Ocean there is the eastern coast of central South America. Even seismically stable zones such as the central part of the Russian Plain, a larger part of Siberia, central Europe, including the Great Britain, are not entirely free of earthquakes. Older chronicles provide evidence. Besides, there are remote effects such as on 4th March 1977, when the earthquake in Bucurest (magnitude 7.2) was felt in Moscow (magnitude 4).

The strong earthquake on 1st November 1755, southwest of the Iberian Peninsula, brought about a strong tsunami which drowned 70 thousand men in Lisbon (Portugal) and neighbourhood. The earthquake was felt in Germany and the Netherlands.

Causes of earthquakes were investigated in old times. Many ancient scientists put forward various hypotheses, but it was Aristotle who explained the problem most comprehensively. He argued that earthquakes were caused by the 'soul'(Greek $\pi\nu\epsilon\upsilon\mu\alpha$) of solid matter, and he sought proofs in volcanic activities. This approach is fully understandable as ancient scientists observed earthquakes on the Balkan Peninsula and numerous islands. Modern investigations seek the primary source of earthquakes in the drift of continents and the resulting processes of the generation of mountains. At many points of the Earth, earthquakes are caused by volcanic phenomena. Man is also responsible for some earthquakes due to atomic explosions, construction of large reservoirs etc.

It was as early as in 1620 that Francis Bacon brought to attention the close resem-

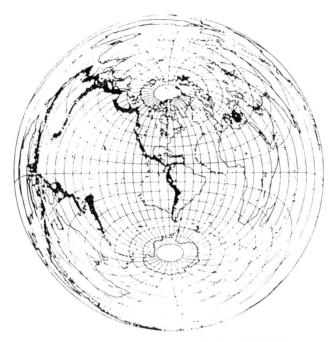

Figure 5.1. General chart of the Earth's seismicity (Bolt 1981). Epicentres of earthquakes from 1963 to 1973, with magnitude equal to or greater than 4.5.

blance of the configuration of opposite coastlines of the Atlantic Ocean. However, it was as late as 1910 that Wegener put forth a hypothesis on the common origin of all continents in a single supercontinent called by him the Pangea. The drift of continents has been measured by instruments fixed in some regions; it reaches several centimetres per year, which is a very high speed in geologic terms. At present one may see that this hypothesis has been supported by the theory of plate tectonics, which is primarily based on the fact that the outer shell of the Earth (lithosphere) consists of some large and strong plates moving with respect to each other. The thickness of these plates is 80 km. At their contact they exert high tectonic forces on adjacent rock massifs so that physical and even chemical transformations occur. The system of plates itself is subject to permanent transformations. At present one distinguishes Euro-Asiatic, African, Pacific, North American, South American, Indo-Australian (including Indian and Arabic Peninsulas) etc plates. The plates can grow due to outflow of magma (African, Antarctic, North and South American), while other decrease (the Pacific with Japanese Islands and Phillipines). If a plate at its contact with another plate is bent downwards, cracks and discontinuites are generated, thus giving birth to shallow-focus earthquakes, and subsequent deformations result in deep earthquakes. If a plate is bent upwards, mountains are generated.

The material of the plates at the depth of 650–700 km is entirely incorporated in deep rock and mixes with the latter. From the above theory it follows that more earthquakes should be produced at the edges of plates than at their centres, but no explanation is given by the theory for earthquakes at the centre of the plates. It seems that these earthquakes are caused by local concentrations of stresses in rock, which give birth to discontinuities from time to time. Generation of these concentrated stresses may be explained in terms of individual properties of their structure.

Most large earthquakes (such as Chilean in 1960 and Alaskan in 1964) resulted from the motion of one plate below another. Permanent mobility of the plates and the time interval between large earthquakes at edges of the plates permit an approximate prediction of the place and time of earthquakes.

5.2 TYPES OF SEISMIC WAVES AND SCALES OF SEISMICITY

Tectonic earthquakes are the strongest variety. They arise if stresses generated in the rock of the Earth's crust are not capable to dissipate and thereby exceed the strength of rock so that the latter breaks and discontinuity occurs as a fault or overfault.

Earthquakes are often accompanied by faults, that is a belt-type discontinuity in the Earth's crust. Fault is an interface in the motion of continents (the San Andreas fault in the USA, Vakhsh fault in the USSR etc). Seismic faults often coincide with geologic ones. Occurrence of faults in rock is associated with release of elastic energy, due to elastic strains (see Section 1.6). A large part of this energy is dissipated for the production of oscillations in the ambient medium.

The plane of discontinuity, together with the surrounding area is referred to as the focus (sometimes centre), or hypocentre of earthquake. The surface of the Earth's crust above the hypocentre is called epicentre.

With respect to the depth of focuses, earthquakes are classified as deep (300–700 km), with the deepest quake recorded at a depth about 720 km under the Flores Sea, intermediate in the range of 60–300 km, and normal, at depths smaller than 60 km.

The most dramatic discontinuities are caused by shallow earthquakes, with hypocentres at depths smaller than 15 km, their contribution to the total energy released in earthquakes being 75%.

The growth of oscillations in the ambient environment due to the release of elastic energy is manifested as propagation of three major types of elastic waves. Among them, one distinguishes only two types of waves propagating inside rock: longitudinal or primary (p) (Fig.5.2a) (cf. Bolt 1981), the motion of which is similar to the propagation of acoustic waves; and slower transverse or secondary waves (s) (Fig.5.2b). The transverse waves do not propagate over the Earth's regions consisting of liquid. The celerity of propagation of these waves depends on the density and other properties of rock through which they travel.

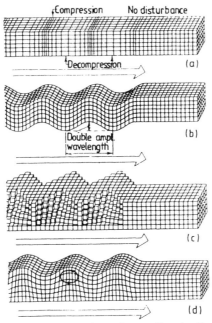

Figure 5.2. Basic types of elastic waves. (a) primary (longitudinal); (b) secondary (transverse); (c) Love waves; (d) Rayleigh waves.

In most cases, longitudinal waves are noticed first. If they reach the the Earth's surface, they can propagate into the atmosphere as acoustic waves felt by people and animals if the frequency of oscillations is above 15 Hz (lower limit of audibility).

The seismic waves of the third type are called surface waves as their propagation is confined to the ground surface. This type of waves is divided into two classes: Love waves (Fig.5.2c), similar to transverse ones though with lateral but not vertical displacement, perpendicular to the direction of propagation; and Rayleigh waves (Fig.5.2d), in which particles move in the horizontal and vertical direction in the vertical plane, that is over ellipses. The surface waves propagate at smaller speed than the three-dimensional waves, and Love waves are faster than Rayleigh waves.

When p- and s- waves reach the Earth's surface they are reflected; superposition of waves occurs which enhances the oscillations so that the amplitude of the transformed waves can be twice as high as that of the incident waves, which in turn affects buildings and structures. It has been observed that oscillations due to underground explosions are weaker than those at the the Earth's surface.

A type of wave is difficult to identify in the case of strong earthquakes because a mixture of seismic waves arises and the discrimination requires a lot of expertise. It is known that the seismic waves are affected by topography, soil conditions etc. For instance, the amplitudes of seismic waves in alluvium can both increase and decrease, and amplification may occur at a top of a ridge or fill.

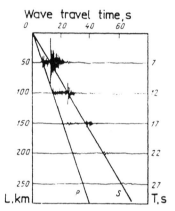

Figure 5.3. Time of travel of p—wave and s—wave (in seconds, horizontal axis) versus distance from earthquake focus.

The difference in the celerity of various types of waves makes it possible to identify the focus of earthquake. The celerity of p-waves is about 8 km per second, and the transverse s-waves travel with only 4.5 km/s, so that a p-wave comes first and an s-wave lags behind it. If one identifies these waves by seismic records and measures the lag time then the distance to the focus is easy to find. Figure 5.3 (Aiby 1982) shows a record which determines the distance from seismic station to the focus of earthquake by the time lag between p-waves and s-waves:

$$T = \frac{L}{v_s} - \frac{L}{v_p}; \quad L = (\frac{1}{v_s} - \frac{1}{v_p})^{-1}T \tag{5.1}$$

in which
  L = distance to focus,
  v_s, v_p = celerity of s-waves and p-waves, respectively (these celerities are often denoted by c_p and c_s),
  T = time interval between arrival of p-waves and s-waves.

It was noted above that the celerity of p-waves provides information on the location of earthquake epicentre. Using the relationships derived in the theory of elasticity one may link the elastic characteristics of the Earth's crust (and mantle; in which the Earth's crust is roughly 30 km thick and the so-called Moho depth demarcates the crust and the mantle) to with the celerities of p-waves and s-waves read

$$v_p = \sqrt{\frac{E(1-\nu)}{\rho(1+\nu)(1-2\nu)}}; \quad v_s = \sqrt{G}\rho = \sqrt{\frac{E}{2\rho(1+\nu)}} \tag{5.2}$$

in which
  E, ν, G = Young's modulus, Poisson's ratio and shear modulus, respectively,

ρ = density of rock.

If v_p and v_s are known one may find E and ν, and vice versa. Taking ν one has $v_p/v_s = 1.73$. For shallow earthquakes this ratio is 1.67 and for deep ones it reaches 1.78.

Approximate celerities of transverse and longitudinal waves in various types of rock are given in Table 1.7 (Krasnikov 1970).

The celerity of surface waves is about $0.9\ v_s$.

The energy of seismic waves is measured by magnitude of earthquake (M), that is the decimal logarithm of the maximum amplitudes of seismic record, in micrometres, obtained on a standard seismic recorder, 100 km from the earthquake epicentre. If displacement is recorded at another distance or by another equipment, the magnitude is computed by analogous records. This characteristic quantity of earthquakes has been proposed by Richter.

The relationship between magnitude and total energy in joules and ergs is given as follows:

$$\log E = 4.8 + 1.5M = K \text{ joules}; \quad \log E = 11.8 + 1.5M = K \text{ ergs} \qquad (5.3)$$

The quantity K is referred to as energy class of earthquake.

Earthquakes quantified by magnitude are referred to by Richter points. The strongest earthquakes occurred in Columbia and Ecuador on 21th January 1906 and in Japan (Sankrikyu) in 1933; their magnitude was 8.9. It is possible that the aforementioned Lisbon earthquake was 9 points by Richter, but accurate information is unavailable.

Earthquakes above 7 points can bring about great catastrophes if they occur close to populated areas. As mentioned above, the magnitude of the earthquake in Bucurest on 4th March 1977 was 7.2. Five-point earthquakes close to populated areas can cause failure of pipelines, linings etc. From Equation 5.3 it is seen that earthquakes below zero are possible — they are the waves which do not induce records on a standard seismic apparatus. The energy of these earthquakes is $E < 10^{11.8}$ erg.

Hence magnitude characterizes the earthquake focus.

In contrast to the total energy computed by magnitude, one may also introduce earthquake intensity, which is computed by MSK, Mercalli, or JMA scales, used in different countries.

Intensity scales characterize occurrence of an earthquake at a point of the Earth, that is the intensity of the same earthquake can be different at various points, for the same magnitude.

The most perfect intensity scale is MSK-64 (denoting the names of Medvedev (USSR), Schponhoyer (GDR), and Karnik (Czechoslovakia), accepted in many countries worldwide. The classification by this scale is based on three clearcut criteria: sensitivity of people and ambient effects, action on structures of different types, and residual phenomena in soil, including groundwater changes. The gradation of MSK-64 is 12 points. The twelve-degree Mercalli scale is widely used in the North America.

The JMA, scale derived in Japan in 1948, has 8 degrees (the maximum is 7 degrees, and 0 is meaningful).

The transition from JMA to Mercalli may be approximated by the following formula:

$$J_M = 0.5 + 1.5 J_{JMA} \tag{5.4}$$

in which

J_M = Mercalli scale (12 degrees),

J_{JMA} = JMA degrees.

Aside from description of consequences of earthquakes, the MSK scale also contains respective degrees for displacement, velocities and acceleration of the Earth's surface. These parameters are particularly constructive in design of different structures. Accelerations are often given in fractions of the acceleration due to gravity.

It can be noted that, in the engineering design enforced by the Soviet standard SNiP II-7-81, the data on acceleration, velocity and displacement is sufficient. Yet the modern methods of stress-strain computations for structures, and even more so for dams, require the knowledge of design acceleration graphs and seismic graphs (for changes in acceleration and displacements of the Earth's surface in time, respectively), which makes much more complex the preparation of input for structural design.

Table 5.1 incorporates accelerations, velocities and displacements by MSK-64. Because the SNiP standard requires seismic design for earthquakes above 5 points, Table 5.1 does not contain weaker earthquakes; yet they may be obtained by extrapolation. The same is true for intensities above 10 points, although permanent structures are forbidden in such cases.

For known intensity of displacements at any given point and the distance to earthquake focus one may determine the magnitude by using Figure 5.4. Reverse transitions are also possible — intensity at points may be determined from earthquake forecast by magnitude, depending on the distance involved.

Construction of dams in seismic regions can induce earthquakes. The primary reason is not explored so far although the role of water in the generation of earthquakes has been realized since very long. It was already Democritus who attributed earthquakes to saturation of the Earth with water. The motion of groundwater during rains, when 'water finds its way' inside the Earth, or during dry periods, when 'ground dries out and entrains water from oversaturated cavities inside, whereupon its falling down into the cavities brings about the Earth's tremor'.

Contemporary engineering activities have quite recently emphasized the fact that water can induce earthquakes.

A series of earthquakes (700 shocks) occurred about Denver (USA), with the magnitude ranging from 0.7 to 4.3. It was exactly at that time that the military works in the Rocky Mountains (Bolt 1981) began pumping of wastewater into

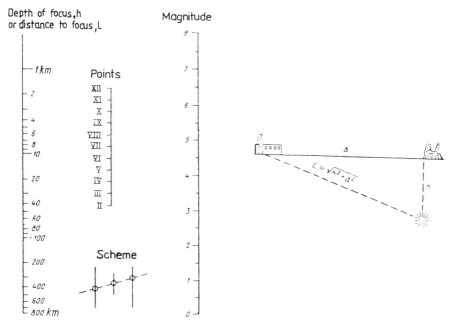

Figure 5.4. Nomograph for transition from intensity to magnitude, and vice versa.

a 3670 m deep borehole. The increased head of water brought about flow of groundwater into cavities and cracks along earlier faults. It was argued that the increased pore pressure in turn caused reduction of effective stresses in rock and clay, which decreased friction and enhanced displacements.

The investigations carried out by the US Geological Survey in 1969 on an oil field have revealed that pumping of water changed pore pressure in seismically active regions, so that the hypothesis on the role of pore pressure in activation of seismic phenomenon has been confirmed. Since pore pressure can provoke earthquakes only if rock is in the condition close to the limit state, it was proposed that pumping of water induce earthquakes in the regions where real earthquakes can cause hazards. Indeed, the pumping brought about a series of earthquakes and thereby prevented one strong earthquake.

In 1958 the Kariba reservoir (the Zambia River; 128 m deep) was impounded. In 1963 the reservoir was filled up to 160 km^3, and at that time more than 2000 shocks were recorded. The strongest quake, 5.8 in magnitude, was recorded in September 1963 and afterwards the seismic activity fell down.

On 11th December 1967 an earthquake occurred about Coyna Dam (India) with the maximum magnitude of 6.5, causing considerable death penalty (177 men killed). The maximum acceleration was 0.63 g. The series of earthquakes coincided with rainfall.

Table 5.1.

Intensity	Acceleration, 10^{-2} m/s^2	Velocity 10^{-2} m/s^2	Deflection of centre of standard pendulum, 10^{-3} m
5	15-25	1-2	0.5-1
6	26-50	2.1-4	1.1-2
7	51-100	4.1-8	2.1-4
8	101-200	8.1-16	4.1-8
9	201-400	16.1-32	8.1-16
10	401-800	32.1-64	16.1-32

Impoundment of a 105 m deep reservoir close to Huan Joy (North Korea) brought about an increasing number of surface earthquakes (shallower than 10 m), and 250 000 of them occurred over 12 years; one of the earthquakes was strong (6.1 in magnitude), and brought about failure of the concrete dam.

Impoundment of Nurek reservoir on the Vakhsh River, 300 m deep, also caused a higher local seismic activity — 200 quakes were recorded, and the maximum intensity was 6.0 points by MSK. These earthquakes did not cause any practical disturbance in structures of the hydraulic engineering scheme.

The above data indicates that the weight of a dam or reservoir could not bring about earthquakes with focuses at a depth of 10 km, as the weight itself is very small. The most probable explanation is that the action of pore pressure and reduction of the bearing capacity of rock in the condition close to limit state causes the earthquakes of the discussed type. Moreover, in the construction of large reservoirs one can also encounter seiches, i.e. standing waves in reservoirs, caused by earthquakes.

The design of an earth dam in a seismically active region is based on seismic data obtained from microseismic prospecting about the construction site. The seismologic practice depends on the applicability of design methods. Considered below are two basic design methods — dynamic and static (or quasi-dynamic). The dynamic method provides an assessment of the stress-strain condition and stability of dams as a function of the duration of acceleration graph for the earthquake. This method involves the design acceleration graph and seismograph trace (diagram of base displacement in time).

The static, or quasi-dynamic, method of design does not take into account the temporal course of events. It is based on the design value of acceleration for the foundation of a structure.

5.3 DYNAMIC METHOD OF STRESS-STRAIN COMPUTATIONS FOR EARTH DAMS (EXPLICIT SCHEME, VARIATION-DIFFERENCE VERSION)

Chapter 1 describes soil model as a hardening elastoplastic body, which obeys the associated flow rule (Zaretskiy-Lombardo model). The model has also been used in the variation-difference formulation, in solution of stress-strain problems for static forces (Section 4.1).

Consider application of the method in the solution of stress-strain problems for earth dams under seismic loading (Zaretskiy & Lombardo 1983).

In solution of dynamic problems one must possess a design acceleration graph, as already noted in Section 5.2. The graph may be constructed if complex seismologic problems are solved, for which the following factors must be known:

1. Location of possible focus of earthquake;
2. Magnitude of earthquake;
3. Intensity of earthquake (points) in dam section;
4. Frequency distribution of acceleration graph etc.

Construction of any acceleration graph (as possible focuses of earthquake can be numerous) may be carried out upon detailed seismologic investigation of the construction site. Since the above investigations require a long time before dam construction, and this is impractical in many cases, the constructed design acceleration graph is usually applied to support the results of earlier computations, with analogous acceleration graphs. The latter must be selected for the earthquakes occurring under similar geologic conditions, for a similar distance from focus of earthquake etc. Since good earthquake records are rather scarce, selection of similar acceleration graphs is simplified at present, and design acceleration graphs are taken as three or more seismic graphs i.e. low, medium and high frequency; and the conclusions are drawn on the basis of the worst design case.

Earth dams can be classified in the following two groups:

1. Plan dimensions comparable with the length of seismic wave (high and very high dams);
2. Plan dimensions much smaller than seismic wavelength (low and medium dams).

For dams of the second group one may assume that all points of the structure are subject to identical displacement at the same time, so that the problem is reduced to computation by the design acceleration graph.

If the motion of dam points is described by the equation

$$F(z) = \rho \frac{\partial^2 z}{\partial t^2} \tag{5.5}$$

in which

$F(z)$ = differential operator,
$z = z(x_i, t)$ = displacement within dam,
i = number of nodes in finite difference grid.

then the boundary conditions at the interface between dam and foundation Ω are given as follows (Zaretskiy & Lombardo 1983):

$$z(x_i, t) = z_0(t); \quad x_i \in \Omega \tag{5.6}$$

in which

$z_0(t)$ = displacement of foundation at its interface with dam (Lombardo 1983).

If the entire surface Ω moves as a whole, then one may introduce a new variable which determines the displacement of the structure with respect to Ω (Zaretskiy & Lombardo 1983):

$$z = \bar{z}(x_i, t) + z_0(t) \tag{5.7}$$

The function $\bar{z}$ on the surface Ω satisfies the condition $z(x_i, T)=0$ for $x_i \in \Omega$ and $0 \le t \le T$, in which T = duration of earthquake.

Substituting Equation 5.7 into Equation 5.5 and assuming that the displacement of the entire structure does not induce strains and stresses one obtains

$$F(\bar{z}) = \rho \frac{\partial^2 \bar{z}}{t^2} + \rho \frac{\partial^2 z_0}{t^2} \tag{5.8}$$

but with inclusion of homogeneity of the boundary conditions of the function $\bar{z}$ on the surface Ω one has

$$F(\bar{z}) = F_0 = \rho \frac{\partial^2 z_0}{t^2} \tag{5.9}$$

in which

$\frac{\partial z_0}{\partial t^2}$ = acceleration at the dam-foundation interface.

Hence the seismic design is reduced to computations by acceleration graph at the dam foundation.

If the plan dimensions of the structure are comparable with the seismic wavelength, the function of foundation displacement $z_0 = z_0(x_i, t)$ depends on the ordinate x_i, $x_i \in \Omega$, $i = 1, 2, 3$ and $0 \le t \le T$. On the strength of $\frac{\partial z}{\partial x_j} \ne 0$ on Ω, the presence of displacements having various phases, for different displacements generated at a given point of foundation at the same time, brings about additional strains and stresses.

Hence one must have not only an acceleration graph but also a seismogram. This information will be sufficient provided celerities of waves in bedrock are known, so that one may compute the mutual accelerations and displacements at the interface.

Comparison of the recorded acceleration graphs or seismograms for a given earthquake does not display an even remote resemblance to the results obtained from double differentiation of a seismogram or double integration of an acceleration graph. This is caused by the fact that a seismogram provides information on low frequencies while the acceleration graph yields high-frequency components of earthquakes.

If the deformabilities of bedrock and body of dam are comparable, it becomes necessary to include a part of bedrock into the design scheme, as the displacement of bedrock in the form of seismogram must not be given. After construction of dam, the deformations of foundation will be different under the action of seismic forces; this will be due to the consolidation.

Lombardo (19780 elaborated another method which permits the consideration of compressibility of bedrock and the associated transformation of the seismogram. In this method, the field of stresses and strains in the foundation due to seismic action in the absence of the structure must be added to additional stress-strain field due to the reaction at the dam foot.

In order to find the reaction stresses over the interface and to determine the accurate stresses and strains in the dam body over the area Ω one formulates the following conditions

$$\sigma_{dam}(x,t) = \sigma_{fnd}(x,t)$$

$$\tau_{dam}(x,t) = \tau_{fnd}(x,t) \tag{5.10}$$

$$u'_{dam}(x,t) = u'_{fnd}(x,t) + u'_0(x,t); \quad x \in \Omega; \quad t > 0$$

in which

$u'_0(x,t)$ = displacement over Ω given by the seismogram in the absence of structure.

By analogy to Equation 5.10 one may elaborate a velocity graph.

The latter approach is reliable if the design seismogram is fixed at the Earth's free surface within the construction site or was computed as a design acceleration graph ('synthetic'); or if a seismogram did not account for the future role of the structure.

Equations 5.10 are utilized in an implicit temporal scheme, as the system is solved in parallel with the stress-strain problem of dam body at each time step for the area occupied by the structure and foundation if the acceleration graph or seismogram is given at the dam-foundation interface.

Zaretskiy & Lombardo (1983) utilized physical equations (soil model) based on the theory of plastic flow with hardening. The primary assumptions of this theory applied to the static problem are given in Chapter 1. One must add that the version based on the associated flow rules by Koiter is used in the following form

$$de_{ij}^p = \sum_\tau d\lambda^\tau \frac{\partial f_\tau}{\partial \sigma_{ij}} \tag{5.11}$$

in which

$d\lambda^\tau > 0$ = parameter,
$f\tau(\sigma_{ij}, e_{ij}^p)$ = smooth functions for loading surface between singular points.

The Koiter flow rule in Equation 5.11 is a generalized foundamental quasi-thermo-dynamic postulate of Drucker (1962) and the principle of maximum dissipation rate for mechanical work (the Mises principle), extrapolated to the case of loading area with finite or infinite number of singular points. Since the loading point is always singular, this is equivalent to the use of the loading surface on which all points are singular. This is exactly the conclusion drawn earlier on the basis of experimental behaviour of the loading surface (Ioselevich et al. 1979).

The major advantage of this soil model consists in the clear formulation of the interface of loading and unloading area, which is of paramount importance to the solution of dynamic problems, in particular seismic ones. The extrapolation of the soil model to dynamic actions is associated with the necessity of including the viscoplastic strains, as their progressive accumulation in time can bring about soil failure.

Zaretskiy & Lombardo (1983) assumed that the strain consists of elastoviscous, e^{ev}, and viscoplastic, e^{vp}, strains

$$e_{ij} = e_{ij}^{ev} + e_{ij}^{vp} \tag{5.12}$$

The elastoviscous strains are given by the following relationship

$$\sigma_{ij} = \left(\bar{E}_0 - \frac{2}{3}\bar{G} \right) e_{ij}^{ev}\delta_{ij} + 2\bar{G}e_{ij}^{ev} \tag{5.13}$$

in which
$\bar{E}_0$ and $\bar{G}$ = integral Volterra operators, other notation being unchanged.

The following properties of viscoplastic strains are notable:
  a) Considerable time lag with respect to the acting stresses;
  b) Cyclic action generates additional strains in comparison with the strains due to equivalent static stresses T_w = max and σ_h = max; in which σ_N and T_N = mean stress and tangential stress intensity under cyclic loading, respectively;
  c) The spatial boundary between elastoviscous and viscoplastic stresses depends on the rate of elastoviscous strains.

The matematical soil model given above has been extended to the case of cyclic loading, through additional introduction of the rate of viscoplastic strains as a loading parameter added to residual strains; this parameter is supplemented by the loading surface creating additional viscoplastic strains under repeated loading.

For multiple loading one introduces secondary plasticity defined as additional accumulation of residual strains under repeated loading after unloading, so that Equation 5.12 is written down as

$$e_{ij} = e_{ij}^{ev} + e_{ij}^{vp(1)} + e_{ij}^{vp(2)} \tag{5.14}$$

in which
$e_{ij}^{vp(1)}$ and $e_{ij}^{vp(2)}$ = strains due to secondary viscoplasticity.

The above process can be described if the model includes a so-called instantaneous primary loading surface (in contrast to the stabilized loading surface, which does

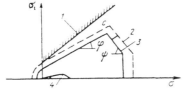

Figure 5.5. Stabilized and instantaneous loading surface of primary plasticity. (1) limit surface; (2) instantaneous surface; (3) stabilized surface; (4) original surface.

not depend on the rate of plastic strain) and additionally instantaneous secondary loading surface, reflecting the lagging strains due to repeated loading.

Figure 5.5 shows the configuration of the limit surface and both loading surfaces.

The configuration of loading surfaces in the space of stresses depends not only on stress components and the accumulated plastic strains but also on the trajectory and time of loading. All these parameters make the model fairly complex, both in implementation and in the search for experimental constants.

Aside from the above circumstances, Zaretskiy & Lombardo (1983) also incorporated consolidation due to dynamic actions, including quasi-two-phase media. The dynamic model of soil was implemented in explicit scheme based on the variation-difference formulation.

Explicit scheme is the one in which solution is obtained for time t on the basis of known solutions at times $t - \Delta t$ and $t - 2\Delta t$, as obtained by reccurential formulae, without solution of the system of equations. The convergence of this solution takes place within Δt.

Lombardo (1983) assumes the following limitation corresponding to the Neumann stability condition:

$$\Delta t \leq \frac{\Delta x}{v_p \sqrt{2}} \tag{5.15}$$

in which

Δt = miminum linear dimensions of grid, other notation unchanged.

Solution of the problem basing on the above variation-difference method of seismic action, with given acceleration graph and seismogram, differs little from the solution of the static problem described in Section 4.1. The difference consists in the relationships between the static and dynamic problems and in the necessity of additional decomposition of plastic strains into primary and secondary. At each step it is necessary to solve nonlinear equations and verify the conditions of existence for secondary viscoplastic strains.

In the solution of seismic problems it becomes compulsory to include the hydrodynamical pressure of water on the upstream side. The motion of water on the upstream side is described by the wave equation

$$\frac{\partial^2 \Phi}{\partial x^2} + \frac{\partial^2 \Phi}{\partial y^2} = \frac{\rho}{K_0} \frac{\partial^2 \Phi}{\partial t^2} \tag{5.16}$$

in which

Φ = velocity potential,

K_0 = bulk modulus of water elasticity; the other notation unchanged.

The hydrodynamic pressure of water is given by velocity potential as

$$P = \rho \frac{\partial \Phi}{\partial t} \tag{5.17}$$

A part of reservoir will be included into the general design scheme. Generation of surface waves is neglected.

One of the chief problems in the analysis of dam operation under dynamic actions, within stress-strain formulation, consists in the transition from the soil state at a point of dam to the stability of the entire system. In this case it is most reliable to assess stability by acceptable displacements. At the same time Zaretskiy & Lombardo (1983) provide an analysis and conditions of occurrence for the limit soil condition basing on the strain parameter (see Chapter 1), which is highly arbitrary and underestimates the data in comparison with the static load.

It has been established that an earth-rockfill dam 300 m high is unstable if the design earthquake has an intensity exceeding 10 points; this is a result of tremendous seismic effects at the foundation and a lower strength of soil. Reduction in the horizontal seismic acceleration down to 0.9 g and of the vertical one down to 0.45 g bring about residual strains of the dam — the crest settles down by 4.1 m (this settlement becomes acceptable if the elevation of the dam crest above normal impoundment datum is 10 m). Further reduction of the seismic action down to 0.64 g and 0.32 g, respectively, causes a residual displacement of crest reaching 2.3 m. If one assumes that the soil strength parameters do not decrease under dynamic actions, then the residual displacement sharply decreases (down to 1.7 m). It must be taken into account that it is exactly in this design case that the initial textural density of the gravel-pebble soil was reduced to 2100 kg/m^3, which caused that residual displacement below 0.9 m should have been expected. This conclusion coincides with the solutions obtained by simpler methods (Rasskazov & Volokhova 1974).

The following conclusions have been drawn from the computation presented by Zaretskiy & Lombardo (1983).

Upon passage of a seismic wave, the pore pressure in the upper prism of a dam increases fairly insignificantly (10–15%), compared with the increasing hydrostatic pressure. This is a very essential effect, as it makes possible optimization of the stability of the upstream prism under extreme seismic conditions. At Nurek Dam (300 m high), the anticipated intensity of seismic action was 9 points, which corresponded to the maximum acceleration of 0.4 g, that is three times below the design value in the given problem. Reduction of the design acceleration to 0.64 g shows that zones of limit state do not arise in dam.

The conclusion made by Rasskazov & Vitenberg (1972) and Rasskazov & Sysoyev (1982) on the transition of soil into limit state in a narrow zone about slip surface, together with small deformations in the soil above this surface, has been confirmed, and the soil is found to move as a whole.

The computations have identified a zone of limit state at the dam crest, in the filters on the tailwater side; this situation favours the formation of slip surfaces with increasing seismic effects.

The horizontal accelerations at the dam crest are below the horizontal acceleration at the foundation. The higher the seismic acceleration at the foundation and the greater the area in limit state, the lower the ratio of the ultimate accelerations $a_{max\,soil}/a_{max\,fnd}$; this points to the different operation of a dam as an elastoplastic body, compared with an elastic dam. Plastic strains, and particularly the zones of limit states, adsorb intensively the seismic energy.

In the fifties Napetvaridze recommended constant acceleration graphs along the dam height, in his seismic design of dams. The field data collected by Seleznev shows that the acceleration graphs under high seismic effects differ substantially from those under low seismicity, when soil works as an elastic body; the accelerations at the crest of high dams are greatly reduced if seismicity is above 7 points.

During the Second Congress on Seismic Construction held in Tokyo in 1960 one of the lectures has provided data on 59 dams of different height subject to earthquakes from 4 to 11 points. Substantial destruction of earth dams was noted only for dams below 20 m.

Hence a bulk of indirect data on seismic stability of earth dams is widely supported by experimental evidence (Zaretskiy & Lombardo 1983).

5.4 DYNAMIC METHOD OF STRESS-STRAIN COMPUTATIONS FOR EARTH DAMS (IMPLICIT SCHEME IN FINITE-ELEMENT METHOD)

Solution of the stress-strain problem arising under seismic loading is possible in FEM formulation by implicit and explicit schemes. Within explicit schemes the method differs little from that presented in Section 5.3; the primary differences consist in decomposition of the area into elements and in the description of strains of elements by displacements of the nodal points.

In solution of stress-strain problems for earth dams in explicit schemes one has a number of advantages — dynamic problems with nonlinear physical equations are solved relatively simply without solving systems of nonlinear algebraic equations; the effect of dam on the design acceleration graph is taken into account; residual displacements at nodal points may readily be determined in time steps; the effect of soil condition on the limit states of dam elements in the course of seismic action is determined easily etc. However, the method requires a long computation time, as the temporal steps are very short (about 0.001 s), while the duration of earthquake is usually 20–120 s so that for $T = 120$ s one has to make arrangements for 120 000 time steps. The implicit scheme has a number of advantages in this respect. The time step in implicit scheme is limited only by the required accuracy. Usually this time step is much longer than that in the explicit scheme given by Equation 5.15. The time step is usually taken from discretization of the acceleration graph and becomes by one order of magnitude higher than the explicit time step. The implicit problems still involve the aforementioned assumptions given in Section 5.3 as to inclusion of the celerity of propagation of elastic seismic waves in the foundation of dam, that is on the construction of design acceleration graph.

In implicit problems, some difficulties arrive in the inclusion of the effect of dam on the transformation of oscillograph traces, as done by Lombardo in explicit scheme; this is due to the fact that the transformation is done in parallel with stress-strain computations. The use of nonlinear equations of physics is also difficult. At the same time, implicit schemes provide a diverse auxiliary information on the performance of structure under seismic effects such as eigenfunctions and eigen frequencies, as explained below; this information gives a deeper insight into mechanical processes.

The stress-strain problems solved by FEM are based on the theory of oscillations.

The condition of dynamic equilibrium for a system with one degree of freedom reads:

$$f_I + f_D + f_S = p(t) \tag{5.18}$$

in which

f_I = inertia force,
f_D = viscous damping force,
f_S = force of elasticity,
$p(t)$ = vector of external force as a function of time.

It is seen from Equation 5.18 that the force $p(t)$ includes various types of forces applied to mass — elastic reaction oriented against the displacements; damping forces resisting the displacements or velocities; and independent external loads. If the inertia force, which also resists acceleration, is taken in addition, then one obtains the equation of motion which expresses the equilibrium of all forces. The d'Alembert principle can be written as follows, in which m is the inertia force proportional to acceleration in the opposite direction:

$$p(t) = -m\ddot{r}_g(t) \tag{5.19}$$

in which

$r_g(t)$ = vector of displacement for foundation.

Then for $f_I = m\ddot{r}$, $f_D = c\dot{r}$, $f_S = kr$ one has

$$m\ddot{r} + c\dot{r} + kr = -m\ddot{r}_g(t) \tag{5.20}$$

in which

c = damping constant,

k = stiffness,
r = vector of deformation (bending) of structure with respect to foundation.

The minus sign on the right-hand side shows that the load is oriented against the acceleration of soil; this sign has no practical meaning in the case of seismic effects because the latter might have an arbitrary orientation.

Equation of motion in matrix form. By using Equation 5.20, that is the general differential equation of motion with damping in the matrix form, for a system of many degrees of freedom, and by decomposing the vector of displacement of foundation one obtains (Zienkiewicz 1975)

$$[M]\{\ddot{r}\} + [C]\{\dot{r}\} + [K]\{r\} = -\{E^x\}\ddot{u}_g^x(t) - \{E^y\}\ddot{u}_g(t) \tag{5.21}$$

in which
$[M], [C], [K]$ = matrices of mass, damping coefficient and stiffness, respectively,
$\{r\}$ = vector of displacement of nodal point with respect to foundation,
$\{E^x\}, \{E^y\}$ = columns of mass showing horizontal and vertical displacement of nodal points, respectively,
$\ddot{u}_g^x(t), \ddot{u}_g^y(t)$ = horizontal and vertical components of foundation accelerations under seismic effects, respectively.

In general, Equation 5.21 can be considered a vector matrix form for dynamic equilibrium in the case of systems with many degrees of freedom; Equation 5.21 is derived in Panovko (1971).

Equation of motion for free oscillations without damping reads:

$$[M]\{\ddot{r}\} + [K]\{r\} = 0 \tag{5.22}$$

In analysis of oscillations for systems with distributed parameters (mass and deformability) one may utilize an approximate method basing on the configuration of a system in motion.

If the oscillating system is stable (which is indeed considered in our case), then the particular solution of the differential equation (Eq.5.22) will read (Fradkin 1973):

$$\begin{aligned}
r_1 &= A_1 \sin(\omega t + a); \\
r_2 &= A_2 \sin(\omega t + a); \\
&\quad\dots\dots\dots\dots\dots \\
r_n &= A_n \sin(\omega t + a)
\end{aligned} \tag{5.23}$$

By substituting Equation 5.22 one obtains a system of homogeneous algebraic equations (with respect to amplitude):

$$-A_1\omega^2 m_{1\,1} - A_2\omega^2 m_{1\,2} - \ldots - A_n\omega^2 m_{1\,n} +$$
$$+A_1 k_{1\,1} + A_2 k_{1\,2} + \ldots + A_n k_{1\,n} = 0$$
$$-A_1\omega^2 m_{2\,1} - A_2\omega^2 m_{2\,2} - \ldots - A_n\omega^2 m_{2\,n} +$$
$$+A_1 k_{2\,1} + A_2 k_{2\,2} + \ldots + A_n k_{2\,n} = 0 \qquad (5.24)$$
$$\ldots\ldots\ldots\ldots\ldots\ldots\ldots$$
$$-A_1\omega^2 m_{n\,1} - A_2\omega^2 m_{n\,2} - \ldots - A_n\omega^2 m_{n\,m} +$$
$$+A_1 k_{n\,1} + A_2 k_{n\,2} + \ldots + A_n k_{n\,n} = 0$$

A trivial solution of the system of algebraic equations arises for the condition $A_1 = A_2 = \ldots = A_n = 0$, that is for no oscillations. We are interested in nonlinear solutions. In this case the following determinant should be zero:

$$\begin{vmatrix} k_{1\,1} - m_{1\,1}\omega^2 \ldots k_{1\,2} - m_{1\,2}\omega^2 \ldots k_{1\,n} - m_{1\,n}\omega^2 \\ k_{2\,1} - m_{2\,1}\omega^2 \ldots k_{2\,2} - m_{2\,2}\omega^2 \ldots k_{2\,n} - m_{2\,n}\omega^2 \\ \ldots\ldots\ldots\ldots\ldots\ldots\ldots\ldots\ldots \\ k_{n\,1} - m_{n\,1}\omega^2 \ldots k_{n\,2} - m_{n\,2}\omega^2 \ldots k_{n\,n} - m_{n\,n}\omega^2 \end{vmatrix} = 0 \qquad (5.25)$$

Upon expansion of the determinant one obtains a polynomial in the power n with respect to ω^2. The number of roots of the polynomial is $n(\omega_1^2, \omega_2^2, \ldots \omega_n^2)$. The frequencies in the increasing order $(\omega_1 < \omega_2 < \ldots \omega_n)$ generate the spectrum of eigen frequencies for oscillations of the system.

Each root $\omega_i (1 \le i \le n)$ corresponds to a particular solution of Equation 5.23 for displacement of the j-coordinate: $r_i = A_{ji} \sin(\omega_i t + \alpha_i)$ (in which the index i shows the respective frequency).

The general equation reads:

$$r_i = \sum_{i=1}^{n} A_{ij} \sin(\omega_i t + a_i), \quad j = 1, 2, 3, \ldots, n \qquad (5.26)$$

If any value ω_i is substituted into Equation 5.25, together with its counterpart a_{ij}, then the number of independent equations will be $n - 1$. These equations permit all amplitudes to be expressed by any of them, for instance the first one. The set of relationships

$$\kappa_{1i} = 1; \quad \kappa_{2i} = A_{2i}/A_{1i}; \quad \kappa_{3i} = A_{3i}/A_{1i}, \ldots, \kappa_{n_i} = A_{ni}/A_{1i} \qquad (5.27)$$

provides relative amplitudes for the i-th eigen oscillation, that is describes the configuration of a system for the highest displacement for oscillations with i-th frequency. This configuration is determined with accuracy to any multiplication factor, that is its scale remains unknown. The set given in Equation 5.27 in the matrix form describes the coefficients of the vector in n-dimensional space. The set of these vectors is referred to as a matrix of eigenforms of oscillations $[X]$. Each eigen frequency ω_i corresponds to i-th eigenform of oscillations $\{i_i\}$.

Determination of eigen frequencies (eigen values) and eigen forms (eigen vectors) is the most difficult problem for oscillations of a system with a large number of degrees of freedom. This problem is usually solved on computers, whereupon some approaches of matrix algebra are utilized, as the aforementioned method is very complex in the case of finding roots for high order polynomials.

Equation 5.22 may be reduced to the following form

$$[1]\{r\} = -[K]^{-1}[M]\{\ddot{r}\} \qquad (5.28)$$

In the Rayleigh method one assumes that the configuration is given with the accuracy of one of the parameters depending on time, whereupon the mechanical model has only one degree of freedom (for instance, free bending of beams).

By the Rayleigh method one obtains:

$$\{r(t)\} = [X]\{y(t)\} \qquad (5.29)$$

in which

$[x]$ = predetermined function of coordinate,
$\{y(t)\}$ = column of unknown time function,

$$r_i(t) = \{x_i\}y_i(t) = \{x_i\}\sin(\omega_i t + \alpha) \qquad (5.30)$$

Upon substitution of Equation 5.30 in Equation 5.29 one obtains the following relationship for the i-th mode:

$$\frac{1}{\omega_i^2}\{x_i\} - [K]^{-1}[M]\{x_i\} = 0 \qquad (5.31)$$

Denoting by $\omega_i^2 = \lambda_i$ one gets the following characteristic equation:

$$\{x_i\} = \lambda_i[H]\{x_i\} \qquad (5.32)$$

If the right-hand side of Equation 5.20 is zero, in order to include attenuation of the oscillations, the above equation may be considered in terms of Zeldovich & Yaglom (1982), for energy in oscillatory processes.

One may assume $r(t) = B\cos(\omega_1 t + a)$. The potential energy at any time of oscillations reads

$$U(r(t)) = \frac{kr^2(t)}{2}$$

as the elastic force is $kr(t)$, the mean rise being $r(t)/2$ and

$$U(r(t)) = \frac{kB_1^2}{2}\cos^2(\omega_1 t + \alpha) \qquad (5.33)$$

The kinetic energy is

$$K(t) = \frac{mv^2}{2} = \frac{m}{2}[-B\omega_1\sin(\omega t + \alpha)]^2 = \frac{mB^2}{2}\omega_1^2\sin^2(\omega_1 t + \alpha) \qquad (5.34)$$

Assuming that for weak attenuation one has $\omega \approx \omega_1$ one obtains $\omega^2 = k/m$ and

$$K(t) = \frac{kB^2}{2}\sin^2(\omega_1 t + \alpha)$$

The total energy becomes

$$U(r(t)) + k(t) = \frac{kB^2}{2}(\sin^2(\omega t + \alpha) + \cos^2(\omega t + \alpha)) = \frac{kB^2}{2} \qquad (5.35)$$

Taking for granted that the friction force is proportional to velocity and is oriented against the oscillatory motion one has $F_T = -cv$, in which v reads

$$v = \frac{dr(t)}{dt} = -B\omega \sin(\omega t + \alpha)$$

Then the momentum of the friction force becomes

$$F_T v = -cv^2 = -cB^2\omega^2 \sin(\omega t + \alpha)$$

This product equals the change in momentum

$$\frac{d}{dt}\left(\frac{kB^2}{2}\right) = kB\frac{dB}{dt} = -cB^2\omega^2 \sin^2(\omega t + \alpha) = -\frac{cB^2\omega^2}{2} \qquad (5.36)$$

as the mean value of $\sin^2 x$ is 1/2.

Indeed one has $y = \sin^2 x$ for x varying from 0 to infinity, and the mean value reads:

$$\sin^2 x = \frac{1 - \overline{\cos 2x}}{2} = \frac{1}{2} - \frac{1}{2}\overline{\cos 2x} = \frac{1}{2}$$

as the mean value of $\cos 2x$ is 0.

To prove it apply the theorem on mean value:

$$\bar{y}(0, b) = \frac{\int\limits_0^b \cos x \, dx}{b - 0} = \frac{\sin x|_0^b}{b - 0} = \frac{\sin b}{b} \qquad (5.37)$$

The maximum possible numerator for $b \to \infty$ does not exceed 1, and the denominator tends to infinity, so that $\bar{y}(0, b)$ goes to 0.

Now one has

$$\frac{dB}{dt} = -\frac{c\omega^2}{2k}B \quad \text{and} \quad \frac{dB}{B} = -\frac{c\omega^2}{2k}dt$$

and accordingly

$$B = B_0 exp(-\frac{c\omega^2}{2k}t) = B_0 e^{-ht} \qquad (5.38)$$

in which

$$h = c\omega^2/2k, \quad \omega^2 = k/m, \quad h = c/2m.$$

Hence the particular solution can be sought in the form $r = e^{-ht}$. The quantity B_0 is found from the initial condition.

If the particular solution is substituted in the homogeneous equation with damping

$$\ddot{r}(t) + c\dot{r}(t) + kr = 0 \tag{5.39}$$

then one obtains

$$(mh^2 - ch + k)e^{-ht} = 0 \tag{5.40}$$

Equation 5.40 is satisfied by the two following values of h:

$$h_{1,2} = \frac{c}{2m} \pm \sqrt{\left(\frac{c}{2m}\right)^2 - \frac{k}{m}} \tag{5.41}$$

For large values of c one has $(c/2m) > k/m$ and the two values h_1; h_2 are real. For small c, which appear in most cases, with inclusion of earth structures, h may assume complex values.

Damping factor. The damping factor c for which the transition from real to complex values occurs is referred to as *critical* and is denoted by c_c:

$$\frac{c_c}{2m} = \sqrt{\frac{k}{m}} = \omega; \quad c_c = 2m\omega \tag{5.42}$$

If the real damping is given in terms of the critical one then the factor reads:

$$\xi = \frac{c}{c_c} \quad \text{and} \quad c = \xi c_c = 2\xi m\omega \tag{5.43}$$

Express $h_{1,2}$ through the relative damping factor ξ, (Clough & Penman 1979):

$$h_{1,2} = \xi\omega \pm \sqrt{(\xi\omega)^2 - \omega^2} = \xi\omega \pm i\sqrt{1 - \xi^2}\omega \tag{5.44}$$

Taking into account that $\sqrt{1 - \xi^2}\omega = \omega_D$ is the damping frequency, Equation 5.44 can be written down as

$$h_{1,2} = \xi\omega \pm i\omega_D \tag{5.45}$$

The solution will assume the form:

$$r(t) = B_1 e^{-\xi\omega t - i\omega_D t} + B_2 e^{-\xi\omega t + i\omega_D t} = e^{-\xi\omega t}(B_1 e^{-i\omega_D t} + B_2 e^{i\omega_D t}) \tag{5.46}$$

in which
B_1 and B_2 = initial amplitudes of oscillations.

The expression in brackets describes a simple harmonic oscillation which is unity by identity for $t = 0$ and integer multiples of π/ω_D. This statement becomes clear if the brackets are shown as a trigonometric form of a complex number (cf. Euler equations) and B_1, B_2 are conjugated complex numbers $B_1 = \bar{A} + i\bar{B}$, $B_2 = \bar{A} - i\bar{B}$. Algebraic transformations yield

$$r(t) = e^{-\xi\omega t}(A\sin\omega_D t + B\cos\omega_D t) \tag{5.47}$$

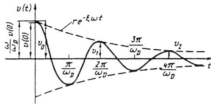

Figure 5.6. Free oscillations of undamped system.

The constants A and B are found from the initial conditions for t approaching zero. By virtue of $A = \frac{\dot{r}(0) + r(0)\xi\omega}{\omega_D}$ and $B = r(t)$ at $t = 0$ one also has the following obvious relationships

$$r(t) = e^{-\xi\omega t}\left[\frac{\dot{r}(0) + r(0)\xi\omega}{\omega_D}\sin\omega_D t + v(0)\cos\omega_D t\right] \tag{5.48}$$

Hence the general solution to Equation 5.39 provides a damped oscillation with the stationary frequency ω_D and gradually decreasing amplitude (Fig.5.6).

The series of the maximum displacements obeys the geometric rule

$$r(t) : r(t + T) = e^{\xi\omega T} \tag{5.49}$$

in which

T = period of oscillations, i.e. time between two adjacent maxima or the same phase.

During time T the argument of sine and cosine changes by 2π, from which it follows $\omega_D(t + T) = \omega_D t + 2\pi$ and $T = 2\pi/\omega_D$.

The natural logarithm of the ratio $r(t) : r(t + T)$ is referred to as logarithmic decrement:

$$\delta = \ln\frac{r(t)}{r(t + T)} = \xi\omega T = 2\pi\xi\frac{\omega}{\omega_D} \tag{5.50}$$

With inclusion of $\omega_D = \omega\sqrt{1 - \xi^2}$ one has

$$\delta = 2\pi\xi\frac{1}{\sqrt{1 - \xi^2}} \tag{5.51}$$

For small ξ (usually below 0.2 for most materials under normal soil conditions, as outlined in Section 1.6), one may assume

$$\delta = 2\pi\xi \tag{5.52}$$

Equation 5.50 can be presented as the series (Newmark & Rozenblatt 1980)

$$\frac{r(t)}{r(t + T)} \simeq e^{2\pi\xi} = 1 + 2\pi\xi + \frac{(2\pi\xi)^2}{2!} + \cdots \tag{5.53}$$

Taking the first two terms of the series, which is fairly sufficient for small ξ, one obtains

$$\xi = \frac{r(t) - r(t - T)}{2\pi r(t + T)} \tag{5.54}$$

Hence the damping factor is approximated.

On comparison with the exact solution

$$\left[\frac{r(t)}{r(t + T)} = e^{2\pi\xi\frac{\omega}{\omega_D}} \right]$$

the error of ξ for $\xi = 0.2$ becomes

$$\frac{\xi_{accur}}{\xi_{appr}} = 0.62 \tag{5.55}$$

so that the accurate ξ is 0.124. The error decreases practically linearly with decreasing ξ. This should be remembered in experimental assessment of ξ.

For small ξ one may take $\omega \sim \omega_D$ and Equation 5.48 transforms into another relationship. For $\omega \neq \omega_D$ the oscillations have the frequency ω_D with respect to the equilibrium condition (Fig.5.6). If the two frequencies are roughly identical, the process will display ω.

Transition to systems with multiple degrees of freedom is implemented in matrix forms. For instance, the damping matrix by Equation 5.43 becomes $2[\xi\omega]$ [M].

Substitute Equation 5.29 into Equation 5.21. From the theory of oscillations one has

$$\begin{aligned}
&[K][M]^{-1} = D(\lambda); \quad [C][M]^{-1} = 2\{\xi_i\omega_i\} \\
&M^* = \{X_i\}^T[M]\{X_i\} \\
&P_n^*(t) = -\{X_n\}^T\{E^x\}\ddot{u}_g^x(t) - \{X_n\}^T\{E^y\}\ddot{u}_g^y(t)
\end{aligned} \tag{5.56}$$

Since the extradiagonal terms $\{X_i\}, [M], \{X_j\}$ are zero for i different from j on the strength of the orthogonality of eigen vectors, the matrix equation Equation 5.21 can be presented as a sum of mutually independent differential equations; for the n-th mode one has:

$$\ddot{y}(t) + 2\xi_n\omega_n\dot{y}(t) + \omega_n^2 y(t) = \frac{P_n^*(t)}{M_n^*} \tag{5.57}$$

Solution of Equation 5.57 can be obtained by different methods, for instance the Runge-Kutta scheme.

Formulation of mass matrix. Mass of the system intervenes in Equation 5.57 in the two manners: as mass matrix $[M]$ of the system and mass column $\{E^x\}$ and $\{E^y\}$, affecting the inertia loading. Moreover, the mass matrix enters the characteristic equation, Equation 5.28, from which one obtains eigen frequencies and modes. The mass columns are specially constructed systems which generate the inertia loading.

Decompose the area into triangular elements by analogy to the solution of the static problem by FEM (cf. Chapter 4). Denote by $M_1, M_2, M_3, \ldots, M_n$ the masses concentrated at nodes, where $M_n = 1/3\Sigma M_\Delta$; (M_Δ = mass of triangular element; m = number of elements converging at the node; n = number of nodes). Taking into account that a concentrated mass displays identical inertia under horizontal and vertical forcing (accelerations) one obtains the following mass columns

$$\{E^x\}^T = \; < M_1 0 \; M_2 0 \ldots M_n 0 >$$
$$\{E^y\}^T = \; < 0 \; M_1 0 \; M_2 \ldots 0 \; M_n > \tag{5.58}$$

The mass matrix $[M]$ can be generated in two ways.

1. By analogy to the mass columns, the mass matrix of a certain triangular element l reads

$$\frac{M_l}{3} \begin{vmatrix} 1 & 0 & 0 & 0 & 0 & 0 \\ 0 & 1 & 0 & 0 & 0 & 0 \\ 0 & 0 & 1 & 0 & 0 & 0 \\ 0 & 0 & 0 & 1 & 0 & 0 \\ 0 & 0 & 0 & 0 & 1 & 0 \\ 0 & 0 & 0 & 0 & 0 & 1 \end{vmatrix} \tag{5.59}$$

The complete mass matrix consists of the nodal superposition of mononomical mass components of the elements of Equation 5.59 converging at the node. This method of distribution of masses and generation of mass matrix is most widespread.

2. The generation of the mass matrix for a single element formulated by Zienkiewicz (1975) is based on the principle of the equivalence of concentrated mass and real distributed mass. The derivation is skipped; the ultimate matrix of masses for an element reads:

$$[m]^l = \frac{M_l}{3} \begin{vmatrix} \frac{1}{2} & 0 & \frac{1}{4} & 0 & \frac{1}{4} & 0 \\ 0 & \frac{1}{2} & 0 & \frac{1}{4} & 0 & \frac{1}{4} \\ \frac{1}{4} & 0 & \frac{1}{2} & 0 & \frac{1}{4} & 0 \\ 0 & \frac{1}{4} & 0 & \frac{1}{2} & 0 & \frac{1}{4} \\ \frac{1}{4} & 0 & \frac{1}{4} & 0 & \frac{1}{2} & 0 \\ 0 & \frac{1}{2} & 0 & \frac{1}{4} & 0 & \frac{1}{2} \end{vmatrix} \tag{5.60}$$

The general mass matrix of the system is formulated similarly to the diagonal mass matrix, Equation 5.59.

One may note in advance that the use of the distributed mass matrix in a plane strain problem using triangular elements increases the accuracy by about 20%.

Solution of the characteristic equation. Equation 5.39 can be solved in a number of ways. The Schwarz method is most widely used; it provides the low eigen frequencies and modes in the early stage of computations but can be used with success only if a limited number of modes and frequencies is needed.

The Schwarz method is applicable if $[H]$ is a symmetric matrix whereas it becomes asymmetric upon use of Equation 5.60. Transform the matrix $[H]$ to the symmetric form by taking $[K]$ as a product of two triangular matrices:

$$[K] = [L]\,[L]^T \tag{5.61}$$

in which

$[K]$ = stiffness matrix constructed by analogy to the static problem,
$[L]$ = triangular form of $[K]$,
$\lambda_i = \omega_i^2$; ω_i = angular frequency of i-th mode.

The matrix $[L]$ is generated from $[K]$ by the method of quadratic root using the following recurrential formulae, with l being an element of $[L]$:

$$
\begin{aligned}
l_{11} &= k^{1/2} \\[4pt]
l_{i1} &= \frac{k_{i1}}{l_{1\,1}} \\[4pt]
l_{ii} &= \left(k_{ii} - \sum_{m=1}^{i-1} l_{im}^2 \right)^{1/2} \\[4pt]
l_{ij} &= \frac{k_{ij} - \sum_{a=1}^{i-1} l_{ia} l_{ja}}{l_{ji}}, \quad 1 < j < i
\end{aligned}
\tag{5.62}
$$

Transform Equation 5.32:

$$
\begin{aligned}
&[L][L]^T\{x_i\} = \lambda_i[M]\{x_i\}; \quad [L]^T\{x_i\} = \{\bar{x}_i\}; \quad [L^T]^{-1}\{\bar{x}_i\} = \{x_i\} \\
&[L]\{\bar{x}_i\} = \lambda_i[M][L^T]^{-1}\{\bar{x}_i\} \\
&\{\bar{x}_i\} = \lambda_i[L]^{-1}[M][L^T]^{-1}\{\bar{x}_i\} \\
&\{\bar{x}_i\} = \lambda_i[\overline{H}]\{\bar{x}i\}; \quad [\overline{H}] \\
&\{\bar{x}_i\} = \frac{1}{\omega_i^2}\{\bar{x}_i\}
\end{aligned}
\tag{5.63}
$$

The matrix $[H]$ is now symmetric.

The quantities $\frac{1}{\omega_i^2}$ and $\{\bar{x}_i\}$ can be obtained by iterations from Equation 5.63 using the initial test vector

$$
v = \begin{vmatrix} 0 \\ 0 \\ 0 \\ \cdots \\ l \end{vmatrix}
$$

The iteration process is continued until the two consecutive terms v_n, v_{n+1} differ by a small predetermined figure, usually 10^{-6}. The convergence has been proved by Newmark & Rozenblatt (1980), where a similar approach by Stodola is presented.

By standardizing x_i one finds the matrix

$$[h_i] = \frac{1}{\omega_i^2}\{\bar{x}_i^0\}\{\bar{x}_i^0\}^T \tag{5.64}$$

in which

$\bar{x}_i^0$ = orthonormal vector.

In order to find the mode $i + 1$ one writes

$$[\overline{H}_{i+1}] = [\overline{H}_i] - [h_i] \tag{5.65}$$

The adequate characteristic equation will read

$$[\overline{H}_{i+1}]\{\bar{x}_{i+1}\} = \frac{1}{\omega_{i+1}^2}\{\bar{x}_{i+1}\} \tag{5.66}$$

The procedure is repeated until the required number of modes is reached. At this point one returns to the initial characteristic Equation 5.32 in accordance with Equation 5.63:

$$x_i = [L^T]^{-1}\{\bar{x}_i\} \tag{5.67}$$

The inverse matrix of the triangular form $[L]$ is found from two recurrence formulae obtained from the formulation for the inverse matrix. Denote by l_{ij} the elements of $[L]$ and by r_{ij} those for the inverse matrix:

$$r_{ii} = \frac{1}{l_{ij}}$$

$$r_{ij} = \frac{\sum_{m=1}^{m=i-1} l_{im} r_{mj}}{l_{ii}}, \quad i \neq j \tag{5.68}$$

The equation of motion (Eq.5.21) is based on the assumption that all points of the foundation of a structure have identical accelerations at the same time, which has been shown to hold true only for short structures, compared with wavelength, or for high celerities of wave in foundation.

Earth-rockfill dams of 300–350 m height are 1200–1700 m at the foot; this corresponds to the time of wave passage of 0.3 s if the celerity is 6000 mps (e.g. in weathered and metamorphic cracked rock such as gneiss, granite etc.

Inclusion of 'running wave'. The celerity of wave in foundation should also be analysed in implicit problems.

The equation of motion incorporating the wave celerity in foundation reads (Clough & Penman 1979):

$$[M]\{\ddot{r}\} + [C]\{\dot{r}\} + [K]\{r\} = -\sum_{k=1}^{N_b}[M]\{b_{2k-1}\}\ddot{r}_{s,2k-1}^b \tag{5.69}$$

in which

$\ddot{r}_{s,2k-1}^b = \ddot{u}^x\left(t - \frac{x_k}{v}\right)$,

N_b = number of nodes at interface of foundation,

x_k = abscissa of k-th node at dam foundation,
v = celerity of seismic wave at dam foundation,
$\{b_{2k-1}\}$ = influence vector, i.e. displacement of k-th node by -1 in the x direction; all horizontal displacements having odd numbers:

$$\{r_s\} = \sum_{k=1}^{N_b} \{b_{2k-1}\} r_{s,2k-1}^b$$

r_s = vector of nodal displacements.

The vertical component of acceleration in this case is not considered for its contribution to the stress state is usually small; if necessary, the term $\{b_{2k}\}$ may be added to the RHS of Equation 5.56.

Equation 5.69 transforms into Equation 5.21 for $\sum\{b_{2k-1}\}$ =<1010...>, which occurs if all points of foundation at time t have the same acceleration $\ddot{u}^x(t - \frac{x_k}{v}) = u^x(t)$, i.e. $v \to \infty$.

By analogy to Equation 5.21 one has

$$\ddot{y} + 2\xi_n \omega_n \dot{y} + \omega_n^2 y = \frac{P_n^*(t)}{M^*} \tag{5.70}$$

in which

$$P_n^*(t) = -\sum_{k=1}^{N_b} \{x_n\}^T [M] \{b_{2k-1}\} \ddot{u}^x \left(t - \frac{x_k}{v}\right)$$

and the remaining notation holds.

The vector $\{b_{2k-1}\}$ can be found as follows. Construct the border part of the stiffness matrix, each vector-column of which (g_k) represents the stresses arising at nodes due to displacement of foundation nodes $r_{sk}=1$. One may write the system of linear algebraic equations:

$$[K]\{b_k\} = -\{g_k\} \tag{5.71}$$

Solving N_b times the system in Equation 5.71, with respect to $\{b_k\}$ one obtains a rectangular influence matrix. Since we are interested in the horizontal effects only, the odd columns q_{2k-1} are chosen for RHS of Equation 5.71, and the respective vector-columns of influence $\{b_{2k-1}\}$ are found.

Hence, in comparison with solution of Equation 5.21, in this case one must solve a system of linear algebraic equations in N_b iterations, in which the right-hand side varies exclusively. To this end one must construct the bordering part of the stiffness matrix.

In solving Equation 5.70 at each nodal point of foundation and any time t, k is determined by the quantity $\ddot{u}_{gk}^x(t - x_k/v)$ given in acceleration graph.

As already pointed out in Section 5.2, upon inclusion of the celerity of seismic waves in the foundation of an earth dam extending over a large zone it is necessary to determine the mutual displacement of foundation points and the reaction of the dam to the displacement. The displacement of nodal points at time t due to mutual displacement of the points caused by the stretch of the dam zone is given as follows:

$$\{r_s^c(t)\} = [b_{2k-1}]\left\{r_{s,2k-1}^b\left(t - \frac{x_k}{v}\right)\right\}$$ (5.72)

The vector r_s^c is added to the vector r_s arising at this time and determined for a given time step t. Stresses and strains are then determined over elements.

The presented method of solution for dynamic problems in FEM extends the possibility of analysis of the operation of earth-rockfill dams. Inclusion of the celerity of seismic waves in the foundation, which is often reffered to as the 'running wave' formulation, enables analysis of very complex aspects of dam performance on various foundations.

Assessment of residual strains. The pulsation of elastic and plastic strains and stress components makes possible assessment of the temporal variation of reliability factor in each element. One or another strength theory can be used for this purpose. The theory including the loading path, for instance the energetic condition of strength in Equation 1.44, which takes into account both plastic strains and loading path, is preferred.

It must be emphasized that plastic strains depend on soil condition, with respect to the limit state, and usual computations of plastic strains in time coincide with assessment of soil condition, i.e. the components of plastic strains and $k_{HI}(t)$ are determined at time t in i-th element by a given value of $k_{Hi}(t - \Delta t)$. The quantity $k_{Hi}(t)$ is computed by k_{Hi} in the static problem.

The greatest complexity arises in the assessment of residual displacements in dam at any time, as computed by the exisiting plastic strains. Solution of the inverse problem (transition from strains to displacement) is impossible without additional assumptions because the number of equations is lower than the number of unknowns. Indeed, in a triangular element one has six unknown components of nodal displacements while the number of equations relating displacements to strains is only three:

$$e_x = \frac{y_m - y_j}{2\Delta}u_{xi} + \frac{y_i - y_m}{2\Delta}u_{xj} + \frac{y_j - y_i}{2\Delta}u_{xm}$$

$$e_y = \frac{x_j - x_m}{2\Delta}u_{yi} + \frac{x_m - x_i}{2\Delta}u_{yj} + \frac{x_i - x_j}{2\Delta}u_{ym}$$

$$e_{xy} = \frac{x_j - x_m}{2\Delta}u_{xi} + \frac{x_m - x_i}{2\Delta}u_{xj} + \frac{x_i - x_j}{2\Delta}u_{xm} +$$

$$+ \frac{y_m - y_j}{2\Delta}u_{yj} + \frac{y_i - y_m}{2\Delta}u_{yj} + \frac{y_j - y_i}{2\Delta}u_{ym}$$

(5.73)

in which

$$2\Delta = (x_m - x_i)(y_j - y_i) - (x_j - x_i)(y_m - y_i).$$

The simplest assumption in a general case can be formulated as equal contribution of displacements at nodal points to generation of strains. The terms with zero numerators, that is the components of nodal displacements which do not contribute to the generation of strain components, must be excluded. The strain component can be decomposed into equal parts within the remaining terms thereafter, and the components of nodal displacement may be determined separately. The total vector of residual displacements is found by summing up at nodes the mononomical displacements obtained in consecutive analysis of all elements belonging to a given nodal point.

The above procedure does not account for the growth of plastic strains in the course of stress fluctuations. This effect may be included in the following iteration procedure.

1. Problem is solved in accordance with the aforementioned inclusion of elastic characteristics for soil deformability.

2. Basing on the above solution for the distribution of k_w^i upon propagation of seismic wave one varies the damping coefficient with respect to the critical coefficient (see Section 1.6), along with the soil stiffness in elements. The solution is repeated thereafter.

The number of iterations will obviously be small; and the obtained solution is approximate.

Operation of dam under seismic effects. Hence the dynamic method makes it possible to analyse quite thoroughly the operation of dams under seismic loadings. It is nevertheless interesting to have a look at some other general qualititive properties of dam operation as a function of design, type of acceleration graphs, formulation of mass, matrix etc.

For the purpose of the above analysis we selected a homogeneous gravel-sand dam, 128 m high, having a symmetric profile with slopes 1:2. In the inclusion of water pressure on the upstream side the counterseepage element was taken in two forms: flexible nonsoil screen and flexible nonsoil diaphragm.

In the assessment of the effect of structural properties of dams, the formulation of mass matrix, effect of damping factor etc, the design acceleration graph was taken as one of California graphs (Parkfield) with the maximum horizontal acceleration of 2.94 m/s^2 occurring at the fourth second (3.947 s) after commencement of earthquake (Fig.5.7). The maximum vertical acceleration (1.572 m/s^2) occurred after the third second (3.094 s).

The solution shows that in the diagonal matrix the displacements and stresses are slightly overestimated and the density of grid strongly affects the accuracy of solution; therefore further analysis was conducted for a distributed matrix.

For comparison of the eigen frequencies for two dams it has become clear that lower values are encountered in dams with diaphragms. The primary effect is due to the added mass of water; and it is this effect which brought about differences in the behaviour of dams.

The presence of water affects considerably the eigen values. The analysis of the damping factor versus displacements and stresses, compared with the critical situation computed (see Section 1.6) has shown that reduction of ξ from 0.2 to 0.1 brought about increase in displacements by roughly 20%, with simultaneous

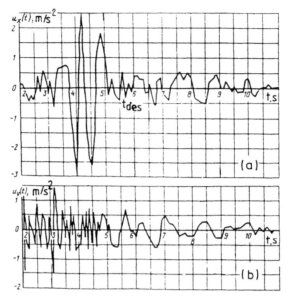

Figure 5.7. Parkfield acceleration graph. (a) horizontal component; (b) vertical component.

increase in stresses, whereupon the type of distribution remained unchanged. Reduction in ξ brought about substantial effect of the primary frequency.

Upon approach of soil to the condition of limit equilibrium the plastic strains grew and the damping capacity of soil increased. A higher damping factor, in relation to the critical one, makes it possible to expose the effect of plastic strains on the generation of stress-strain conditions (Section 5.3). It must be stressed that an increase in damping factor reduces the time of free oscillations of dam.

In order to identify the effect of acceleration graph on the working capacity of the dam we compared results for various acceleration graphs — Parkfield (Fig.5.7b), Eureka and El-Centro (without 'running wave'). The acceleration graphs were scaled to smooth out the maximum peak of horizontal accelerations, which was taken as 0.3 g.

The Parkfield acceleration graph has one strong group of shocks (up to 3 m/s^2) about 3.9 to 4.5 s after beginning of earthquake. The Eureka acceleration graph is characterized by the period $T = 0.45$ s for primary frequency and a group of shocks (up to 3 m/s^2) all over the third, fifth and seventh second (Volokhova & Rasskazov 1974).

The results of computations are compared in Figure 5.8. If the Parkfield acceleration graph is used the peaks $\Delta\tau_{max}$ are lagging behind the maximum shocks, so that $\Delta\tau_{max}$ arises during the shocks of two times smaller in intensity. For the Eureka acceleration graph $\Delta\tau_{max}$ appears during the third group of shocks (first second), and the lag behind the acceleration peak is about 0.35 s; it coincides with the shock having intensity about 1 m/s.

The quantities $\Delta\tau, \Delta\sigma_x$ and $\Delta\sigma_y$ are increments in stress components due to exclusive seismic effect. Total stresses are obtained by summation of static and seismic components. The stress pulsation brings about additional plastic strains.

The difference in stress increments $\Delta\tau_{max}$ reaches 30% at the crest, for equal intensity of peak shocks.

The El-Centro acceleration graph with peak shocks of the same intensity as for Parkfield and Eureka is characterized by the period of primary frequency T = 0.57 s and strong shocks at the beginning of earthquake. Intensive shocks occur at the end of the third second and the intensity of shocks is already two times lower about the seventh and ninth second; strong shocks occurred again at the origin of the twelfth second.

The maximum increments in stresses $\Delta\tau_{max}$ occur during the first and second series of shocks.

Values of $\Delta\tau_{max}$ are found close to those obtained by the Eureka acceleration graph although they are slightly lower. If the El-Centro acceleration graph is used there is practically no lagging of $\Delta\tau_{max}$ versus the time of primary shock (the lag is below 0.1 s).

Figure 5.8 depicts stress pulsations about dam crest for Eureka acceleration graph. Oscillations of all stress components are shown for the most dangerous time.

Stresses $\Delta\sigma_y$ and $\Delta\tau_{xy}$ oscillate in phase, while $\Delta\sigma_x$ are opposite. The stress pulses have a period close to that of the primary frequency of acceleration graph ($t \approx 0.6-0.7$ s), and it is only after the eleventh second that the period of pulses becomes close to the period of the primary frequency of the structure.

The eigen frequencies of the first six forms of dam are 5.488; 8.589; 9.638; 10.590; 11.918; and 12.489 rad/s. The primary eigen frequency of the Eureka acceleration graph is 14 rad/s, that is close to the frequency of the higher forms of eigen frequencies of dam, which are low. The El-Centro acceleration graph has the primary eigen frequency of 11 rad/s and does not fall into resonance with any eigen frequency of structure or its multiple. This seems to be due to the lower effect of the acceleration graph on stresses although the frequencies in the El-Centro graph are lower than those in the Eureka graph.

Hence a high earth-rockfill dam is a structure the primary eigen frequencies of which are below the eigen frequencies of earthquakes. From this it follows that the lower are the eigen frequencies of structures the further away their spacing from resonance zone and the more reliable the condition of the structure, what should obviously be utilized in the design of dams.

The type of acceleration graphs affects stresses in the lower and central parts of dam. The effect of acceleration graphs about slopes is small, which is of particular importance.

The type of acceleration graphs in dams with central core and screen (for which analogous assessments were also made) displays relatively low effect in the zone of the most crucial parts of dam — crest and slopes, where $\Delta\tau_{xy}$ are close to

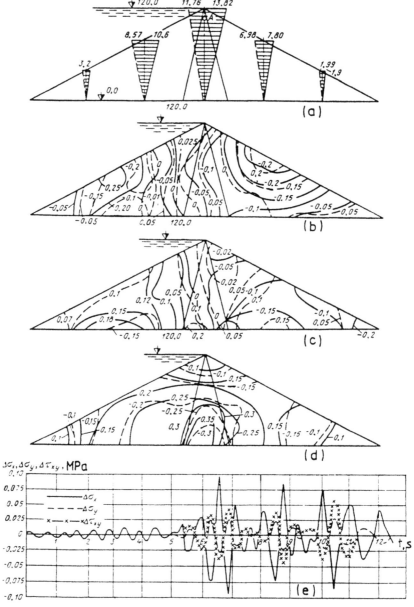

Figure 5.8. Effect of design acceleration graph on stress-strain condition of dam with core, $E_{pr} = 500$ MPa, $\xi = 0.31$, $E_{core} = 800$ MPa, $\nu_{pr} = 0.33$, $\nu_{core} = 0.2$. (a) increments of horizontal displacements Δu, cm; (b) increments of horizontal stresses $\Delta \sigma_x$, MPa; (c) increments of vertical stresses $\Delta \sigma_y$, MPa; (d) increments of tangential stresses τ_{xy}, MPa; (—) Eureka acceleration graph, $\omega_1 = 5.488$ rad/s, $t_{des} = t_1 = 10.435$ s; (- - -) Parkfield acceleration graph, $\omega_1 = 5.488$ rad/s, $t_1 = 5.201$ s; (e) variation of stress increments at point A in time (Eureka graph).

static stresses. The zone is fairly narrow in the dams above 100 m, and for 9-point earthquakes it surrounds the body of dam (δ zone).

The above conclusion may be commented as follows: if pulsating stresses $\Delta\sigma_1$ are about one-half, or more, of the static stress, one encounters the considerable increments in plastic strains, and residual displacements appear in this case.

This occurs mostly in the δ-zone having a thickness about 10 m (that is 10% of dam height), and indeed in this zone the primary part of residual displacements occurred in the cases studied.

The δ zone may embody entire dam and bring about its destruction in low dams, when the period of primary eigen frequency of dam is close to the primary period of earthquake for 9-point earthquakes.

The comparison of stress-strain analysis for various designs shows that dams with central diaphragm and core are identically subjected to earthquake effects, for their elastic characteristics of gravel-pebble and water-saturated cores are similar.

Eigen frequencies in dams with screens are slightly lower, and their horizontal displacements are considerably lower, which is due to higher mass of dam. If low tangential stresses are taken into account in seismic circumstances, then a dam with soil screen or sloping core is preferred. However, it must be borne in mind that a clayey soil of screen is less resistant to shear, and configuration of screen at the slope might cause landslides, even for smaller tangential stresses.

Increased compactness of fill of dam body brings about a higher primary eigen frequency, together with higher elastic displacements and lower residual plastic strains, hence lower residual displacements.

If the dry weight of gravel and pebbles is reduced from 2200 to 2100 kg/m^3, the residual displacements are increased by a factor of 2.

Dams of different height display response to identical seismic actions. For two dams, 300 and 30 m high, with central core and an identical material, one observes the following features.

1. The maximum displacements at the dam crest and the stresses in the dam body in both cases occur at the time corresponding to the maximum of these parameters in the primary form.

2. Eigen frequencies of dams are inversely proportional to the dam height in all design forms.

3. The maximum horizontal displacements of the dam crest for the 300-m dam are roughly 20 times higher than those for the 30-m dam.

4. Seismic stresses in the 300-m dam are only two times higher than for 30 m.

The above conclusions emphasize the earlier finding that seismic effects are more hazardous for lower dams because the background static stresses in them are roughly ten times lower and the hazardous δ-zone enlarges, and embodies the major part of the dam.

In higher dams, the static stresses are greater and therefore the dry unit weight of soil is also greater, for identical initial compactness of fill. This gives rise to increasing dynamic modulus of elasticity, so that the primary period for a 100-m dam is 0.9–1.0 s and that for the 300-m dam becomes 1.1–1.2 s.

Inclusion of 'running wave'. Since the above procedure was presented for 120 m dam, the same dam was considered with a 'running wave'. The solution in Figure 5.9 is presented for $v \to \infty$ which is in agreement with the above procedure. Figure 5.10 shows the solution for $v = 1500$ m/s, while Figure 5.11 depicts the solution for $v = 500$ m/s.

The maximum displacements at crest decrease with the growing v (from 14.2 cm to 12.18 cm and 7.89 cm, respectively). The distribution of stresses is transformed substantially. If for $v \to \infty$ one faces a symmetric distribution of Δ_x (as much as symmetry is possible for asymmetric distribution of masses), then for $v = 1500$ m/s, and particularly for $v = 500$ m/s, one observes a clearcut symmetry, although the absolute stresses differ little (there occurs a reduction in stresses about the slopes if the celerity decreases). Isolines of stresses σ_x follow the form of the 'running wave'.

Stresses $\Delta\sigma_y$ increase, primarily at the foundation of core, where they reach 0.4 MPa, (for $v = 500$ m/s), which corresponds approximately to 30% of static stresses. With increasing celerity of wave the tangential stresses are reduced, particularly at slopes.

The most dangerous moment may be identified for each point of a dam (element). This is most clear for $v = 500$ m/s. For instance, the element at the dam crest has the design time $t_1 = 5.654$ s, and its neighbour is $t = 4.7774$ s while that at the slope has $t = 5.734$ s, whereupon the stress peak does not coincide with the peaks represented by forms of oscillations. Morover, the phases in stresses practically do not coincide at all (Fig.5.10e).

Computations for a dam with screen have provided a qualitatively similar picture. Hence the effect of wave celerity at dam foundation brings about reduction in tangential stresses at the dam crest, so that the δ-zone shrinks.

Computations for the 300-m and 30-m dams, with inclusion of the wave celerity in foundation ($v = 1500$ m/s) have shown that the effect of 'running wave' is manifested in reduction of the maximum tangential stresses by 50%, as compared with the case $v \to \infty$. The dimensions of the δ-zone embodying the external contour of dam have been found two times smaller than for $v \to \infty$.

The residual displacements for $\ddot{u}_{x_{max}} = 0.3\ g$ in the 300-m dam are about 50 cm (see Section 5.3). They were assessed from triaxial tests. The static stress σ_1 and σ_3 created the background stress condition. The sample was then subjected to design stress pulses and the residual displacements were assessed, from which residual displacements in the dam were calculated.

If the above procedure is applied, there arises the question of the number of oscillation modes to include. This criterion has not yet been established. This is due to the fact that some higher forms yield a considerable contribution to displacements (this does not concern the primary low frequency). In the solution of

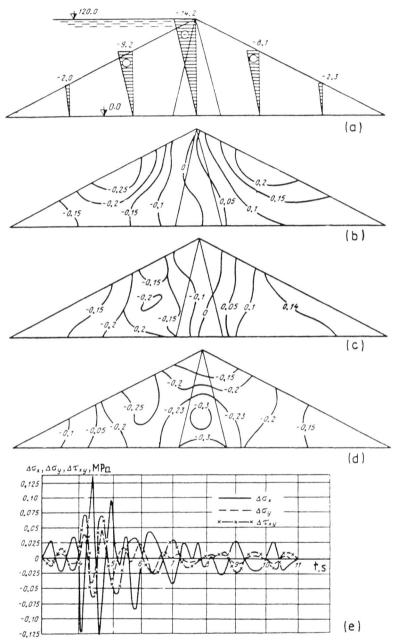

Figure 5.9. Solution of stress-strain problem for dam upon instantaneous propagation of seismic wave in foundation ($v \rightarrow \infty$); Parkfield acceleration graph; $E_{pr} = 500$ MPa; $\nu_{pr} = 0.31$; $\xi = 0.1$; $E_{core} = 800$ MPa; $\nu_{core} = 0.33$; $\omega_1 = 5.488$ rad/s; $t_{des} = t_1 = 5.201$ s; (a) increments of horizontal displacements Δu, cm; (b) increments of horizontal stresses $\Delta \sigma_x$, MPa; (c) increments of vertical stresses $\Delta \sigma_y$, MPa; (d) increments of tangential stresses τ_{xy}, MPa; (e) variation of stress increments at point A in time .

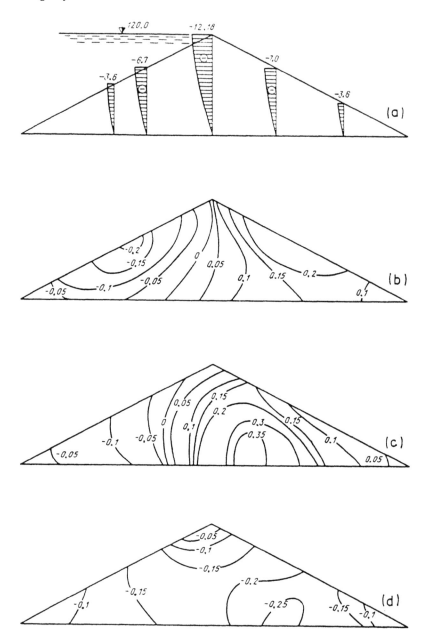

Figure 5.10. Solution of stress-strain problem for dam in which a seismic wave propagates at v=1500 m/s; Parkfield acceleration graph; E=500 MPa; ν =0.31; ξ=0.1; ω_1=5.631 rad/s; $t_{des} = t_1$=5.361 s; (a) increments of horizontal displacements Δu, cm; (b) increments of horizontal stresses $\Delta\sigma_x$, MPa; (c) increments of vertical stresses $\Delta\sigma_y$, MPa; (d) increments of tangential stresses τ_{xy}, MPa.

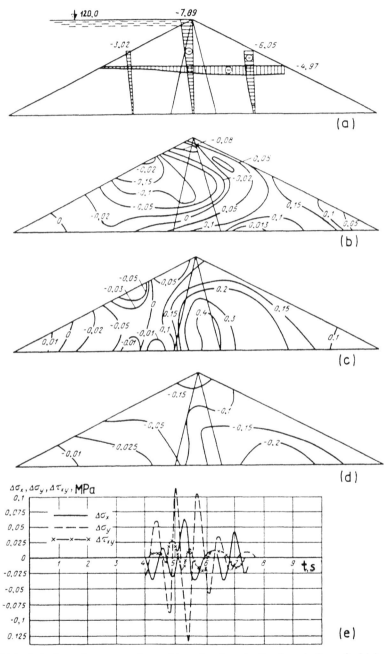

Figure 5.11. Solution of stress-strain problem for dam in the foundation of which a seismic wave propagates at $v=500$ m/s; Parkfield acceleration graph; $t_{des} = t_1 = 5.734$ s; (a) increments of horizontal displacements Δu, cm; (b) increments of horizontal stresses $\Delta \sigma_x$, MPa; (c) increments of vertical stresses $\Delta \sigma_y$, MPa; (d) increments of tangential stresses τ_{xy}, MPa; (e) variation of stress increments $\Delta \sigma_x$, $\Delta \sigma_y$, $\Delta \tau_{xy}$ in time.

dynamic problems by FEM in implicit schemes it is advisable to include not less than 25–30 modes. These modes which give the highest contribution may be identified by experiments. It follows that the modes obtained by experiments do not coincide with those found by design procedures.

5.5 STATIC (QUASIDYNAMIC) METHOD OF SEISMIC COMPUTATIONS

As already noted in Section 5.2, the static or quasidynamic method of seismic computations does not take into account the temporal variation of forces. It is assumed that the seismic forces are acting as static ones after they arise, no matter how long. Like in the implicit dynamic methods, the seismic forces are determined by modes of oscillations. The design should be based on the known forms of oscillations, which is facilitated by spectral analysis; therefore the method is referred to as spectral. It was postulated and elaborated by Zavriyev in 1928, and at present it is included in the Soviet standard SNiP II-7-81 as the compulsory procedure for design of structures in seismic zones.

The method is based on the equilibrium of inertia and external forces (Newton's second law):

$$P(t) = m\ddot{r}(t), \quad \text{or} \quad P(t) - m\ddot{r}(t) = 0 \tag{5.74}$$

in which

$m\ddot{r}(t)$ = inertia force.

Equation 5.74 shows that the seismic force for the i-th eigen frequency related to a certain point k is determined on the assumption of elastic strain:

$$S_{ikj} = k_1 k_2 Q_k \frac{a}{g} \beta_i k_\psi \eta_{ikj} \tag{5.75}$$

in which

Q_k = weight of structure related to point k,

g = acceleration due to gravity,

A/g = relative acceleration taken as fraction of g equal to 0.1; 0.2; 0.4, respectively for seismicity of 7, 8 and 9 points,

η_{ikj} = displacement at k-th point for i-th frequency in direction j.

If weight is divided by g and A/g is multiplied by g one obtains the product of mass and acceleration; g can be reduced and one again obtains Equation 5.75. The coefficients k_1, k_2, k_ψ and β_i, reflect various properties of the operation of structure under seismic conditions.

The coefficient k_1 accounts for the possible damage of buildings and structures and a reduction of acceleration due to plastic strains or cracking in structure. In hydraulic engineering structures, including earth dams, k_1 is assumed 0.25. Hence k_1 reflects the effect of energy absorption upon the growth of plastic strains, as noted in Section 5.3. One must take into account that residual deformations are not allowed for $k_1=1$, and $k_1=0.12$ is assumed in the design of temporary minor structures the destruction of which does not bring about the loss of human life.

The coefficient k_2 accounts for the height of dam and is taken $k_2=0.8$ for dams below 60 m and $k_2=1$ for dams above 100 m. Linear interpolation is suggested between the two values.

The coefficient k_ψ accounts for the damping, although the damping as such does not intervene in computations. Together with k_1, the coefficient k_ψ characterizes the reduction of acceleration due to damping with growing plastic stains. The coefficient of reduction in acceleration in earth dams increases with growing action: $k_\psi=0.7$ for seismicity of 7–8 points and $k_\psi=0.65$ for seismicity of 9 points (in the design of concrete dams one takes respectively $k_\psi=1$ and $k_\psi=0.8$).

The coefficient β_i is referred to as the dynamic coefficient reflecting the growth of acceleration due to properties of the system. This can be explained as follows. In a strictly static theory for seismic forces (without dynamic effects) one has $S = A/g{\cdot}Q$. On the strength of $A/g = (\ddot{u}_g^*/g)$ Equation 5.57 becomes analogous to $S = Q/g{\cdot}\ddot{u}_g^x$, i.e.

$$S = m\ddot{u}_g^x \qquad (5.76)$$

The dynamic property of system was first included by Mononobe in 1920 in his computations of forced oscillations of a system with one degree of freedom without damping. Consider this equation:

$$m\ddot{r} + kr = -m\ddot{r}_s(t) \qquad (5.77)$$

divide both sides of Equation 5.77 by m and take the forcing as $m \cdot B \cdot \sin \omega t$, with $k/m = \omega_0^2$; hence

$$\ddot{r} + \omega_0^2 r = B \sin \omega t \qquad (5.78)$$

in which

> ω_0 = eigen frequency of system (this condition is obtained from the equation of free oscillations without damping)
> ω = frequency of forced oscillations.

It must be stressed that for the forced oscillations $r_s(t) = B_0 \sin \omega t$ as harmonic oscillations one has $\ddot{r}_s(t) = -B_0\omega^2 \cdot \sin \omega t$. On the strength of $B_0 \cdot \omega^2 = const$, which can be denoted by B, upon substitution in Equation 5.77 one obtains the right-hand side effect in $B \cdot \sin \omega t$.

Seek solution in the form

$$\begin{aligned} r(t) &= C\sin(\omega t + \alpha) \\ -\omega^2 C \sin \omega t + \omega_0^2 C \sin \omega t &= B \sin \omega t \end{aligned} \qquad (5.79)$$

and assume that there is no time lag, along with $\alpha = 0$, i.e. that the oscillation begins from the quiescent condition:

$$C(\omega_0^2 - \omega^2)\sin \omega t = B \sin \omega t \qquad (5.80)$$

from which one has

$$C = \frac{B}{\omega_0^2 - \omega^2} = \frac{B}{\omega_0} = \frac{1}{1 - \omega^2/\omega_0^2} \tag{5.81}$$

Hence one has

$$r(t) = \frac{B}{\omega_0^2} \frac{1}{1 - (\omega/\omega_0)^2} \sin \omega t \tag{5.82}$$

or

$$r(t) = \frac{B}{\omega_0^2} \frac{1}{1 - (T_0/T)^2} \sin \omega t \tag{5.83}$$

The seismic inertia force, with inclusion of progressive and elastic displacements for mass m, becomes:

$$S = -m(\ddot{r}_s + \ddot{r}) \tag{5.84}$$

which follows from Equation 5.20.

   Upon substitution into Equation 5.84 of $r(t)$ and $r_s(t)$ one has

$$\begin{aligned} S &= m \left[\frac{B\omega^2}{\omega_0^2[1 - (\omega/\omega_0)^2]} + B \right] \sin \omega t = mB \sin \omega t \left[\frac{\omega^2 + \omega_0^2 - \omega^2}{\omega_0^2(1 - \omega_2/\omega_0^2)} \right] = \\ &= mB \sin \omega t \frac{1}{1 - (T_0/T)^2} \end{aligned} \tag{5.85}$$

The maximum value $S = S_{\max}$ is reached for $\sin \omega t = 1$. In this case one has $B = A_0 = $ amplitude of acceleration at foundation. Then one has

$$S_{\max} = \frac{A_0}{g} gm \frac{1}{1 - (T_0/T)^2} = AQ \frac{1}{1 - (T_0/T)^2} = AQ\beta \tag{5.86}$$

in which

$$\beta = \frac{1}{(1 - (T_0/t)^2}$$

If damping occurs in the system one has

$$\beta = \frac{1}{\sqrt{(1 - (T_0/T)^2)^2 + (2\xi T_0/T)^2}} \tag{5.87}$$

Equation 5.87 is obtained from the equation of forced oscillation with damping, Equation 5.20. The particular solution of homogeneous equation reads

$$r(t) = e^{-\xi\omega t}(A \sin \omega t + B \cos \omega t) \tag{5.88}$$

and the particular solution for harmonic load is

$$r(t) = A \sin \bar{\omega} t + B \cos \bar{\omega} t \tag{5.89}$$

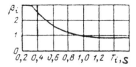

Figure 5.12. Spectral curves.

Upon substitution of Equation 5.89 into Equation 5.20 and division of the coefficient at $\sin \bar{\omega} t$ one obtains two equations which are satisfied separately as $\sin \bar{\omega} t$ and $\cos \bar{\omega} t$ are zero at various times. Two unknowns, A and B intervene in each of these equations; adequate combination yields their values. Upon subsition of A into B in Equation 5.89 one obtains the dynamic coefficient with inclusion of damping (Clough & Penman 1979).

From comparison of Equation 5.86 and Equation 5.76 one may see that the dynamic effects give rise to the coefficient β which is referred to as dynamic coefficient or gain factor of harmonic action. It must be stressed that in order to include the initial phase of earthquake, when the oscillations are superimposed on the oscillating system one takes $r_s(t) = B \cdot \cos \omega t$ instead of $r_s(t) = B_0 \cdot \sin \omega t$, which gives $\beta = \frac{2}{1-(T_0/T)^2}$, which is two times greater than that computed from Equation 5.87. Hence the initial conditions have important effect on the dynamic operation of structures. This result will be contained in the general solution, and not the particular one.

The primary shortcoming of the quasidynamic theory in this version is its complexity and insufficient accuracy of the laws of the motion of soil and structure in real earthquakes. Accordingly, a new method was developed, referred to as linear-spectral (Biot method), which is based on spectral curves. The method predicts maximum values due to dynamic stresses at various points of a structure, in particular earth dams, as related to modes of oscillations. Stresses are determined for a system with one degree of freedom (linear oscillators) for various eigen frequencies and damping coefficients, and are characterized by the dynamic factor β.

An analysis of the maximum seismic accelerations, velocities and displacements of oscillators with various eigen frequencies and damping factors, basing on the recorded earthquakes, has provided spectral curves which relate the maximum accelerations (velocities or displacements) to the period of eigen oscillations (Figure 5.12). It must be remembered that the spectra of accelerations C_w, velocities C_v and displacements C_u are interrelated as follows:

$$C_w = \frac{2\pi}{T}C_v = \left(\frac{2\pi}{T}\right)^2 C_u \tag{5.90}$$

The Soviet standard SNIP II-7-81 provides guidelines for the dynamic coefficient as a function of not only eigen frequencies, by the form of T_i, but also types of soil in foundations, which may be categorized as follows.

Category I: rock and coarse grained dense soil with less than 30% of fine fractions (see Chapter 1) etc. The coefficient β_i is below three in this case

$$\beta_i = 1/T_i \ \text{ at } T_i > 0.15s$$
$$\beta_i = 1.5 + 10T_i \ \text{ at } T_i \le 0.15s \tag{5.91}$$

Category II: rock, weathered and strongly weathered coarse grained dense soil containing fine fractions above 30%, sand, clayey soil with liquidity index $I_L \le 0.15$ (see Section 1.1) and void ratio $e < 0.9$ for clay and clay loam, and $e < 0.7$ for sandy loam. For these soils one has $\beta_i \le 2.7$:

$$\beta_i = 1.1/T_i \qquad \text{at} \quad T_i > 0.2s$$
$$\beta_i = 1.5 + 8T_i \qquad \text{at} \quad T_i \le 0.2s \tag{5.92}$$

Category III: all other soils; the coefficient β_i is not greater than 2.

$$\beta_i = 1.5/T_i \qquad \text{at} \quad T_i > 0.2s$$
$$\beta_i = 1.5 + 2.5T_i \qquad \text{at} \quad T_i > 0.2s \tag{5.93}$$

Moreover, the condition $k_\psi \beta_i \le 0.8$ should be satisfied, and if it is greater than 0.8 it must be taken as 0.8.

For category I structures, their particular reliability requires that the design acceleration be 1.2 A.

For a system with n degrees of freedom such as dams, the dynamic coefficient β should be supplemented by the coefficient of oscillation mode η_{ik}. The nature of the latter factor is clear from Equation 5.21 or Equation 5.62, if the right-hand side of the differential equation for forced oscillations with damping in the i-th mode is taken instead of mass, as it should be for an equation with one degree of freedom replaced by a certain set of eigen functions. One can assume that this function corresponds to the type of acceleration A.

In agreement with Equation 5.21 or Equation 5.62 for $b_{2k-1} = \ <1010....> \ $ one may take

$$\eta_i' = \frac{\{x_i\}^T \{E^x\}}{\{x_i\}^T [M] \{x_i\}} \tag{5.94}$$

in which

E^x = mass column in determination of inertia forces along x-axis, generated from the mass matrix by its multiplication by the vector $<1010....>$,
η_i' = distribution parameter for soil acceleration over eigen mode, which can be considered the coefficient of contribution by each mode (Okamoto 1980).

Taking into account that the diagonal mass matrix is used only, and more precisely the mass column (one-dimensional mass block), one obtains by multiplication of the numerator and denominator by g

$$\eta_i' = \frac{\sum\limits_{m=1}^{m=n} Q_m x_{im}}{\sum\limits_{m=1}^{m=n} Q_m x_{im}^2} \tag{5.95}$$

Solution of Equation 5.57 or Equation 5.70 yields only $y(t)$ from Equation 5.29, and transition to the displacements $r(t)$ requires additional x_i, so that at the point k one has

$$\eta_{ik} = x_{ik}\eta_i' = x_{ik}\frac{\sum\limits_{m=1}^{m=n} Q_m x_{im}}{\sum\limits_{m=1}^{m=n} Q_m x_{im}^2} \tag{5.96}$$

In the spatial case Equation 5.96 reads

$$\eta_{ikj} = x_{ikj}\frac{\sum\limits_{m=1}^{m=n} Q_m \sum\limits_{j=1}^{J} x_{imj}\cos(x_{\overparen{imj}}, \bar{u}_0)}{\sum\limits_{m=1}^{m=n} Q_m \sum\limits_{j=1}^{J} x_{imj}^2} \tag{5.97}$$

in which

j = projection of displacement on the coordinate axis (1,2,3),
$\cos(x_{\overparen{imj}}, \bar{u}_0)$ = directional cosines between directions of the vector $\bar{u}_0$ of seismic effect and displacements x_{imj},
Q_m = mass of element related to point m (includes added mass of water).

In usual computations for earth dams in one- or two-dimensional schemes (shear wedge method in the former case) one merely includes the horizontal component of seismic action, although the computations are not more sophisticated with inclusion of the vertical component, as follows from Equation 5.96 and Equation 5.97.

In the computations in three-dimensional schemes one is recommended to include the oblique seismic effect at an angle of 30° to horizontal plane. With this effect as the design acceleration one assumes the absolute magnitude of the vector, and its components along coordinates will be smaller.

The design load at a given point k in the space S_{ikj}, with inclusion of n design formulae in the j-th direction is given by

$$S_{kj} = \sqrt{\sum_{i=1}^{n} S_{ikj}^2} \tag{5.98}$$

where S_{kj} as given by Equation 5.75 represents the maximum possible stress at a given point by each mode i. These stresses are different at various times. It is shown (Okamoto 1980) that reaction parameters are overestimated if the dynamic analysis includes spectra with higher modes of eigen frequencies.

Another method for the resulting force at a given point of dam is also possible by simple summing up of the values $S_{pkj} = \sum\limits_{i=1}^{n} S_{ijk}$ for the modes of forces. It is shown by Okamoto (1980) that even for structures with a narrow spectrum of modes this method provides overestimated forces. The former method is recommended in the standard SNiP II-7-81.

Earth dams below 25–30 m usually have primary eigen frequencies below primary eigen frequencies of earthquakes (0.2–0.4 s). Medium size earth dams, high and very high dams have these frequencies much above the primary frequency of earthquakes (0.8–1.2 s).

It has already been mentioned that the effect of upstream water can be described with the concept of added mass.

Added mass load m_B included for horizontal seismic actions is

$$Q_B = m_B g = \gamma y \mu \psi \qquad (5.99)$$

in which

> y = depth of water at a given point of structure at slope (the added mass loading should be applied to slope),
> μ = dimensionless coefficient of added mass,
> ψ = dimensionless coefficient accounting for limited length of reservoir, taken as 1.0 for $l/h \geq 3$,
> l = length of reservoir,
> h = depth of water in reservoir.

Practically in all cases one takes $\psi = 1$. In the rare cases with $l/h < 3$ one should follow Table 12 of SNiP II-7-81.

The coefficient of added mass for water is computed by the formula

$$\mu = R \sin^3 \theta \qquad (5.100)$$

in which

> θ = slope of upstream thrust edge,
> R = coefficient depending on the ratio of submergence on the thrust edge below water level to the depth of water in reservoir (the origin of y is taken at the upstream datum):

y/h	0.1	0.2	0.3	0.4	0.5	0.6	0.7	0.8	0.9	1.0
R	0.23	0.36	0.47	0.55	0.61	0.66	0.70	0.72	0.74	0.74

Upon inclusion of the vertical component one obtains additional seismic pressure of water on the inclined thrust edge

$$P_{add} = 0.5 \gamma_0 y \frac{A}{g} k_1 \sin \theta \qquad (5.101)$$

The load due to added mass is multiplied by the respective area associated with a given point k, and this force is added to the other forces. By analogy, one determines the total vertical force at point k and arranges for the summation of S_{pkj}.

The weight of the element of dam below water datum is computed for full saturation.

Waves can be generated in reservoir under earthquake conditions. The overtopping of dams above normal impoundment datum is given by SNiP II-7-81 as follows:

$$\Delta h = 0.4 + 0.7(J - 6) \tag{5.102}$$

in which

    J = intensity of design earthquake in points of 12-point scale.

5.6 SEISMIC EFFECTS IN DESIGN SCHEMES AND SEISMIC PREVENTION MEASURES

Inclusion of seismic effects in the design of earth dams is possible in two basic versions:

    1. Explicit or implicit dynamic scheme;

    2. Quasidynamic (static) method.

Both versions have advantages and shortcomings.

The dynamic method is highly potential in the exploration of the performance of an earth dam, which is a complex structure due to inhomogeneity of dam material (core soil, transition soil, thrust prism, foundation etc.) and elastoviscoplastic properties of soils. The method provides not only qualitative but also quantitative data for the analysis of earth dams. However, its application in design practice is difficult because of the following factors: complex input (design seismograms and acceleration graphs); absence of assessment methods for dam working capacity in traditional form, that is by reliability coefficient, coupled with the necessity of the assessment of working capacity by residual displacements; the necessity of using high-speed computers with large memory; and execution of a large number of complex dynamic soil tests.

In addition, the dynamic analysis requires a highly skilled personnel specializing in this type of computations. All these factors create a situation in which the dynamic method is used only in specialized laboratories or workshops.

Nevertheless, one should remember that dynamic computations involving implicit schemes, even incomplete, provide the necessary information on the form of eigen vectors and eigen values for dam design, much more relevant than these in the simple quasidynamic method. Therefore one should stipulate application of the dynamic methods in design practice, particularly in the context of improving seismic investigations and growing amount of statistical material for earthquakes.

The use of the quasidynamic method also encounters difficulties due to the necessity of determining the modes of oscillations and eigen values. The latter are usually given for high dams about 100 m in height (Figure 5.13).

The transition to dams having different heights is possible by the following Cauchy modelling criterion:

$$T'' = T' \frac{H''}{H'} \sqrt{\frac{\rho'' E'}{\rho' E''}} \tag{5.103}$$

in which

    T', H' and ρ' = oscillation period, dam height and density of material, respectively (H' in this case is 100 m),

T'', H'', ρ'' = counterparts for dam under design.

In more accurate computations of eigen oscillations one should solve the characteristic equation (Eq. 5.32); the method is given in Section 5.4.

Other methods, for instance the rotation method (Jacobi method) are also possible. In the latter, the cross-section of dam is decomposed into elements, the matrices of stiffness and mass are constructed etc. Once the modes and eigen values are obtained, the values of seismic forces at various points of dams are sought. At this stage it is again necessary to determine the quantity of modes so that the solution be sufficiently accurate. In the use of the modes determined by experiments it is usually sufficient to take three to six modes as the experimental determination of resonance is possible only for modes with considerable energy, for the sixth experimental mode sometimes corresponds to 25th-30th mode. In experimental modes one usually defines clearly the mode corresponding to mere horizontal and mere vertical oscillations. This distinction is inadequate in the design schemes, with the exclusion of the first mode corresponding primarily to horizontal oscillations and the second mode corresponding to vertical oscillations (Figure 5.13). Therefore, a high number of modes (30 or even more) is required, but this in turn stipulates the use of high speed computers.

In rough estimates of seismic forces one may use the eigen modes by Figure 5.13, having in view that they have been constructed for a 120 m dam. In accordance with Equation 5.103, one recalculates the periods by the modes, whereupon the value ρ'' and respective E'' in dams higher than 120 m is taken against the prototype (model), which results in a certain increase of T compared with T' (see Section 5.4). The prediction of the growth of density and deformation modulus is impossible by experiments on soil compressibility (see Sections 1.3 and 1.7), whereupon the modulus should be determined on the unloading branch. In the design of dams above 120 m, the reduction of the density and the modulus of elasticity must not be taken into account if these characteristics (in soil fill) are close to those for 120-m dams.

In the determination of the resulting seismic modes, corresponding to the mode i at a certain point k in the direction j, one may be guided by the following circumstances (Lombardo 1983).

1. Structure interacts with foundation, and the inclusion of the interaction brings about increasing periods of oscillations in individual modes, but the majority remains practically unchanged if the modes with the highest energy contribution and their earthquake counterparts are removed. Such a first mode is usually in the range $0 < f_1 \leq 0.8$ Hz, accompanied by the second frequency $1.5 < f_2 \leq 3.0$ Hz.

2. The modal coefficient is recommended as

$$\eta_{ijk} = x_{ijk} \frac{\sum\limits_{m=1}^{m=n} Q_m x_{ijm} d_{ij}}{\sum\limits_{m=1}^{m=n} Q_m x_{ijm}^2} \tag{5.104}$$

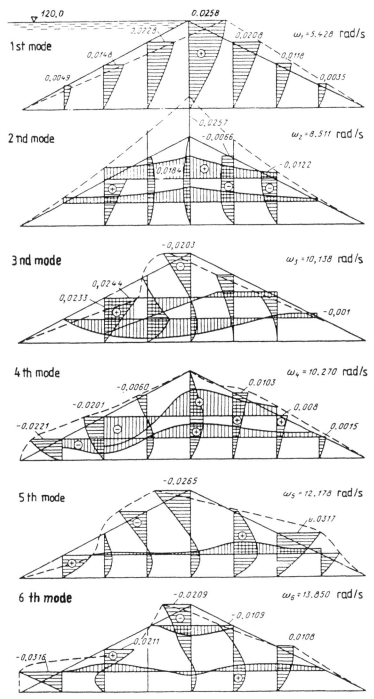

Figure 5.13. Oscillation modes upon inclusion of added mass of water on the surface of upstream slope.

in which

> $d_{ij} = 1$ for all j of the structure and $d_{jj} = 0$ for all j of foundation.

This relationship alters principally the analysis of the contribution of individual modes to the energy. For instance, the modes with frequencies close to the frequency of foundation provide very minor contribution as the denominator increases. At the same time, the modes with frequencies close to those of the structure provide the highest contribution. It is therefore recommended that the range of foundation be considered by the depth 2–3 H, in which H = height of dam. This is to be adhered to if the foundation is compressible. In the case of bedrock it is sufficient to include the structure alone if an earth dam is designed.

3. The presence of a large number of modes analogous by form and close in frequency to their cantilever modes make Equation 5.98 inadequate, so that the following summation of loads for close eigen frequencies is recommended:

$$S_k = \sqrt{\sum_i \sum_j \frac{|S_{kj}||S_{ki}|}{1 + q_{ij}^2}} \tag{5.105}$$

in which

$$q_{ij} = \frac{\omega_i - \omega_j}{\xi_i' \omega_i + \xi_j' \omega_j}; \quad \xi_i' = \xi_i + \frac{2}{\omega_i l}$$

in which

> ω_i and ω_j = respective eigen frequencies of i-th and j-th mode (more strictly, the quantity q_{ij} in the numerator of the formula should be replaced by ω_{Di} and ω_{Dj}, cf. Equation 5.50),
> ξ_i, ξ_j = respective damping factors related to the critical values for i-th and j-th modes,
> l = equivalent length of the white-noise band to simulate acceleration graph taken 20 s for rock, 30 s for semi-rock and 40 s for lightweight soil.

4. For a similar inclusion of the effect of foundation one must consider about 100 modes which are in frequency range of the first 4–5 modes of the cantilever rod.

5. Equation 105 for the summation of loads and inclusion of 100 modes requires a very large bulk of computations and is applicable only to high-speed computers.

It must be noted that Equation 5.105 can be transformed to

$$S_{kj} = \sqrt{\sum_i S_{ikj}^2 + \sum_{i \neq j} \sum \frac{|S_{ik}||S_{jk}|}{1 + q_{ikj}^2}} \tag{5.106}$$

The second term under the root tends to zero if the eigen frequencies tend to the values sufficiently differing from each other, if the damping factors go to zero or if the ratio l/T_i goes to infinity, in which $T_i = i$-th period of damping eigen oscillations.

Hence if the modes are determined by FEM, then the inclusion of six modes by Equation 5.106 provides a slightly higher accuracy of solution. Since the fundamental eigen frequency of dam carries about 60–80% of energy, preliminary estimates can be found from the first mode.

Estimation of the period of the fundamental mode can be facilitated by the following relationship

$$T_1 = 0.7 \frac{B}{B + B_0} H \sqrt{\frac{\rho}{G_y}} \tag{5.107}$$

in which

B_0 and B = width of dam at crest and foundation, respectively,
H = dam height,
ρ = density of dam soil,
σ_y = shear modulus of elasticity.

For a very rough assessment of the stability of dam slopes consisting of loose soil and subject to seismic effects one may provide the following formula, which overestimates the stability of stone on revetments under horizontal seismic forces

$$k_s = \frac{\tan \varphi}{\tan(\alpha + \delta} \tag{5.108}$$

in which

$\tan \varphi$ = soil friction factor at the surface of revetment,
α = angle of slope to horizontal,
$\delta = \arctan \frac{A}{g} \cdot k_1$ (notation in Equation 5.75).

Generation of pore pressure in coarse soil (see Chapter 3 and Section 5.3) due to seismic factors can reduce substantially the stability of dam slopes. The growth of pore pressure can be controlled if slopes are constructed of strongly permeable material — rock or homogeneous pebbles. Drainage of sorted pebbles (see Shimen Dam) can be arranged inside thrust prisms (mostly upstream one) if they are made of sand and gravel and if the dam has a core. The downstream drainage is provided to dissipate the pore pressure generated by watering of the tailwater prism due to infiltration under seismic condition.

An analysis of the operation of dam under seismic conditions, in terms of circular cylindric failure surfaces, shows that the most hazardous curves are generated in the upper parts of dams where the seismic forces are most important. Therefore the dam stability factor can be increased if the slopes in the upper one-third of a dam are made more shallow, if the crest is wider, and if pebbles (of which dam is constructed) are replaced by stones having a higher coefficient of internal friction (see Section 1.5). Moreover, a higher compactness of soil fill can also increase the stability of a dam as a direct outcome of the growth of the internal friction.

There is data (Latkher et al. 1984) concerning armouring of the upper part of dams by the use of antiseismic reinforced-concrete belts (see Figure I.1); however, these measures are hardly recommendable as the design methods and simulation of the system soil-reinforced-concrete belts are very unreliable. If the belts are to be included into the operating scheme of dam one requires substantial deformations in the upper part of dam as the dam crest under static loads is subject to high compressive strains. Substantial bending also occurs in the steel rods of the belts above dam core under static load. Arrangement of these seismic belts brings about higher stiffness of the dam crest, which in turn might give rise to concentration of stresses at the crest and some other unfavourable effects.

In the construction of earth dams in seismically active regions, the effectiveness of the dam–canyon bank transition is provided by thorough preparation of the banks at their interface with the dam core. To this end one must eliminate the crushed parts, provide sputtering of concrete on the banks, remove the fractions above 10–20 mm from the contact zone, thoroughly select grain size distribution of filler with regard to the self-filling of cracks (as cracking is possible) etc.

Selection of the most adequate dam design in seismically active zones is a technological and economic problem that must be solved individually for specific site conditions. General guidelines are impossible as various dam designs are available for earth dams in seismic regions.

CHAPTER 6

Slope stability

6.1 STABILITY COMPUTATIONS FOR CIRCULAR CYLINDRIC SLIP SURFACE

Stability computations of earth slopes are the least explored and the most important chapter in the design of earth structures. A first step towards solution of this problem was done by Coulomb, who (upon his analysis of retaining walls) established that the failure surface was a plane, and derived a method for soil pressure computations.

In 1916 Swedish engineers Petersen and Glutin found in their analysis of bulkheads that the failure surfaces in clayey soil were curvilinear and could be regarded as cylindric. A circular arc is an approximation for the cross-section of the failure surface. From this time on, theoretical works on circular cylindric slip surface have been intensified. In numerous versions, the method is an approximated engineering tool for assessment of slope stability. The major condition underlying the method is the necessity of satisfying three equations of statics for the slip body, having in view that the number of unknowns is greater than three, and therefore requiring other assumptions, which have given rise to a large variety of methods derived to date.

There have been attempts giving up the concept of a circular cylindric surface and replacing it by a logarithmic spiral or any other smooth curvilinear surface. However, in all such cases the basic controversy remains — the number of unknowns is greater than the number of equations; in addition, new difficulties accompany those new approaches.

Terzaghi method. Developments in the computations of circular cylindric failure surface have incorporated a method with fairly simple assumptions postulated by Terzaghi in the thirties. This is well substantiated as a large number of tests and investigations carried out by various methods in many countries (including the works by Chugayev in the USSR in the sixties) have shown that the difference in accuracy of solutions does not arise from the method itself but rather from input for the majority of the structures tested.

Sometimes one faces the slope stability problem for a predetermined plane slip

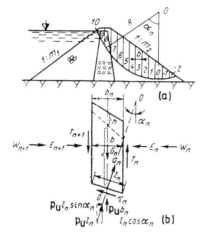

Figure 6.1. Stability computations for downstream slope. (a) circular cylindric slip surface; (b) forces acting on slip slice.

surface (a dam with a clayey screen, interlayers of weak soil at foundation etc) and in this case one also has engineering guidelines at hand.

Assume that the sliding soil volume in the body of an earth dam is bounded by a circular cylindric surface and can be decomposed into slices by vertical planes shown in Figure 6.1a.

The computations are based on the assumption of plane deformation in a dam slice 1 m thick. Divide the failure body into vertical columns of width b and express the reliability factor as a ratio of the moment of maximum possible reaction M_r to the moment of active forces M_a. The moment is taken with respect to any centre of a circular arc bordering the failure body. The moments are computed as a sum of moments for all forces acting on each slice:

$$k_s = \frac{M_r}{M_a} = \frac{\int_l R\tau_r \, dl}{\int_l R\tau_a \, dl} = \frac{\sum_n \tau_r \Delta l_n}{\sum_n \tau_a \Delta l_n} \tag{6.1}$$

in which

τ_r = tangential reaction stresses,
τ_a = active tangential stresses,
l = failure arc.

Apply to the n-th slice (Fig.6.1b) the following forces:

G_n = weight of slice,
T_n, and T_{n+1} = friction forces at side surfaces of slice,
W_n and W_{n+1} = seepage pressure at sides,
E_n and E_{n+1} = soil pressure from adjacent slices on the sides of slices,
σ_n and τ_n = normal and tangential stresses at failure surface, respectively.

If the slice is in its limit state, τ_r reaches its maximum for the following normal stress σ_n:

$$\tau_r = \sigma_n \tan \varphi_n + c_n \tag{6.2}$$

in which

φ_n, c_n = angle of internal friction and cohesion in n-th slice, respectively.

Taking projection of all forces on the axis 0-0, normal to the failure surface, and considering $T_n = T_{n+1} = 0$; $E_n = E_{n+1}$; $W_n = W_{n+1}$ (these internal forces are in mutual equilibrium) one obtains

$$(G_n - pwb_n) \cos \alpha_n - \sigma_n l_n = 0$$

or

$$\sigma_n l_n = (G_n - p_w b_n) \cos \alpha_n \tag{6.3}$$

in which

p_w = pore pressure at foot of n-th slice.

Pore pressure can be generated by the soil pressure of the overlying strata, pressure of water, seepage or dynamic effects (Section 3.3); one also has: $l_n = b \times \cos \alpha_n$.

The condition $T_n = T_{n+1} = 0$ shows that vertical planes are the principal planes, i.e. there is no mutual movement of slices:

$$G_n \sin \alpha_n = \tau_a \tag{6.4}$$

Equation 6.4 stipulates $\sigma_y = \gamma H$, in which γ = unit weight of soil (with water in pores); H = height of slice.

The above condition holds true for a homogeneous dry slice. If the slice consists of heterogeneous soil and incorporates depression line (Fig.6.2), then the weight must be computed as follows:

$$G_i = b_i(\gamma' h^I + \gamma'_{sub} h^{II} + \gamma''_{sub} h^{III} + \gamma'''_{sub} h^{IV}) \tag{6.5}$$

in which

γ', h^I, h^{II} = unit weight of soil and thickness of layers at water content of filling, respectively,

$\gamma_{sub}, \gamma^{II}_{sub}, \gamma^{III}_{sub}, h^{III}, h^{IV}$ = unit weight and thickness of saturated layers, respectively (soil below depression surface),

b_i = width of slice, usually taken as $\frac{1}{10}$ or $\frac{1}{20}R$,

R = radius of failure arc.

Accordingly, Equation 6.1 reads

$$k_s = \frac{\sum\limits_n (G_i - p_w b_i) \cos \alpha_i \tan \varphi_i + \sum\limits_n c_i l_i}{\sum G_i \sin \alpha_i} \frac{m}{n_c} \tag{6.6}$$

in which

m = structure operation factor,

n_c = load combination factor, which is n_c = 1 for primary combination of

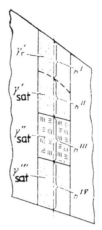

Figure 6.2. Scheme of computations for *n*-th failure slice.

loads, 0.9 for particular combinations, and 0.95 during construction.

The coefficient m incorporates the type of limit state, approximation of design schemes, particular features of structure and construction etc. In the computations of slope stability for primary and particular combinations of loads in earth dams one usually takes $m=1$.

Inclusion of pore pressure in Equation 6.6 was proposed by Florin (1961). The quantity $G_i \sin \alpha_i$ includes the shear component of seepage forces, which is indirect as the numerator embodies bouyant soil (pore pressure p_w equals vertical component of seepage force W), whereas the soil in the denominator is water-saturated.

Hence Equation 6.6 is derived on the assumption of 'hardening failure body', which stipulates that the latter move along failure surface as a whole, without changes in its shape.

The quantity p_w is determined as specified in Section 3.3. If the stability at time $t \to \infty$ is considered one has $p_w = W$ and

$$W = h_p \gamma_w \tag{6.7}$$

in which
  h_p = piezometric pressure at the centre of slice.

On the tailwater slope one has $h_p \cong h_w$, in which h_w = depth of water at slice considered, as the equipotential lines are almost vertical.

Applicability of Equation 6.7 in stability computations for the upstream slope with the above condition $h_p \cong h_w$ depends on the inclination of equipotential lines (Fig.6.3).

In stability computations for a failure slice it is convenient to include water on the upstream and tailwater sides as a material with zero internal friction and cohesion,

Figure 6.3. Quantities h_w and h_p in n-th failure slice in stability computations for upstream side of homogeneous dam.

i.e. the failure surface should be continued in water with inclusion of the weight of the slices consisting exclusively of water (Figs 6.3–6.5).

The seismic forces acting on the structure can be determined as specified in Chapter 5. Therefore we confine ourselves to the description of their inclusion in the discussed method. One computes the vertical and horizontal seismic forces, S_v and S_h, acting at various points of the structure. The vertical components reduce the weight of failure slices, and therefore are incorporated in the numerator of Equation 6.7, while the horizontal forces generate the overturning moment. One has

$$k_s = \frac{m}{n_c} \frac{\sum\limits_{n}(G_i - p_w b_i - S_{vi})\cos\alpha_i \tan\varphi_i + \sum\limits_{n} c_i l_i}{\sum\limits_{n} G_i \sin\alpha_i + \sum\limits_{n} S_h \frac{r_i}{R} - \sum\limits_{n} S_{vi}\sin\alpha_i} \qquad (6.8)$$

Vertical and horizontal seismic forces within the failure slice checked are included in Equation 6.8. If the method of failure wedge is used, the vertical component is excluded, and the horizontal component is decomposed into the normal and tangential parts, $S_h \sin\alpha$ and $S_h \cos\alpha$, together with the seismic force acting along its line on the slip curve, then one obtains

$$k_s = \frac{m}{n_c} \frac{\sum\limits_{n}[(G_i - p_w b_i)\cos\alpha_i - S_h \sin\alpha_i]\tan\varphi_i + \sum\limits_{n} c_i l_i}{\sum\limits_{n}(G_i \sin\alpha_i + S_h \cos\alpha_i)} \qquad (6.9)$$

The seismic force is considered in the slices in which it is incorporated upon its transposition along a line of action towards the slip curve. The same procedure is applied to the component S_v.

The reliability factor of slope can be determined in a number of procedures if the failure curves are presented and the curve with the minimum $k_{s.min}$ is selected.

In preliminary computations for the critical configuration of failure surface one may be guided by the following circumstances:

a) If foundation soil is stronger than dam soil, the failure surface must not incorporate the foundation;

b) If foundation soil or soil interlayers in foundation are weaker than dam soil then one must secure maximum inclusion of failure surfaces in the foundation.

The following procedure may be suggested for the selection of the most hazardous failure curve. A few curves are drawn from each centre, and the one with the minimum reliability factor is chosen. Graphs of k_s are drawn for various centres as shown in Figure 6.4. The quantity k_{min} is selected from these graphs. The

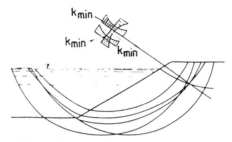

Figure 6.4. First scheme of determination of minimum reliability factor.

standard SNiP II-50-74 under normal conditions of operation: 1.25 for class I; 1.2 for class II; 1.15 for class III; 1.1 for class IV. For particular combination of loads one may take $k_s n_c < 1$ for class IV structures, as the loads are temporary in this case.

One must remember that the standard reliability factors are small. This stems from the fact that the computation methods (incorporating overestimating assumptions) and the factors intervening in soil characteristics (most often mean minimum values) do already have some safety margins.

Landslide processes are fairly slow, so the assumption $k_s \leq 1$ for temporary loads is well substantiated for particular combinations of loads.

Another method of searching for the most hazardous centre of failure curve is also available. Take the area of possible hazardous centres and mark the centres of slip curves in this area with a predetermined step. Draw a few curves with various radii from each centre. The local minimum is taken for k_s corresponding to the radius having $k_{s.min}$. Next draw the line of equal $k_{s.min}$ from which one finds $k_{s.minmin}$ (Fig.6.5). It must be stressed that the centre corresponding to $k_{s.min.min}$ should be inside the selected area of centres. If the most hazardous centre is at the boundary of the chosen region of centres, then the latter must be expanded. This procedure provides good results but the bulk of computations is larger. The method is implemented on computers.

Cray's method. If the assumption on 'hardening failure slice' is rejected then one faces a large class of methods which incorporate the deformation of failure slice; the widest application has been found in the methods basing on Cray's method. The latter assumes that the integral of tangential stresses along the boundaries of slices is zero. Accordingly, the horizontal forces acting between the slices of failure body remain active. In this case the following relationships for the reliability factor and stability hold true (Fedorov 1962)

$$k_s = \frac{m}{n_c} \frac{\sum_n (G_i - p_B b) + \sum c_i b}{\sum_n G_i \sin \alpha_i} \frac{1}{\sum_n \left(1 + \frac{\tan \varphi \tan \alpha_i}{k_s} \cos \alpha_i \right)} \tag{6.10}$$

Equation 6.10 contains k_s on both sides so that the iteration processes must be applied.

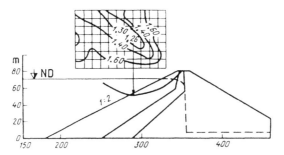

Figure 6.5. Second scheme of determination of minimum reliability factor.

All stability computations for the tailwater slope are usually conducted for the water level at NPU (normal impoundment datum) and FPU (forced impoundment datum) on the upstream side and respective minimum and maximum water level on tailwater side.

Stability computations for the upstream slope are conducted for various water datum on the upstream side: NPU, UMO (maximum datum), and $\frac{1}{3}h$ above foundation, if the latter is possible during operation or construction.

If the level $\frac{1}{3}h$ occurs only during the impoundment, and no longer, then the criterion of extraordinary conditions of operation must be applied.

In stability computations for the upstream slope, the case of rapid fall of water level is of interest. In this case the unsteady groundwater flow from dam body to reservoir must be considered. In approximation, soil above water level in reservoir may be regarded as saturated.

6.2 ASSESSMENT OF LOCAL SLOPE STABILITY

Hazard of washing and creep of soil particles on the dam slope. If water emerges on a dam slope, the velocity of seepage at the exit point A (Fig.6.6a) is oriented along the tangential line of the depression line and equals $v = k \sin \theta$, in which k is the coefficient of permeability. At various points in the exit zone AB the direction of the outlet velocity varies, and becomes normal to the slope at point B. The tangential component of the velocity $v_1 = k \sin \theta_1$ remains constant while the normal velocity $v_2 = k \sin \theta \tan \alpha$ increases and becomes theoretically infinite at B (curve AB in Figure 6.6), whereupon it is practically finite. Below water level along the slope section BC the streamlines are oriented along the normal to the slope, whereupon the velocities are lower than those at the exit section, so that the stability of soil grains in the exit zone is lower than that on the underwater section.

The unit volume of loose soil at point M (Fig.6.6a) under the action of gravity γ and seepage pressure $\gamma_w J$ oriented at angle α to the slope is maintained on the slope because of friction forces. The equilibrium of all forces projected on the

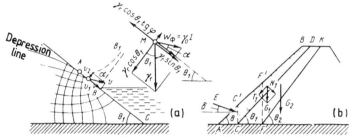

Figure 6.6. Schemes of slope stability computations for washing (a) and soil screen and protection layer (b).

slope line is given as follows

$$\gamma \sin \theta_1 + \gamma_w \sin \theta_1 = (\gamma \cos \theta_1 - \gamma_w \sin \theta_1 \tan \alpha) \tan \varphi \tag{6.11}$$

from which the coefficient of internal friction can be found as

$$\tan \varphi = \frac{(\gamma + \gamma_w) \sin \theta_1}{\gamma \cos \theta_1 - \gamma_w \sin \theta_1 \tan \alpha} = \frac{\gamma + \gamma_w}{\gamma \cot \theta_1 - \gamma_w \tan \alpha} \tag{6.12}$$

At the exit point A the angle α is zero and accordingly

$$\tan \varphi \geq \frac{\gamma + \gamma_w}{\gamma_1} \tan \theta_1 \approx 2 \tan \theta_1 \tag{6.13}$$

or

$$\tan \theta_1 \approx \frac{1}{2} \tan \varphi \tag{6.14}$$

Upon introduction of the standard coefficient for working conditions m and combination of loads n_c one obtains

$$\tan \theta_1 \approx \frac{1}{2} \tan \varphi \frac{n_c}{m} \tag{6.15}$$

If a dam is constructed on a permeable foundation, the angle $\tan \theta_1$ should be taken slightly smaller, as suggested by Shestakov (approximately by 15%).

If the angle α below point A increases, the value of $\tan \varphi$ should also increase, but along the line AB there will occur not only seepage flow but also open channel flow, with different intervening characteristics.

There is no clear-cut solution for these conditions; one may use Equation 6.15, with a certain margin of safety.

In order to prevent washing of slope it is sufficient to apply a sloping drainage. A more radical measure is the arrangement of drainage within the dam body, which will bring about a lower depression curve and its disappearance on the dam slope (see Chapter 2).

The stability of cohesive slope in the seepage exit zone is better secured.

Stability computations for dam screens, if a dam is made of cohesive soil, include a check on possible slip of the protection loose-soil layer; this sliding can take place along the screen or otherwise the screen can move together with the protection layer.

The above check on the protection layer and the screen combined with the protection layer can be performed by the method of circular cylindric slip surfaces inside the protection layer and inside the screen.

In the case of thin screens the check can be carried out by the method of planar surfaces.

The design of the protection layer is based on the assumption that its part $C'CDB$ in Figure 6.6b, having the weight G, can creep along the inclined plane CD under the action of the weight component $T_1 = G_1 \sin \theta_1$, which is counteracted by the friction forces:

$$S_1 = N_1 \tan \varphi = G_1 \cos \theta_1 \tan \varphi \tag{6.16}$$

in which φ, the angle of internal friction for the screen material, and the reaction forces E for the prism ACC', should also be taken into account.

The dam foundation is usually taken as strong enough. The equilibrium condition for the protection layer is given as

$$E \cos \delta + G_1 \cos^2 \theta_1 \tan \varphi - G_1 \sin \theta_1 \cos \theta_1 = 0$$

from which one has

$$E = G_1 \frac{\sin \theta_1 \cos \theta_1 - \cos^2 \theta_1 \tan \varphi}{\cos \delta} \frac{n_c}{m} \tag{6.17}$$

in which

δ = angle of the reaction E to horizontal; in particular cases $\delta = 0$.

The computation of the reaction E_r of the prism $AC'C$ should be greater than E or one has $k_s = E_p/E \geq 1.2$–1.5.

In these computations one should make a check on the most dangerous case of fast evacuation of water from the reservoir; the weight of the protection layer above the lowered water level on the upstream side should be taken as saturated or semi-saturated with water, and not as dry weight.

The screen is designed like the protection layer, but the total weight of the part of the protection layer and the screen in the body $FF'BK(G_2)$ is taken in Equation 6.17; and the angle θ_2 is taken instead of the angle θ_1, whereupon the reaction E is found for the prism AF':

$$E = G_2 \frac{\sin \theta_2 \cos \theta_2 - \cos^2 \theta_2 \tan \varphi}{\cos \delta} \frac{n_c}{m} \tag{6.18}$$

6.3 MOZHEVITINOV'S METHOD OF OBLIQUE FORCES

A couple of methods have been derived for the computations of branching solidified slices, in attempts to fully represent the equilibrium conditions on any arbirtary shear surface. From among the procedures proposed by Janbu, Morgenshtern, Price, and Mozhevitinov we are taking the most convenient method derived for practical computations by Mozhevitinov.

As above, we are considering the body bounded by non-deforming vertical slices shown in Figure 6.7 in the limit state on the slip surface.

The slice shown in Figure 6.7 is subjected to the following forces:

$rds =$ resultant force of reactive normal stresses and ultimate tangential stresses (only due to the friction component $\sigma \tan \varphi$),

$cds =$ the same due to cohesion,

$gdx =$ resultant of weight, buoyancy, seepage or seismic forces and loading on the elementary surface,

$E =$ interaction between elements (with the components E_x and E_z).

The soil strength properties φ and c are in general variable along the slip surface. An arbitrary shape of the surface is given, so that its length s and the inclination of each elementary section ds to horizontal α are known functions of the abscissa x.

The equations of equilibrium for the slice of failure body can be written down most conveniently in the following form (Fig.6.7).

1. Projection of forces on the direction of the component of soil reaction

$$rds - g\cos(\alpha + \delta - \varphi)dx - \sin(\alpha - \varphi)dE_x + \cos(\alpha - \varphi)dE_z +$$
$$+c\sin\varphi ds = 0 \tag{6.19}$$

2. Projection of forces on the line perpendicular to the component r

$$-g\sin(\alpha - \delta + \varphi)dx - \cos(\alpha - \varphi)dE_x + \sin(\alpha - \varphi)dE_z +$$
$$+c\cos\varphi ds = 0 \tag{6.20}$$

3. Moment of forces about the foot of the element (slip surface)

$$mdx - E_z dx - E_z \cdot 0.5dx + E_x(a' + 0.5dz) -$$
$$(E_x + dE_x)(a + da - 0.5dz) = 0$$

Taking into account that the terms $dE_z dx$, $dE_x da$ and $dE_x dz$ are second order infinitesimal quantities one has

$$mdx + E_x dz - E_z dx - dM = 0 \tag{6.20a}$$

in which

$m = gb\sin\delta =$ moment of the force g

$M = E_x a =$ moment of horizontal component of interaction force.

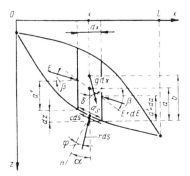

Figure 6.7. Explanation diagram for slope stability computations by Mozhevitinov's method.

In addition, for the sliding body as a whole one should also observe the equilibrium conditions, for which one assumes the natural conditions at the edges of the body in the form of interaction forces and moments, which both are zero:

$$x = 0 : \quad E_x = 0, \quad E_z = 0, \quad M = 0 \tag{6.21}$$

$$x = l : \quad E_x = 0, \quad E_z = 0, \quad M = 0 \tag{6.22}$$

As already noted, these equations are insufficient to determine k_s. Therefore additional physical relationships or hypotheses, sometimes very artificial and arbitrary must be harnessed. As in a number of other methods, we are assuming the constant angle β of inclination of the interaction E, for all slices:

$$E_z = E_x \tan \beta \tag{6.23}$$

in which
$$\beta = \text{const.}$$

In contrast to other methods, in this procedure the value of β is not given but instead found from the conditions of equilibrium for the entire sliding body. By integrating Equation 6.20a one obtains an expression for $M(x)$, and with inclusion of Equation 6.22, i.e. $M(l)=0$ the angle β reads

$$\tan \beta = \frac{\int\limits_0^l (E_x \tan \alpha + m)dx}{\int\limits_0^l E_x dx} \tag{6.24}$$

Equation 6.24 shows that the angle β can be found by iterations; Mozhevitinov proposes the following approximate relationship

$$\beta = \frac{\sum (\alpha_i + \delta_i)\Delta x}{l} \tag{6.25}$$

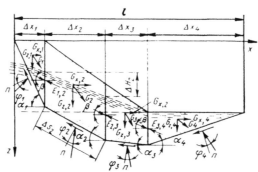

Figure 6.8. Decomposition of failure slice for broken failure surface.

By integrating Equation 6.19 and using Equation 6.23 one obtains a general expression for the interaction forces E which intervenes in the equilibrium conditions (Eqs 6.21–6.22), i.e. $E(0)=0$, $E(l)=0$:

$$E(x) = \int\limits_0^x \frac{g \sin(\alpha + \delta - \varphi)dx - c \cos \varphi ds}{\cos(\alpha - \beta - \varphi)} \tag{6.26}$$

As a result of transition in Equation 6.26 from integration to elementary summation along the slice l of the sliding body, and upon separation of the terms with $\tan \varphi$ and c one obtains the following relationship for the reliability factor:

$$k_s = \frac{R}{A} = \frac{\Sigma G_i \cos(\beta + \delta_i) \tan(\varphi_i + \beta - \alpha_i) + \Sigma \frac{c_i \cos \varphi_i \Delta s_i}{\cos(\varphi_i + \beta - \alpha_i)}}{\Sigma G_i \sin(\beta + \delta_i)} \tag{6.27}$$

in which
$\quad G$ = resultant forces acting on the slice (Fig.6.8).

The search for the location and form of most dangerous sliding surface (the one with $k_{s.min}$) is carried on like in the case of circular cylindric slip surfaces, that is by trials and errors.

Comparison of results by Mozhevitinov's method and other data shows that it practically coincides with the method derived by Cray, and is equivalent to the 'rigorous' solution given by Taylor in the case of homogeneous slopes.

The use of Mozhevitinov's method for smoothly varying groundwater flow in the slope often permits the assumption that the lines of equal heads are vertical; the distribution of pressures at any vertical below the depression curve will then be hydrostatic, and the seepage forces will be horizontal. In this case the consideration of seepage forces is reduced to the determination of φ_z with inclusion of submerged weight, while φ_x encompasses the component equal to the seepage force:

$$\varphi_{xw} = \gamma_{sub} h_w \Delta x \Delta H / \Delta x = \gamma_{sub} h_w \Delta H$$

in which
$\quad h_w$ = height of saturated sliding prisms, averaged for each slice,

ΔH = decrement of depression curve along each slice Δx (Fig.6.8).

6.4 RECOMMENDATIONS ON THE USE OF SLOPE STABILITY ASSESSMENT METHODS

Design of slope stability by methods of circular cylindric slip surfaces can be acceptable without practical limitations. This means that one may compute not only slopes but also screens, slope revetments, slopes with weak laminae etc; in all these cases one should inscribe a circular cylindric surface passing basically over weakened soil. Special methods can accelerate computations.

The Terzaghi method remains most widely accepted despite its limitations; it yields results which are fully acceptable from the practical point of view.

Cray's method, and similar methods basing on the acceptability of deformations in sliding body can be used, but for the sake of simplicity one usually rejects iterations for determination of k_s, as the right-hand side of Equation 6.10 incorporates the standard value of k_s and real value of k_s must be found thereafter.

CHAPTER 7

Dam-foundation transitions

7.1 SEEPAGE PREVENTION MEASURES IN DAM FOUNDATION

Geological structure of foundations determines the choice of seepage prevention measures. The following configurations are possible.

1. Small thickness of permeable deposits (or no deposits at all, for instance at flanks); the strata are underlain by almost impermeable bedrock (such cases are extremely rare). On such occasions special seepage prevention measures are not required in bedrock, and the design is confined to embedment in rock of a dam cut-off. The foundation soil is herein referred to as permeable if its coefficient of permeability is close to or greater than its counterpart for dam or a seepage prevention measure.

The bedrock-core transition, much as any interface, is the most responsible constituent of a structure, because its low quality can bring about irreversible seepage deformations of fine grained soil in the core (screen).

A reliable interface can be provided if the weathered layer of rock is removed. The depth of removal can be several metres. At present there is a tendency towards reducing the volume of removed rock at the cost of a strengthening grout. Coarse cracks are cleaned up and grouted. Concrete is also used to smooth out the surface of bedrock, so that large troughs are filled up. The surface of the excavation for core is cleaned up and washed. The surface of rock is sometimes covered with gunite, which can smooth out the sputtered area. One also uses gunite applied to a reinforcing mesh.

If the bedrock is weak, and semi-rock foundation occurs, subject to rapid weathering, one must provide anti-weathering measures immediately upon exposure of rock; they include guniting or covering with concrete, bitumen or core soil.

Steep slopes of flanks are covered with a more plastic material to improve the operation of the core; bentonite clay up to 30 mm thick or fine soil of core up to 500 mm (with more water and limited additions of material having $d \leq 10-20$ mm) are applied, followed by regular soil of dam core (Fig.7.1). Usage of a more plastic material at the transition secures a higher strength of the interface and higher deformations of the core material without cracking.

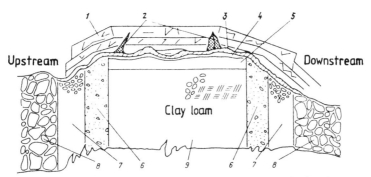

Figure 7.1. Possible design of core transition in large dam on bedrock, at steep canyon flanks. (1) bedrock; (2) concrete-filled cracks; (3) concrete gunite up to 100 mm thick; (4) bentonite clay layer up to 30 mm thick (can be abandoned if the core soil is sufficiently fine and plastic); (5) soil layer up to 500 mm, with inclusions of $d \leq 10\text{-}20$ mm; (6) primary filter layer; (7) secondary filter layer; (8) coarse grained soil of fills (thrust prisms); (9) regular core soil with inclusions.

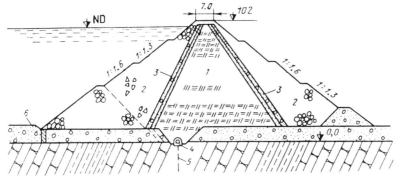

Figure 7.2. Globocica Dam (Yugoslavia). (1) lacustrine clay core; (2) rubble mound; (3) transition filters of sand (first layer) and pebbles (second layer); (4) grouting gallery; (5) grout curtain; (6) concrete diaphragm wall.

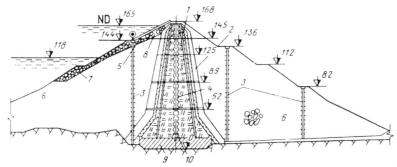

Figure 7.3. Charvak Dam. (1) loam core; (2) double-layer transition zones of fine and coarse mixtures of sand and gravel; (3) inclinometer pipe for displacement measurements; (4) inspection shaft; (5) concrete revetment 0.5 m thick; (6) rubble revetment in zones by size; (7) coarse stone revetment; (8) cobbles; (9) concrete cushion (in channel); (10) grouting hole.

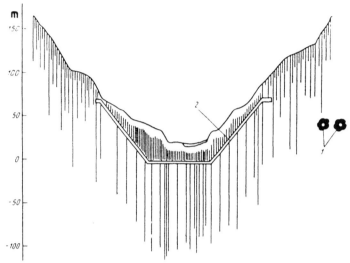

Figure 7.4. Grout (cementation) curtain in the foundation of El-Infiernillo Dam (Fig.7.10).
(1) tunnels of bank spillways; (2) holes.

At the transition at steep slopes on the tailwater side or screen, the filter should be selected on the condition of self-filling of cracks. If canyon flanks are sharply configurated they must be smoothed out so as to reach an even form although it is not always economical; in this case the dam is designed on the condition of possible cracking (see Chapter 4).

The presence of holes in the foundation and the application of surface and deep grouting (for relief of groundwater flow at the transition) do not bring about substantial changes in the design of the transition between dam core (screen) and foundation, but can reduce the thickness of removed rock.

If a dam is constructed on a nonrock foundation with permeability close to that of the seepage prevention measure, the latter is connected to the foundation by analogy to the case of a practically impermeable bedrock, with preceding removal of the upper vegetation layer or weak clayey sediment. Additional seepage prevention facilities in foundation are not required in this case.

2. Small thickness of permeable strata (10–20 m). In this case various measures may be taken for the foundation. In dams with cores one may try to remove the permeable alluvium and achieve a direct connection of the core with the water-bearing stratum through a tooth or cut-off made of core material (Fig.7.2), or a concrete cushion (Fig.7.3).

3. Permeable strata do not exist or have small thickness, but the underlying rock is strongly cracked. In this case the seepage prevention constituent of dam body is connected with the permeable cracked bedrock. In the case of strongly weathered rock the foundation is covered with bitumen upon removal of the overfill (the

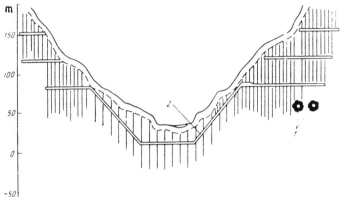

Figure 7.5. Drainage in the foundation of El-Infiernillo Dam (Fig.7.10). (1) tunnels of bank spillways; (2) holes.

foundation can be semi-rock or rock with semi-rock interlayers); sputtering with concrete or clayey soil may also be provided. The seepage prevention member in the foundation can have a form of grout curtain.

The curtain can be necessary in combination with drainage which must be substantiated by geological exploration and seepage computations. The curtain should prevent removal by seepage of the crack filler. Drainage serves the same purpose as it intercepts the groundwater flow behind the curtain and provides orderly patterns for its evacuation into tailwater.

Screens and drainage boreholes are often executed from holes arranged in a concrete cushion (Fig.7.3), cut-off, trench, or at the surface of the foundation of core or screen. If a curtain is constructed by this scheme, organization of the construction works is often impeded because of the laying of counterseepage soil as the works at flanks are very laborious and cumbersome (removal of rock for borehole, concrete works, impermeable coating etc). The construction of the curtain and drainage for rock and earth dams from galleries arranged at dam foundation (Fig.7.4) may therefore be regarded as a successful solution. Transition between the curtain and the core or screen is provided by up-going boreholes.

Drainage holes are often constructed from the above galleries serving the same purpose as the inspection shaft (Figs 7.5, 7.6), although independent drainage galleries are sometimes provided.

The above method of construction of seepage prevention facilities in foundations is additionally convenient because it makes possible a more accurate investigation of bedrock upon construction of the holes, and facilitates the works in connection with galleries and grouting operations, quite independently of the primary works on the dam. It remains to control one deadline for all works (generally for all types of works) i.e. filling of reservoir.

4. Thickness of permeable alluvial strata permits cutting through with sheet

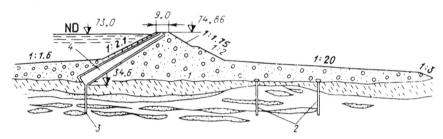

Figure 7.6. Nehranice Dam (Czechoslovakia). (1) gravelly sand; (2) relief drainage holes; (3) seepage prevention wall; (4) loessial loam screen.

piling (up to 40 m), soil sheet (up to 80 m) or grout curtain (up to 100–200 m). In this case the construction of seepage prevention facilities is possible for dams with soil core (screen) or homogeneous dams made of practically impermeable fine soil; and also for dams with nonsoil screens or diaphragms.

Draglines or special machinery are used for construction of counterseepage soil walls in dam foundations. A trench, 0.2–0.3 m wide (depending on type of machinery) is filled up with a clayey solution in colloidal form (the best one is bentonite clay) which provides stability of trench walls.

The density of the suspension is 1050–1200 kg/m³. Next the trench is filled with the removed soil (Figs 7.7, 7.8) mixed with clay (clayey concrete), clayey material or a mixture of clayey material and cement (with 7–10% of cement, and more, by soil mass). The following composition of mixture, in kilograms per 1 m³, was used at the transition of the dam diaphragm ('soil wall') and foundations of Formitz Dam (FRG): 1300 sand, 150 clay, 90 cement, 30 bentonite clay and 400 water. This soil-cement mixture is sufficiently plastic. Additional plastification of the material in trench is not required as a rule.

At small depths of walls (up to 8–10 m), they can be made more rigid — composite reinforced-concrete slabs are lowered and the joints are made monolithic. The diaphragm in the foundation is constructed by putting concrete in the trench.

A reinforcement framework is sometimes arranged prior to concrete works so as to configurate a reinforced-concrete diaphragm. The units can be constructed section by section.

In the design of seepage prevention facilities in highly permeable foundation (gravel and pebbles, sand-gravel-pebbles etc), particular attention must be paid to the transition between the seepage confinement measures and dam body. It is at the interface that high gradients of groundwater flow arise, because the penetration of sheet piling and 'wall in soil' in the case of dam (screen) reaches not more than 2–4 m, which gives rise to seepage path of 4–9 m, if additional measures are not taken. In order to reduce the gradients (increase the seepage path) one may use concrete galleries connected with the wall (if the latter is needed for other purposes), concrete slab at the core-foundation transition, and clayey cut-off in the

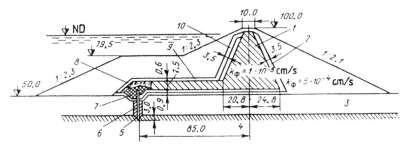

Figure 7.7. Peneos Dam (Greece). (1) clay core; (2) filter; (3) alluvium; (4) clayey marl; (5) zone of spreading of bentonite solution; (6) 0.6-m armoured-concrete wall; (7) plastic clay; (8) clay; (9) upstream puddle blanket; (10) transition filter.

upper part of the wall. If additional measures for elongation of seepage path in the foundation have not been foreseen, one must provide sufficient seepage strength of the core soil at its transition to the foundation so as to increase the hydraulic gradients by arrangement of inverse filter or other means.

Drainage holes behind the above seepage prevention facilities are sometimes purposeful to control the effectiveness of these measures in various periods of operation of structure.

A dam with sheet piling and grout curtain at the foundation is shown in Figure 7.8. Clay-concrete walls up to 100 m deep (Fig.7.9) can be constructed in gravel and pebbles with boulders, when other types of curtains are troublesome; in this case the curtain consists of piles intersecting in plan view. The filler of the holes is similar to that used in the construction of 'wall in soil'. The material of the clay and concrete wall of the dam in Figure 7.9 had the following composition, in kg per 1 m^3: 160 of cement (class 300), 132 of bentonite clay, 100 of gravel up to 20 mm, 500 of sand with gravel up to 5 mm, and 345 of water. The compressive strength of the material was 3 MPa, and its coefficient of permeability reached 10^{-5}–10^{-6} cm/s.

If the clay of a core has the coefficient of permeability of 10^{-8}–10^{-9} cm/s, then such a diaphragm (wall) becomes a drainage for the core (Fig.7.9b). The exit gradients are close to unity, and the seepage deformations of the core-clay material at its interface with the pebble foundation are not hazardous.

For soil with high permeability, the exit gradient at the distance of 0.5 m from point A reaches 11 (Fig.7.9c). Actually, the gradient at point A goes to infinity. In this case one must increase the clayey cut-off about the wall, apply a filter beneath the core on the tailwater side of the wall, or use any other structural measure (Fig.7.14).

If the permeabilities of a core material and diaphragm are similar (Fig.7.9d) the exit gradients at the distance 0.5 m from point A are about 8. This case is simpler than the preceding one but remains more complex than the first case. The necessary structural elements are the same as in the second case.

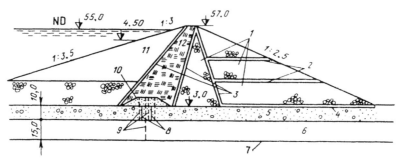

Figure 7.8. Shek-Peak Dam (Hongkong). (1) clay-filled stone revetment; (2) drainage layers for removal of infiltration and seepage water; (3) filter; (4) drainage mattress; (5) alluvium; (6) strongly broken granite; (7) quartz porphyrite; (8) piles; (9) grouting holes; (10) sand overfill; (11) sand; (12) loam core.

A dam with grout curtain in an alluvial stratum of foundation is depicted in Figure I.2. In this case the thickness of the alluvial layer increases in the course of construction upon filling of the very first prism of gravel and pebbles in the river bed above the water level in excavation.

In general, the above prism may be implemented as a cofferdam, so the filling of coarse material of the upstream prism of the cofferdam is accomplished in water (at low water stages).

5. Considerable thickness of permeable soil; water-confining stratum at a depth practically inaccessible for sheet piling, counterseepage soil wall, or even grout curtain. In this case, if the dam incorporates a core or diaphragm, a suspended grout curtain is constructed in the foundation; an alternative is a curtain reaching the water-confining stratum in which the depth of the curtain is 0.5–1.0 or even $5H$, in which H = head on structure. Construction of the curtain is possible from galleries or holes arranged in the core (Fig.7.13) or in foundation (Figs 7.10 and 7.11). A horizontal or deep drainage is often arranged behind the curtain, upstream of flow exit into tailwater; it facilitates the operation of the structure and permits control of the effectiveness of the curtain.

If a dam is homogeneous or incorporates a screen, a convenient counterseepage measure can be implemented as an upstream puddle blanket (Fig.7.12). In some cases one may simultaneously use various curtains or walls, but this requires special substantiation. The upstream puddle blanket is sometimes combined with a core (Fig.7.13), but in this case the blanket slightly reduces the stability of the upstream fill (thrust prism); the prism is usually made of a stronger material, compared with the core, and the foundation of the prism consists of a soil that enhances the generation of a failure surface. In such a design, the upstream puddle blanket and core are therefore used in extremely important structures if the seepage stability of foundation soil must be increased. An example is provided by High Assuan Dam (Fig.7.13).

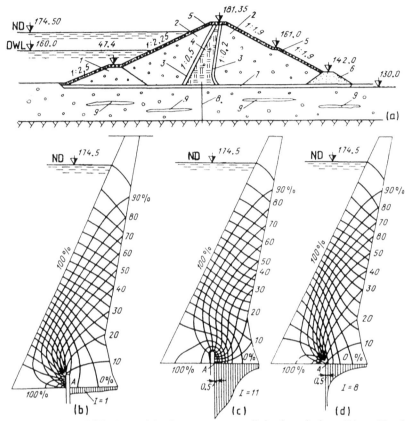

Figure 7.9. Earth-rockfill dam with clay-concrete wall in foundation, River North Kibir (UAR), Soyuzgiprovodkhoz design. (a) design dam profile; (b) flow net for the ratio of the coefficients of permeability (wall versus core) $k_w/k_c \approx 10^{-2}$; (c) for $k_w/k_c \approx 10^2$; (d) for $k_w/k_c \approx 1.0$; (1) upstream cofferdam of gravel and pebbles; (2) thrust prism of gravel and pebbles; (3) transition layer; (4) clay core; (5) stone revetment; (6) drainage prism; (7) levelling layer of gravel and pebbles; (8) seepage prevention curtain of clay concrete; (9) clayey lenses in foundation.

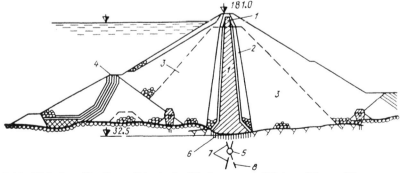

Figure 7.10. El-Infiernillo Dam (Mexico). (1) dam core; (2) transitions; (3) zones of stone material (conglomerate); (4) cofferdam screen; (5) gallery in dam foundation; (6) area grouting; (7) grouting holes; (8) drainage holes.

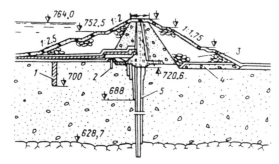

Figure 7.11. Silverstein Dam (FRG) with grout curtain and horizontal drainage. (1) concrete cut-off (to increase seepage path); (2) well points; (3) river pebbles; (4) horizontal drainage; (5) grout curtain.

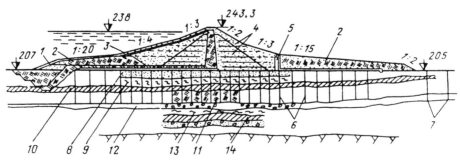

Figure 7.12. Derwent Dam (Great Britain) with core and upstream puddle blanket. (1) clayey puddle blanket; (2) sandy clay; (3) concrete blocks; (4) drain interlayers; (5) drainage wall; (6) vertical drainage boreholes; (7) distant holes; (8) sandy gravel; (9) mud sand; (10) moraine clay; (11) layered clay; (12) sand and gravel; (13) mud; (14) varved clay.

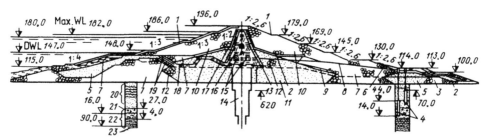

Figure 7.13. High Assuan Dam (Egypt). (1) facing on cobbles; (2) three-layer filter (1 m of 40–150-mm cobbles, 0.5 m of 5–35-mm shingle, and 1.5-mm coarse sand); (3) quarry fines; (4) drainage holes; (5) dune sand; (6) compacted dune sand; (7) stone above 50 mm covered with sand; (8) depression curve; (9) 40–150-mm cobbles; (10) hydraulically compacted rubble mound; (11) sand-filled rubble; (12) rock; (13) compacted shellac sand; (14) cement-clay curtain; (15) grouting and inspection galleries; (16) clayey core; (17) cobbles; (18) upstream clay blanket; (19) 40–150-mm cobbles; (20) diverse sand fractions; (21) shingle and boulders with sand filler; (22) interlayers of loam, sandy loam, sand and sandstone; (23) magmatites.

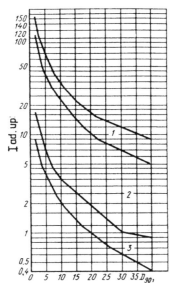

Figure 7.14. Critical gradient based on contact uplift condition ($I_{up.cr}$) as a function of D_{90} of soil beneath upstream puddle blanket. (1) clay; (2) loam; (3) sandy loam.

The length of the upstream puddle blanket is determined through seepage computations and depends on acceptable exit gradients with inclusion of inverse filters at the exit to the foundation drainage.

The upstream puddle blanket must be covered with a protecting layer along the sections susceptible to freezing and scouring due to considerable velocities of water. The upstream puddle blanket is sometimes combined with drainage and curtains.

The thickness of the upstream puddle blanket is determined by a strength of its soil and the grain size distribution of the underlying soil. The acceptable thickness can be computed by methods of seepage strength (see Chapter 2). The minimum acceptable thickness is 0.2 m, and depends on conditions of construction. For approximate selection of this thickness one may use the graph given in Figure 7.14. The known D_{90} of the underlying layer and of upstream puddle blanket material one finds the allowable gradient $I_{all.upl}$. The thickness of the upstream puddle blanket becomes $\delta = H/I_{all.upl}$. The upstream boundary of each branch depicted in the drawing corresponds to a fatter soil while the lower limit is attributed to leaner soil. Hence, knowing the plasticity of the upstream puddle blanket soil and D_{90} of the underlying soil one may find the critical gradient $I_{cr.upl}$ and the corresponding δ.

If large areas are screened, the thickness δ becomes variable because of changing H and perhaps D_{90}. The assumed thickness of the upstream blanket can be checked by computation and experiments (Section 2.3, for a class of structure).

6. Permeable foundation with anisotropic seepage properties (usually higher permeability in the horizontal direction). In this case the combat of seepage losses

by construction of an upstream puddle blanket is ineffective; one must apply vertical walls in the form of sheet piling, 'wall in soil', grout curtains etc. The drainage for interception of groundwater flow must be vertical. The above remarks also hold in the case of layered foundation (Fig.7.13), which is actual for a curtain and deep drainage.

7. Highly compressible clayey foundation. In this case one must enhance consolidation of foundation, and the stability of dam should be determined with consideration of pore pressure in the foundation. The consolidation can be accelerated by construction of vertical sandy drains (Fig.7.12). The diameter of the drains is usually 35×40 cm. The drains are arranged in staggered order and are equally spaced from each other. The zone of operation of the drainage is assumed convieniently as a circle in plan view. The drains are effective for the coefficient of permeability $k \leq 10^{-3}$ cm/s and $c_v = k \frac{1+\epsilon_m}{\alpha \gamma_w} \leq 0.1$ m³/day (Eq.3.10a).

Transitions between fills (thrust prisms) and foundations of earth dams. They should provide maximum strength (stability) of the dam. The following cases are encountered.

1. Weaker soils (often mud) than those in the thrust prisms, occur in foundation. In this case the stability of the structure should be determined by the strength characteristics of the foundation soil. Many examples of such structures may be provided.

First of all, one must design measures to increase the strength and decrease the deformability of foundation, which is implemented by drainage, if possible (see Section 3.4). Next, aside from strengthening of foundation, one must check the soil of foundation on possible growth of seepage deformation at the transition of the foundation and the thrust prism. If seepage deformation occurs, a filter between the foundation and the thrust prism must be arranged in accordance with the guidelines given in Chapter 2. The filter of High Assuan Dam was arranged under water to the depth of 20 m; it was very laborious so that a new design unknown by that time was applied — underwater dumping of rock with sand (Fig.7.13).

Thrust prisms of stronger material (characterized by φ and c), compared with foundation, are not always recommendable, and they should be used only in the absence of weaker and cheaper material.

2. Thick layer of soil as strong as the fill (thrust prism) arranged at the level of foundation and covered above with a thin layer (1–3 m) of weaker soil. The necessity of removal of weak soil must be substantiated by technological and economic analysis and comparison of various dams on different foundations.

3. Soil of equal strength is arranged in the foundation; in this case it is only the uppermost vegetation layer which must be removed.

4. Bedrock covered completely or partly with loose sediment of small thickness. The properties of the deposits should be determined in laboratory on undisturbed samples. If the strength characteristics are below those of the fill (thrust prism), the loose material should be removed. However, if the characteristics of sediment

and thrust prism are close to each other, and the topography does not facilitate a smooth failure surface, the removal is not necessary and the loose material can be left behind at the foundation of the thrust prisms.

7.2 GROUT CURTAINS IN DAM FOUNDATION

Natural phenomena in rock (daily and seasonal variation of temperature, tectonic effects, varying water content, etc bring about generation of cracks, so that rock can lose cohesion and transform to smooth granular material. Cracking rock has very low strength and a higher permeability. It is the latter factor which dictates the necessity of using grout curtains.

The following classification of cracks in rock can be borrowed from (Cambefort 1971).

Width of crack, mm	< 1	1–5	5–20	20–100	> 100
Type of crack	fine	small	medium	coarse	very coarse

Cracks can be filled or unfilled with loose material. The permeability of cracked rock is determined by the degree of opening and spacing of cracks, which can vary from several millimetres to several (sometimes tens of) metres.

The presence of soluble inclusions in rocks brings about chemical suffosion, that is leaching of some components of rock (limestone, sandstone, rock salt, potash salt etc), so coarse cracks and cavities arise; prediction of chemical suffosion is outlined below.

Permeability can be reduced and the strength properties of large rock bodies at dam foundations may be improved by cement grouting, silicatization, cold or hot bitumization, and pressure filling in deep boreholes with clay or other materials. Surface improvements of the quality of water-bearing strata of rock and semi-rock soil can also be implemented by grouting through shallow boreholes (5–10 m). Deep and surface pressure techniques are usually combined in one structure — deep pressure filling is used to generate an almost impermeable curtain (Section 7.1), which elongates the seepage path, whereas the surface filling over the entire area of core (screen) transition generates rock foundation for displacement of groundwater flow from the transition and elimination of seepage deformations at the interface.

Grouting (cementation) increases strength and reduces permeability of bedrock; this is the most widely used technique for improvement of rock and non-rock foundation. The use of grouting depends on the size of cracks and the velocity and composition of groundwater.

The grain size of cement varies from 0.001 to 0.12 mm, with the mean diameter $d_{50} \approx 0.04$ mm. If cement grains are to penetrate pores or cracks, their size must be two or more times smaller than the size of cracks (pores). Accordingly, grouting can be used in cracks with opening range above 0.1 mm. Successful grouting requires relatively low velocity of water in cracks; it must be below $v_D \approx$ 0.6 cm/s (Eq.1.16). If water-soluble minerals are contained in rock or a semi-rock foundation is grouted, the above conditions must be softened to $v_D \leq 0.3$ cm/s.

Grouting of rock with fine, medium and coarse cracking employs thick thixotropic solution injected almost without pressure (0.1–0.2 MPa), which reduces the penetration of the solution and requires that the boreholes be spaced by not more than 1–2 m. The solution consists of cement, clay, silicate (which enhances thixotropic properties) and fine sand. The consumption of cement sometimes reaches 450 kg per 1 m³ of solution. The recommended solution may be established by laboratory tests on site, because quality of water and other factors affect the properties of grout solution. Little cement and a lot of clay is used in clay-cement solutions. If bentonite clay is applied, it decelerates sedimentation of cement particles.

Grouting in gravel and pebbles without sand additions can be used for permeabilities k of 0.1–0.2 cm/s at pressure of 1–3 MPa, which corresponds to turbulent flow of clay-cement solution in soil pores. Grouting is usually applied under pressure. Pressure can be reduced or increased, even by a factor of 2 and more. Tests show that curtains can be generated in soil with coefficients of permeability k ranging from 10^{-3} to 10^{-5} cm/s.

Bitumization is applied to provide temporary impermeability, as organic bonding agents do not possess structural strength. Impermeability can also be reached due to viscosity. Light bitumen heated to 200–220° can be injected into coarse cavities; in fine ones it cools down very rapidly and grouting is arrested. Filler with grains $d < 1$ mm, in the quantity up to 60% can be added to the bitumen. Cold bitumization, with bitumen emulsion in water, can also be used for grouting.

Sodium silicate and reactants (electrolytes, acid or other colloidal solutions) can be used for grouting in soil. The cost of these solutions can be reduced if fine powder (cement or bentonite clay) is added as suspended matter.

Some other materials are also used in the construction of grout curtains.

Permeability of rock is assessed by specific water absorption i.e. flow rate of water in the borehole per 1 metre of length under head of 1 metre:

$$q = \frac{Q}{Hl} \tag{7.1}$$

in which

    Q = flow rate of water absorbed in the tested segment of borehole at head H, litres/min;

    l = length of tested borehole segment.

The water absorption is investigated under heads of 20–50 m.

A grout curtain can be constructed for $q < 0.1$ l/min, or roughly a permeability of 2×10^{-5} cm/s. In approximation, the permeability in cm/s is 500–1000 times smaller than the specific water absorption in l/min.

The allowable gradients in grout curtain in rock depend on specific water absorption and vary from 10 (for $q=0.05$ l/min) to 20 ($q=0.01$ l/min) (Churakov 1976). In sand, the allowable gradient varies from 2 (fine sand) to 3 (coarse sand).

Injection curtains can be arranged in one or many rows. The number of rows depends on permissible hydraulic gradient and structural requirements. For instance, for grout curtains in alluvial foundation beneath a core (screen), the thickness of the curtain is usually taken as the thickness of core (screen) at foundation.

The thickness of a single-row curtain is

$$\delta_c = (0.7 \div 0.8)R \tag{7.2}$$

in which
  R = spacing of boreholes in row.

The thickness of multiple row curtain becomes

$$\delta_c = (n-1)R + 0.4R \tag{7.3}$$

in which
  n = number of rows.

The spacing of boreholes is usually taken 2–3 m, but one also encounters smaller spacings (1 m at Irkutsk Dam) and greater ones (6 m at Bear Creek in the USA). The spacing between boreholes is specified more accurately under construction upon implementation of a test grouting scheme.

In both rock and non-rock soil employing multiple-row curtains (which are customary), the external rows incorporate thicker and stronger solution; these rows are implemented first. Fine cracks sometimes require thin-dispersion colloidal bitumen (this was the case for the upstream row in Mingechaur Dam curtain; for the tailwater road one used grouting). The grouting row of Ortotokoy Dam consisted of cement solution in the upstream row and clay-cement solution on the tailwater side. Intermediate rows incorporate clay-cement or other solutions which can easily penetrate fine cracks. The number of rows in multiple row schemes of grout curtains gradually decreases with depth (Figs 7.11 and 7.13).

7.3 DISSOLUTION OF SALTED SOIL IN BODY AND FOUNDATION OF HYDRAULIC ENGINEERING STRUCTURES

Earth dams are often constructed on salted soil, or the construction material can also contain salts. Structural properties of salted soil vary in the course of construction and operation because the composition of soil incorporates various soluble minerals which can be dissolved by groundwater and removed by seepage (Verigin & Oradovskaya 1960); this is referred to as chemical suffosion. The processes of salt removal from soil have been investigated by Oradovskaya guided by Verigin (Verigin & Oradovskaya 1960).

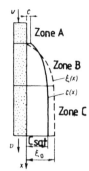

Figure 7.15. Generation of leaching zones upon seepage in gypsum soil.

Salts in soil can be classified in three groups as to their solubility: easily soluble ($NaCl$, $MgCl_2$, $Na_2SO_4 \cdot 2H_2O$, K_2SO_4, $MgSO_4$, Na_2CO_3) of medium solubility ($CaSO_4 \cdot 2H_2$), and of low solubility ($CaCO_3$, $MgCO_3$).

The usual content of easily soluble salt in soil is about a few percent. Complete removal of these salts from soil is fast if the volume of the water passing through the soil exceeds three times the volume of soil pores. The process of dissolution (leaching) thus occurs at the beginning of the operation of a structure, while the leaching of hydraulic-fill earth dams occurs during the construction period. The salts of low solubility undergo slow processes (with exception of very aggressive groundwater), so such processes of leaching are not usually considered. A soil incorporating carbonates does not change its properties over long spans of time.

Salts of medium solubility (gypsum, anhydrite), the contents of which can be quite considerable, pose most hazards to hydraulic engineering structures. Even for low contents of gypsum in soil (several percent), its gradual solution and removal will increase the permeability of soil (in particular, if the foundation is rock or semi-rock) which can amount to catastrophic proportions. If gypsum is contained in loose soil, its dissolution can undermine structural bonds in soil, and the properties of soil will change considerably — shear strength can be reduced and additional settlement will occur. Quantitatively and qualitatively, these processes are controlled by initial properties of soil and the conditions of its operation in the foundation or structure. The assessment of the properties of soil with gypsum should be performed in laboratory.

The intensity and duration of dissolution of salts in soil can also be assessed by theoretical procedures (Virigin & Oradovskaya 1960). The theoretical solutions are applicable to salted soil with dispersed salts. Layers of gypsum and other water soluble rock are not considered.

The process of leaching and removal of salt from soil due to groundwater belongs to problems of physico-chemical hydromechanics. In general, it is described by the equation of convective diffusion with additional terms accounting for dissolution of salts due to seepage and an equation standing for leaching as a function of groundwater flow.

Table 7.1. Dissolution of salts in water at 20°C.

Salt formula	Contents[**]	Salt formula	Contents[**]
$CaCO_3$	$14.5 \cdot 10^{-3}$	$MgSO_4 \cdot 7H_2O$	261.9
$CaSO_4 \cdot 2H_2O$[*]	2.1	$Na_2CO_3 \cdot 10H_2O$	176.9
$FeSO_4 \cdot 7H_2O$	209.5	$NaCl$	264.7
KCl	253.7	$Na_2SO_4 \cdot 10H_2O$	162.5
K_2SO_4	100.0		

[*] at 30°C.
[**] Contents of anhydrous compound in 1000 g of saturated solution.

In the derivation of respective equations (Verigin & Oradovskaya 1960) the diffusion far away from the surface of soluble salts was neglected, and so was the variation of salt concentration in time. For one-dimensional groundwater flow (Verigin 1957) the equations of motion and conservation of mass of soluble salts, for linear flow along the Ox-axis yield

$$v\frac{\partial c}{\partial x} + \rho\frac{\partial \xi}{\partial t} = 0 \qquad (7.4)$$

The equation for dissolution of salt under seepage conditions reads:

$$\rho\frac{\partial \xi}{\partial t} = -\kappa(c_{sat} - c_0)\sqrt{\xi_0} \qquad (7.5)$$

in which

$$\gamma = \delta\left(\sqrt{\frac{\nu}{\eta}} + 1\right), \qquad (7.5a)$$

ξ_0 = initial unit volume of gypsum at $n = 0$, i.e. before inicipient washing due to seepage (in fractions of unity),
ρ = density of soil in g/cm³,
d = grain diameter of gypsum, cm,
c_0 = salt concentration in water at inflow into salted soil g/cm³,
c_{sat} = saturation concentration for salt in groundwater flow (Table 7.1),
η = coefficient of convective leaching, cm/s,,
δ = coefficient of diffusive leaching, 1/s,
κ = general coefficient of leaching, 1/s.

The coefficients δ and k depend on η_0, gypsum grain diameter, diffusivity D, viscosity of water ν and porosity of soil n. In each case they should be determined by laboratory experiments (Verigin & Oradovskaya 1960). The solution can be simplified under certain assumptions whereupon the coefficients δ and ν do not appear.

It may be assumed that a simplified solution for salt dissolution fully satisfies the design requirements for general hydraulic engineering structures. More accurate solutions will be required in the forecast of leaching of easily soluble salts or for the design of special hydraulic engineering structures (Verigin & Oradovskaya 1960).

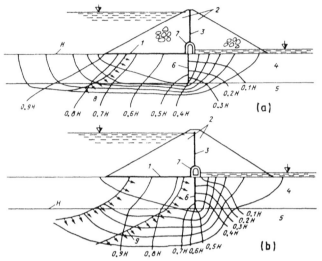

Figure 7.16. Propagation of leaching front in dam foundation. (1) asphalt-concrete upstream puddle blanket; (2) fills (thrust prisms); (3) asphalt-concrete diaphragm; (4) highly
permeable gypsum-saturated foundation zone; (5) almost impermeable foundation zone; (6)
impearmeable wall in foundation; (7) gallery; (8) leaching front after Δt; (9) leaching front
after $2\Delta t$.

In one-dimensional problems (Fig.7.15) three zones of leaching may be distinguished in the direction of groundwater flow:

Zone A — region of rock already without salt with $\xi = 0$ and $c = c_0$ in groundwater
flow.

Zone B — 'active' zone in which leaching takes place as gypsum is removed from
soil, with gradual saturation of groundwater with gypsum. In this zone ξ increases
from zero at the boundary of zone A to ξ_0 at the boundary of zone C, while c
varies from $c = c_0$ at the boundary with A to $c = c_{sat}$ at boundary with C. The
interface with zone C is usually taken conditionally at the cross-section $c = 0.9c_{sat}$,
as the accurate boundary for $c = c_{sat}$ is very difficult to find.

Zone C — area of unleached rock, in which groundwater flow saturated with
gypsum is practically depleted of its dissolving capacity and one has $\xi = \xi_0$ and
$c = c_{sat}$.

For constant seepage rate v the zone A grows, the zone B (partial leaching)
does not vary, and the zone C shrinks. For varying seepage rate in the flow, the
zone B is variable. If the seepage rate is high and the content of gypsum in soil
is low, the time T_o of full leaching of salt at the inflow section is insignificant, and
the leaching process is practically determined by zone B alone.

For small seepage rates and high contents of gypsum, the length of zone B (x_B)
is reduced. Zones B and C can be identical. For small T_o and x_B the problem is
solved in a simpler manner, so the coefficients η, δ and κ are not required; one has

$$\frac{dx}{dt} = 1.26\frac{(c_{sat} - c_0)\nu(x,t)}{\rho\xi_0} \tag{7.6}$$

Equation 7.6 is referred to as an equation of the celerity of complete leaching front i.e. it describes the boundary of zone A for unsteady, non-uniform groundwater flow.

The problem may be solved by Equation 7.6 if the method of finite differences is applied along streamlines of groundwater flow. The leaching time t along a certain segment of foundation (dam body) can be computed if one knows the following characteristics of salted soil and groundwater flow: specific volumetric content of $CaSO_4 \cdot 2H_2O$ is soil (ξ_0), and initial (c_0) and ultimate (c_{sat}) content of $CaSO_4$ in flowing groundwater. Determination of these characteristics belongs to the problem of general identification of engineering and geologic parameters in the design of hydraulic engineering structures.

Solution of soil leaching problems (simplified version) proceeds as follows:

1. Determine design values k, ξ_0, ρ, c_0 and c_{sat} for a sorted soil (soils) of foundation or dam body. All above constants must be determined as a result of investigations.

2. Construct flow net by the electrohydrodynamical analogy or any other method (first approximation, Figure 7.16a).

3. Determine mean velocity of seepage (Chapter 2) along streamline.

4. For a given small increment Δt (from 2 to 10 years, depending on the length of impermeable dam contour), find the length of segment along design streamlines for fully soluble gypsum by using Equation 7.6:

$$\Delta x = 1.26\frac{(c_{sat} - c_0)\nu(x,t)}{\rho\xi_0 M_b}\Delta t \tag{7.7}$$

in which

M_b = coefficient accounting for deviation of mean seepage velocity along streamline from the real one, along each section between potential isolines.

If a flow net has been constructed by hydroeletrodynamic analogy or graphically one has

$$M_b = \frac{n}{b^2}\sum_{k=1}^{k=n}(\Delta x_k)^2$$

in which

n = number of segments along streamlines,
Δx_k = spacing of adjacent potential isolines,
b = complete length of streamline.

If one must determine the time of leaching along a streamline having the length l (b being a multiple of l) by the use of Equation 7.6, then one has

$$t = t_b M_b = t_b\frac{\sum_{k=1}^{k=p}(\Delta x_k)^2}{\sum_{k=1}^{k=p}(\Delta x_k)^2}$$

in which

$$l = \sum_{k=1}^{k=p} \Delta x_k$$

5. Combination of the obtained flow rates yields the boundary of fully dissolved salts (leaching front) after the time increment Δt (Fig.7.16b).

6. If it is known that leaching is accompanied by a sharp increase in k in the leaching zone, then the problem is reduced to the construction of a flow net (hydrodynamical grid) in second approximation. Since the quantities of gypsum contained in rock upon leaching can reach as much as 1–2%, and the permeability amounts to 1 cm/s, the losses along the leaching segment can be neglected and full head may be applied along the leaching front, as k in the original condition is smaller by 2 or 3 orders of magnitude (Fig.7.16b).

7. The second step of solution, associated with higher flow rates along streamlines, brings about higher Δx during the same time internal Δt.

8. Next one finds the leaching front in second approximation, which corresponds to $2\Delta t$. The third approximation is found thereafter by analogy etc until the leaching front reaches tailwater.

Hence the time of full dissolution of salts in dam foundation along the designed impermeable contour (upstream puddle blanket, grout curtain or other complex measures for leaching prevention) is $n \cdot \Delta t$, in which n = number of required approximations, ussually 5–10 in practical applications.

9. If permeability does not change upon leaching or if these changes may be neglected, and the foundation is regarded homogeneous, leaching simplifies as one can assume that the time of leaching along streamline is

$$t = 0.79 \frac{\rho \xi_0}{c_{sat} - c_0} \frac{x}{v(x)} M_b \tag{7.8}$$

in which

x = seepage length along the considered streamline.

Some other techniques can be used for solution of the leaching problem; they depend on particular conditions of the problem.

A strict solution of the leaching problem requires full scale investigations of the properties of salted soil in laboratory so that δ, η and κ are known. If δ and η are known then κ is determined by Equation 7.5a with v standing for seepage rate at inflow cross-section.

The duration of complete leaching along the original segment T_o is determined as follows:

$$T_0 = 1.78 \frac{\rho \sqrt{\xi_0}}{\kappa (c_{sat} - c_0)} \tag{7.9}$$

The length of fully leached soil x_A at any time t, counted from the time of incipient seepage, for constant seepage rate v=const is determined as follows:

$$x_A = 1.26\frac{(c_{sat} - c_0)v}{\xi_0\rho}(t - T_0) \tag{7.10}$$

For $v = v(x, t)$ Equation 7.6 holds true.

The length of the active zone B reads:

$$x_B = 5.18\frac{v}{\kappa\sqrt{\xi_0}} \tag{7.11}$$

Equation 7.11 determines dissolution of salts for $v=$const, but it can also approximate the conditions with $v \neq$ const (Verigin & Oradovskaya 1960); upon substitution of v_{max} one obtains the longest zone B.

In soil in which dissolution of salts does not bring about structural failure or consolidation (intruding rock or semi-rock soil), the coefficient of permeability k upon complete leaching can be approximated (with overestimation) by the following formula:

$$\frac{k}{k_0} = \frac{(n_0 + \xi_0)^3(1 - n_0)^2}{(1 - n_0 - \xi_0)^2 n_0^3} \tag{7.12}$$

in which

k, k_0 = permeability prior to and after complete removal of salt,
n_o = real porosity of soil (after removal of salts),
ξ_o = unit volume of soluble salts.

If soil structure is disturbed due to removal of salts, the maximum additional settlement of soil due to leaching can be determined as follows:

$$\Delta s = \frac{\epsilon - \epsilon_0}{1 + \epsilon}h \tag{7.13}$$

in which

$$\epsilon = \frac{n_0 + \xi_0}{1 - (n_0 + \xi_0)} \tag{7.14}$$

$$\epsilon_0 = \frac{n_0}{1 - n_0} \tag{7.15}$$

h = thickness of soil layer from which salts have been completely removed,
ϵ_o, ϵ and n_o, n = original and ultimate (after leaching) void ratios and porosities, respectively.

If soluble salts occupy a thick layer of foundation, the primary design problem of earth dams is reduced to the separation of salt from groundwater flow. For instance, a stratum of rock salt (NaCl) the leaching of which could bring about displacements of rock bodies in the dam foundation is encountered beneath the upstream prism of Rogun Dam (Fig.8.20). This could have brought about cracking on the upstream side, with possible stability failure. Therefore the accepted solution included prevention of salt removal. The design incorporated a concrete plug on the upstream side where the stratum outcrops into river bed, and a hydraulic curtain established on the tailwater side where groundwater flow occurs. The latter measure was constructed from special boreholes into which a saturated solution of salt was pumped under pressure. The role of this curtain is to stop the flow of salt solution. The curtain should arrest the leaching process; in the end it can slow down. It is realized that the celerity of propagation of the leaching front into the stratum is fairly low (1 cm/year).

7.4 SELECTION OF UNDERGROUND CONTOUR

The design of seepage prevention measures in dam foundation is fairly complex. An adequate selection of measures is impossible without consideration of dam design; according to Section 7.2 one must consider the conjugation of these structures. For instance, if a dam incorporates a core then upstream puddle blanket should not be applied.

One must endeavour to design such seepage prevention facilities in dam foundation which permit simultaneous construction of dam and counterseepage measures. This goal may be reached by construction of grout curtains and drainage from holes in foundation, from the ground surface of canyon banks, by creation of soil upstream puddle blanket upon construction of dam with soil screen, implementation of clay-concrete wall; sheet piling, or 'wall in soil', in combination with screen.

One must bear in mind that grout curtains have permeabilities $k < A \cdot 10^{-5}$ cm/s, and if cracking is so fine as to make impossible or very expensive grout curtains, and seepage prevention measures are still needed, then one may construct a clay-concrete wall whereupon the effectiveness of seepage prevention measures increases. One must strive for removal of as low as possible a volume of rock beneath seepage prevention measures and for surface grouting of rock, to make it more rigid.

All types of seepage prevention measures may be constructed in the practice of non-rock foundation. The major guideline in this case is to refrain from excavation and to use in the foundation a wide grout curtain partly from a centrally filled prism, which can play a role of cofferdam, and partly from a hole generated at the top of this prism; one may also use the well only, especially if they are several. Nonrock foundation can be cut by a 'wall in soil'.

The selection of a seepage prevention measure is dictated not only by economic factors but also by potentials of cheap and fast maintenance and repairs. A grout curtain constructed from a gallery can be renovated if the operation is disturbed. Repairs of 'walls in soil' are also associated with grouting techniques, but this requires a gallery (which is not necessary upon construction of a 'wall in soil'); or alternatively injection works upon repairs must be conducted from dam crest (if dam has a central core). Special means on the upstream slopes can also be used if the core is inclined. Upon construction of dams with screens, in the absence of holes at the connection of the screen and the 'wall in soil', repairs are possible if the reservoir is emptied. Hence a well is often required if a clay-concrete wall or 'wall in soil' are used, although this is not linked to construction proper.

Continuous seepage prevention structures in bedrock are recommended if the foundation soil includes water soluble compounds; small permeability of such facilities should be secured.

In order to extend the leaching time, the foundation can be protected with upstream puddle blankets of soil, asphalt and concrete (Chapter 8) or man-made materials. The upstream puddle blankets are to be connected with cores, screens or diaphragms, depending on the volume of reservoir and the height of dam. For instance, if a reservoir provides seasonal or multi-year control, and repairs of seepage prevention facilities in the dam body for foundation are impossible in an emptied reservoir, then structures with seepage prevention members should be preferred, as they are more convenient for grouting works to execute during repairs; accordingly the upstream puddle blanket will be connected with the core or diaphragm. In this case the material of the upstream puddle blanket can be selected upon consideration of technological and economic factors, because a cheaper upstream puddle blanket of soil can bring about considerable shallowness of the upstream slope. In any case, the seepage prevention constituents of dam foundation should guarantee a sufficiently long time of leaching, not less than 50 years counted from the incipient seepage to the time of the exit of leaching front into tailwater; this time depends on the class of permanent structures.

In vertical screens that cut into upper, more permeable layer, continuous seepage prevention measures are preferred such as clay-concrete wall or 'wall in soil', which are characterized by lower permeability, as noted above.

One should remember that the simplest design is secured by upstream puddle blankets made of soil. Grouting and other types of works in nonsoil foundations are usually conducted by specialized firms.

CHAPTER 8

Dam design using earth materials

8.1 GENERAL DESIGN GUIDELINES

In the design of a hydraulic engineering system (hydroelectric, agricultural or combined one), the selection of dam for impoundment purposes should stem from an analysis of technological and economic factors for various versions of dams. The following issues should be examined:

1. Availability of local construction materials for dam body. One should try to make maximum possible use of the materials produced in situ by mining and excavations. In the presence of coarse grained soil such as stone, cobbles, pebbles and gravel, or cohesive soil (clay or loam), one must first consider dam versions with water confining constituents such as cores or screens, with water confining prisms of coarse soil — such dams have the most compact cross sections.

2. Engineering and geologic properties of the dam area, in particular permeability of soil or rock at the foot of the dam, and the strength characteristics. Bedrock may be used practically for any type of dam. If weak soil is available at the foundation, the dam itself must be much wider.

3. Hydrologic conditions and the capacity of passing flood discharges, for instance over the crest of a dam still under construction, inclusion of cofferdams in the dam body etc.

4. Climatic conditions of site (extended periods of negative temperatures, high precipitation, long period of high temperatures). In the dam construction in arctic regions, the activities involving soil are seasonal, therefore the dam must be surrounded by non-soil seepage prevention constituents. If precipitation is high, large quantities of cohesive soil are impossible to lay down without excessive moisture, which requires a dam with minimum volume of seepage prevention components.

5. Regional seismicity, as dam design should secure reliable performance under earthquakes. In view of seismic stability requirements one prefers dams made of coarse grained soils with high permeability and seepage prevention components, capable of resisting considerable deformations.

6. Quality of construction, available machinery, experience, management etc.

As a result of analysis of technical and economic factors, along with the inclusion of expert's assessment, one selects a certain type of dam and solves the following problems:

a) Determination of the overall dimensions of dam, i.e. upstream and downstream slopes, width of dam crest, and the location and width of berms on the slopes, if necessary.

b) Selection of the type and design of slope revetments and dam crest.

c) Selection of the type and design of seepage prevention measures within the dam.

d) Selection of the type and design of drainage.

e) Design of underground dam contour and transitions to the above seepage prevention and drainage measures.

f) Determination of the type of transition between the dam body, foundation and banks, depending on the design of dam and its underground contour.

The above factors are mutually dependent. For instance, the seepage prevention and drainage components are often determined by the type of dam. The seepage prevention and drainage units in the foundation are connected to their counterparts in the dam body.

In the course of dam design one should consider different versions of passing the discharges during the construction period and should select the optimum one based on technological and economic factors.

Dam design also includes a section on construction practices and another one on instrumentation for surveyance of the condition of dam during construction and under operation.

The selection and dimensioning of a dam are the most important issues of the design of a hydraulic engineering scheme, linked to the complex objectives of the latter, often of key importance to all other activities.

8.2 EARTHFILL DAMS

Dam profile. Earthfill dams are the most common type of earth dam designed for hydraulic engineering schemes with low and medium head — in principle they have heights not greater than 50–60 m. They are usually homogeneous, but inhomogeneous designs with screens, cores and diaphragms are also practised. The dam cross-section depends on the height, strength of soil within dam body, and the technology of construction. The cross-section is determined by the upstream and tailwater slopes, presence of berms, and the datum and width of dam crest.

In dam design it is the crest datum which often becomes the first quantity to determine; it is found from the datum for normal and back-water level in the reservoir, and also from computations of wind waves and construction properties of dam crest. The crest datum should be above the static water level in the reservoir given as follows

$$d = \Delta h + h_H + a \tag{8.1}$$

Figure 8.1. Determination of dam crest. (1) design static datum; (2) mean wave set-up level.

in which

Δh = wind wave set-up,

h_{rup} = run-up height,

a = freeboard taken as not less than 0.5 m.

The design values of Δh and h_{rup} are determined by the Soviet standard SNiP 2.06.04-82. The value of d determines the design crest datum. In order to avoid the subsidence of dam crest below the design datum under operation it is necessary to consider a certain construction freeboard equal to the anticipated magnitude of subsidence. The scheme for determination of the crest datum is shown in Figure 8.1.

Dam crest width is determined by factors of construction, operation and transportation. If traffic is not allowed on the crest, its width should not be less than 3 m; this value becomes 6 m for high dams. If a highway and/or railway exists, then the dam width depends on the overall dimensions of the traffic means, determined by appropriate standards. The design of the dam crest depends on the design of highway or railway, cable passages, storm sewers etc. If the dam body or its core consists of clay, then the dam crest is covered with a protecting layer of sand or gravel, the thickness of which should exceed the frost thickness in the given region; these measures prevent frost swelling of clay and protect the pavement from destruction. In seismic regions, the design crest width and its datum are taken slighly greater than those determined for usual conditions. This stems primarily from the possibility of considerable deformations and seismic activities and is due to generation of waves in the reservoir caused by landslides.

Dam slopes are determined through stability computations and depend on the design of dam, soil strength, operation of dam, geologic structure of foundation etc. In preliminary design the slopes are found from experience and by analogy to the existing dams. The analogs should be selected as modern designs under similar natural conditions, wih soils similar by their grain composition, strength and deformation properties. The experience gained so far for sandy and clayey dams permits the following recommendations for preliminary design:

a) Low dams, 5–15 m high: upstream slope factor (cotangent) of 2–3, tailwater slope of 1.5–2.5;

b) dams 15–50 m high: upstream slope of 3–4, downstream 2.5–4.0;

c) dams higher than 50 m: upstream 4–5, downstream 4–4.5.

Shallower upstream slopes are used if tailwater sections have to be loaded or if weak soil is used for the screen. Shallower tailwater slope is necessary if the general stability of the dam has to be secured. In the presence of weak soil in the foundation, dams must be much shallower than it is shown above.

Berms on the slopes of a dam should be constructed with a width not less than 1 m at a spacing of 15–20 m; they are designed for inspection and repairs. Wide berms are constructed for service vehicles, construction of highways and railways, special pipelines etc. Precipitation water is evacuated in berms via channels or side drains. Berms are also employed where the dam body is connected to draining bank or cofferdam, included into the dam profile.

Dam slopes are protected from waves by adequate revetments. The revetment can consist of rubble, concrete or reinforced-concrete slabs, asphalt or bituminous concrete. In the last two cases the revetment not only protects from destruction but also serves seepage prevention purposes. The revetment is constructed in the zone of water level oscillations. The upper boundary of the primary revetment should coincide with the datum of run-up, or practically with dam crest. The lower boundary is located two-third wave heights bellow DVL (dead volume level). The lightweight revetment (secondary revetment) is arranged below.

A rubble revetment is designed for waves up to 2.5 m high; it is either rubble or a stone layer on inverse filter. The thickness of inverse filter should not be less than 0.15 m, and in responsible cases it reaches 0.5–0.6 m.

Revetment material should be of considerable strength ($\sigma_t \geq 150$ MPa), frost-resistant (more than 50 cycles of freezing and thawing) and of high unit weight ($\gamma > 24$ kN/m). The design weight of rubble capable of resisting wave action is found by the formula:

$$Q = \frac{\mu \gamma_{st} h L}{(\gamma_{st}/\gamma_w - 1)^3 \sqrt{1 + m^3}} \tag{8.2}$$

in which

μ = coefficient equal to 0.021 for concrete and 0.025 for stone,
γ_{st} = unit weight of stone,
m = slope factor (cotangent),
h; L = design wave height and length, respectively.

It is most convenient to characterize the rubble material by its geometrical dimensions. The mean diameter reduced to the diameter of an equivalent sphere is given as

$$D_{sph} = \sqrt[3]{\frac{Q}{0.524 \gamma_{st}}} \tag{8.3}$$

The approximate diameter can be found from $D_{sph} \approx \epsilon h$, with wave height h and the coefficient ϵ from 0.25 to 0.35.

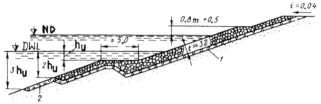

Figure 8.2. Rubble slope revetment. (1) inverse filter; (2) gravel or pebbles.

The thickness of rubble layer should provide protection from erosion of the filter and dam body and is required to be not smaller than $3D_{sph}$ for unsorted rubble and $2.5D_{sph}$ for sorted rubble. Unsorted rubble is recommendable for revetments where the maximum dimension is not limited; still more than 50% of total quantity should contain a rubble with a design size or greater. Rubble is a reliable revetment as its strength is not affected by dam deformation. Rubble is constructed upon use of machinery at any time of the year, and can be easily maintained and repaired.

A rubble cover is constructed on a layer of gravel or pebbles and in this case requires less rock. The thickness of rubble cover is given by

$$\delta = 1.7 \frac{\gamma_w}{\gamma_{st} - \gamma_w} \frac{\sqrt{1 + m^2}}{m(m + 2)} h \tag{8.4}$$

A rubble cover is chosen for low waves up to 1–1.2 m. The construction is more difficult than for a typical rubble revetment and is less reliable.

The selection of inverse filters under cover layers should be based on the condition of nonsuffosion, lower layers of structure should not be removed through the upper layers.

One of the earth dams on the Volga-Don canal, under reconstruction of its damaged revetment, was repaired by pouring cement mortar on the cover layer (3×3 m segments with unfilled bands 0.2–0.3 m wide). Further operation has not brought about any damage.

Figure 8.2 shows revetment design using rubble cover layer 30 cm thick (not less than 3 diameters of the maximum rubble size) underlain by inverse filter.

Concrete and reinforced-concrete revetments are made as monolithic, cast in situ, or composite slabs. Monolithic slabs are used for protection of upstream slopes under waves 2 to 4 m high. Their thickness is 0.15–0.5 m and their plan dimensions are 5×5 up to 20×20 m, or even more. The joints between slabs are open or sealed. In the case of waterproof joints it is necessary to provide a filter at the foundation of the slabs. The slope revetment with monolitic reinforced-concrete slabs is very sensitive to soil deformation at the foundation, and can involve cracking in the slabs. The experience gained from the Volga cascade shows that the dimensions of weakly reinforced concrete slabs having a thickness of 0.4–0.5 m should be not below 10×10 m. Before the construction an inverse filter should be thoroughly consolidated. If slabs are arranged on a cohesive soil slope,

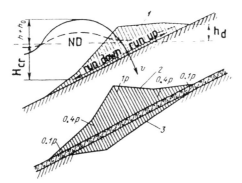

Figure 8.3. Wave action on slope. (1) velocity distribution graph; (2) wave pressure p upon wave breaking; (3) wave-induced seepage pressure W_s; (v) direction of water elements in plunging wave.

a 1.0–1.5 m layer of sand becomes necessary.

Overall dimensions of slabs and their reinforcement are determined by computations. The design scheme is a slab on elastic foundation under dynamic loading. Under conditions of breaking waves the wave pressure on slope p is determined by wave height. In the run-down situation, counterpressure W_s is induced by seepage which tends to detach revetment from the slope. The graphs of wave-induced pressure and counterpressure are shown in Figure 8.3. The maximum value of W_s is found approximately at $z=0.9$ h/m counted from still water line (more details in SNiP 2.06.04-82):

$$W_s = 0.46\gamma_w h\sqrt{1 + m^2} \tag{8.5}$$

The design of slabs on elastic foundation under wave action makes it necessary to find moments and shear forces. The strength of a slab depends on the maximum bending moment. The reinforcement percentage is usually 0.4–0.6%. Thickness of a monolithic slab is determined from its resistance to uplift pressure:

$$\delta_{upl} = 0.07kh\frac{\gamma_w}{\gamma_{conc} - \gamma_w}\frac{\sqrt{1 + m^2}}{m}\sqrt[3]{\frac{L}{B}} \tag{8.6}$$

in which

    k = coefficient ranging from 1.25 to 1.3,
    B = slab length normal to water line,
    γ_{conc} = unit weight of concrete,
    h, L = wave parameters.

The uplift pressure is considerably affected by aeration of water. For aerated flow (γ_w=2.5–3.0 kN/m) the thickness of slab should be determined by the formula derived by Shaytan:

$$\delta_{upl} = 0.6\frac{h_{50\%}^2\sqrt[4]{B_{rel}^3}}{B\cos\alpha}\frac{\gamma_w}{\gamma_{conc} - 0.3\omega\gamma_w} \tag{8.7}$$

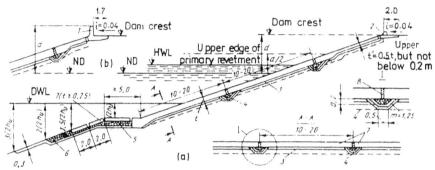

Figure 8.4. Slope revetment with concrete slabs (a) and design version for dam crest (b). (1) parapet; (2) 0.2-m openings in parapet every 10 m; (3) 0.1-m layer of tamped cobbles; (5) belt drainage; (6) 0.2-m layer of gravel or pebbles; (7) slabs; (8) 2.5-cm planks.

in which

$h_{50\%}$ = 50-% wave height,
B_{rel} = relative length of slab side (ratio of B to wave height),
α = angle between wave direction and normal line,
ω = coefficient of loading fullness (for B from 1 to 6 it is 1 to 0.5).

Slab revetments on slopes prevent ice loading. In this case the strength of slabs is checked on stresses exerted on the ice-covered side of reservoir. The computations are prescribed by standards which give ice loading for different varieties of ice and under various conditions of wind, temperature, expansion of ice and freezing of ice cover in the reservoir.

Figure 8.4 shows a design of a slope revetment with concrete slabs. The slabs are arranged on a 0.1-m stone layer. The joints between slabs are filled with tarred planks, and belt drainage is arranged at the foot of each joint. The upper part of the slope above the boundary of primary revetment encompasses slabs which are two times thinner. Below the boundary of primary revetment, ending on the berm, an additional load of small-size slabs with fine stone is arranged. The components of revetment with monolithic concrete slabs on slopes of earth dam are shown in Figures 8.5 and 8.6. The effectiveness largely depends on the permeability of joints. In a number of cases, unloading openings are implemented to reduce the uplift forces; those openings are filled with coarse material which prevents destruction under waves and upon rapid emptying of reservoir.

Composite slabs are 8 to 20 cm thick and have dimensions 1.5×1.5 up to 5×5 m. The slabs are arranged on a continuous inverse filter and are hinged on each other. The joints between slabs are open or cast in situ. Sometimes the joints of a group of slabs are made monolithic so that a composite monolithic cover is created. Metallic joints between slabs are subject to corrosion, which strongly affects the durability of the structure. In order to make them stronger, the joints between

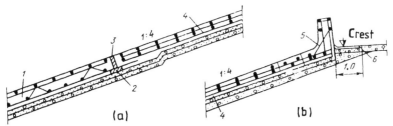

Figure 8.5. Revetment units with monolithic concrete slabs on slope of earth dam, Lenin Hydropower Plant on the Volga River. (a) transition between slabs; (b) slope revetment at crest; (1) 0.5-m reinforced-concrete slab on 0.1-m layer of cobbles and 0.15-m cover layer of cobbles; (2) 0.08×0.3 m reinforced-concrete slabs with bituminous mats at periphery; (3) tarred plank; (4) 0.35-m reinforced-concrete slabs; (5) parapet; (6) 0.3-m fine grained asphalt-concrete on 0.13-m layer of cobbles.

slabs are filled with bituminous concrete or resins, which ensure a better flexibility. The design of composite reinforced concrete slabs having sizes 2.5×2.5 and 3×3 m with monolithic joints is shown in Figure 8.7.

Monolithic slabs are widely used; they provide better protection from damage, but still are subject to considerable cracking if the dam body is deformed non-uniformly. Composite revetments are much better under variable settlement of the slope and if the filter below slabs is eroded. The selection of reinforcement also depends on conditions of construction.

Concrete, bituminous concrete and metallic cover layers of the upstream slope, playing simultaneously the role of seepage prevention units (screens) are also used. These types will be considered in Section 8.5. Selection of revetment design involves comparison of technological, economic and other factors.

Shallow dissipating slopes. Since rubble and concrete revetments are expensive, earth dam with shallow dissipating slopes can be more economical.

With increasing slope shallowness the dynamic action of waves becomes less important and the slope can be stable without a revetment. Field observations on slopes under waves carried out by Pyshkin have yielded the graph shown in Figure 8.8 for preliminary design of slope, for different waves and varieties of soil. This graph makes possible the design of a low or medium dam with unprotected upstream slope (or with a lightweight revetment). The dissipating slope is designed from the datum of NIL (normal impoundment level) plus run-up height, up to DVD (dead volume datum) i.e. H_{cr}.

Dam tailwater slopes subject to wave action are protected similarly to the upstream slopes, depending on wave height on the downstream side. A layer of gravel or rubble is designed above tailwater level in the zone of anticipated seepage, while above this zone the slope is protected with either a layer of gravel or grass sods or turf in a 0.2–0.3-m layer.

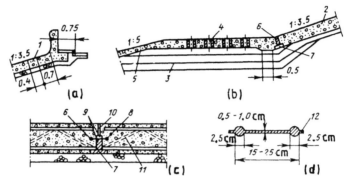

Figure 8.6. Joints of monolithic concrete slabs in slope revetment in river dam of 22nd CPSU Assembly Hydropower Plant on the Volga River. (a) at transition to dam crest; (b) at berm; (c) joint of slabs; (d) cross-section of rubber seating; (1) closing casting unit; (2) & (5) armoured concrete slabs on 0.15-m layer of cobbles, 0.5 and 0.25 m thick, respectively; (3) three-layer filter; (4) drainage holes; (6) tarred timber planks; (7) reinforced-concrete platens; (8) & (12) rubber seating and fastening wire eye; (9) wool; (10) two layers of roofing felt; (11) reinforcement.

The construction technologies for earth dams are described by Moiseyev & Moiseyev (1977).

8.3 HYDRAULIC-FILL EARTH DAMS

Hydraulic-fill dams differ from rockfill dams by the method of construction. In this section we consider hydraulic transport methods. The material for construction is mined either in a quarry using a hydraulic jet or in river bed with a dredger. Soil is transported via pipelines or channels. Water and soil produce the so-called slurry. In general, the consistency of slurry is 1/7 to 1/10, so that one part of transported soil corresponds to 7 up to 10 parts of water. The motion of slurry in open channels is free-surface type, and takes place at low velocities. In this case the concentration of slurry is lowest. The flow under pressure in pipelines makes it possible to increase the concentration of slurry, so this method is more economic. The theoretical background for hydraulic filling is given by Melentev et al. (1973).

Construction schemes for subaerial and subaqueous hydraulic filling are useful in various stages of construction. The initial stage embodies subaqueous filling of the lower part, after which the remaining part above water level is filled. In subaqueous filling, in contrast to the former case, one of the following two schemes is used:

a) Water level in settling basin or soil level at the extremity of filling slope is raised at a speed equal to the rate of soil sedimentation at the slurry outlet;

b) Constant water level or soil level at the edge of filling slope.

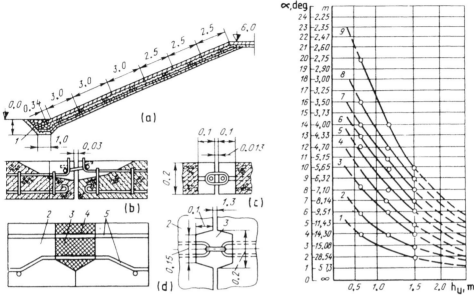

Figure 8.7. Slope revetment using reinforced-concrete slabs (left). (a) revetment cross-section; (b) articulated connection of slabs; (c) connection of slabs by cable-fastening rods; (d) joint cast in situ; (10 thrust prism; (2) slab; (3) asphalt concrete in joint; (4) weld; (5) reinforcement.

Figure 8.8. Graph for determination of slope cotangent m for inerodible upstream slope and variety of soils (right). (1) clay; (20 loess; (3) loam; (4) fine sand; (5) medium sand; (6) loam with pebbles and boulders; (7) coarse sand; (8) gravel; (9) pebbles; extrapolation by dashed lines.

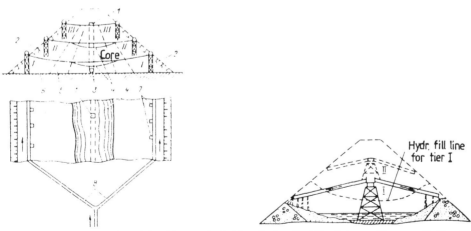

Figure 8.9. Scheme of double-side trestle-bridge hydraulic fill of dam with core (left). (1) pit; (2) dyke; (3) collecting pipe; (4) settling basin; (5) transition zone; (6) side prism; (7) outlet; (8) slurry line.

Figure 8.10. Hydraulic filling of dam from central slurry line (right). (I) trestle bridge for tier I fill; (II) layout for tier II fill.

The mean surface slope for subaqueous filling is given by

$$i_m = 0.15c^{1/3}(\frac{d_{50}}{h_*})^{1/6} \tag{8.8}$$

in which

c = initial consistency of slurry by mass (in fractions of unity),
h = depth of water corresponding to non-eroding velocity of clear water.

The profile of ground surface under subaqueous filling at any time, for constant water level, is given as:

$$y = \sqrt{\frac{q}{\gamma_{dry}}}\sqrt{1.33i_m}(1 - x_0)^{4/3}\sqrt{t} \tag{8.9}$$

in which

q = unit discharge of solid matter.

The coordinate $x=0$ corresponds to the outlet of slurry, while $y=0$ is the datum of water surface.

The profile of ground surface under subaeral filling for sandy soils becomes

$$y = h[1 - exp(-2\sqrt{\frac{w_m}{v_m}}\frac{\alpha}{h})] \tag{8.10}$$

in which

h = depth of water in reservoir,
v_m = mean velocity of water inflow into the reservoir,
w_m = mean settling velocity of soil.

Hydraulic-fill dams of non-uniform soil are characterized by sorting of different fractions of soil along the slurry line, which makes the dam body inhomogenenous at the outlet of the slurry line. Such dams have soils with the coefficient of uniformity $\eta > 3$–4, in which the quantity of fine particles ($d < 0.05$ mm) and the coarse fraction (from 0.5–1 to 50–60 mm) is sufficient but does not exceed 15–20% for soils of first to third category. If non-uniform soil is used for filling on both sides, then the central core of the dam can consist of finer fractions. It is merely fine particles which reach the settling basin and it is these particles that generate the dam core. The lateral parts of the dam are made of coarser particles with high permeability. A transient zone is generated between the core and the lateral parts. The evacuation of clear water takes place through the filtering lateral prisms, or partly through temporary discharge pits (trays) which are arranged at the foot of the pipeline. Water is sometimes evacuated by floating pumping facilities.

The width of the settling basin should be obviously contained in the range of the design width of the dam core, that is 15-20% of the dam width. The basin width is controlled by water discharge, so that the discharging channels must incorporate controlling gates. The spacing of the pits should not be smaller than the minimum width necessary for settling of particles coarser than 0.005 mm. Finer particles are evacuated together with water. In order to accelerate the hydraulic filling process one may enforce deposition of coarser fractions, for instance below 0.05 or 0.1 mm. In this case the required length of slurry travel in the settling basin should be as follows

$$l_0 = \frac{v_m}{w_m}h$$

In approximate computations one may assume the following mean settling velocities w_m (cm/s): 0.173 for sand with $d=0.05$ mm; 1 for $d=0.13$ mm; and 3 for $d=0.3$ mm.

A dam with core (Fig.8.9) can also be constructed from a central slurry line on a trestle bridge (this method was widely accepted in the fifties, but is now abandoned), with slurry discharge on both sides (Fig.8.10).

One-side filling (Fig.8.11) can produce a dam with screen. Slurry is discharged on the side of the rubble toe (a) and the filled inverse filter (b), and sorting of soil on the tailwater side brings about deposition of coarser particles in the layers c_1, c_2.... The finest particles, which generate the screen, settle on the upstream side about the small filtering dykes d_1 and d_2.

Mingechaur hydroelectric plant (Fig.8.12a) on the Kura River, in the region of 8-point seismic activities is the highest Soviet hydraulic-fill dam (80 m). This earth dam was constructed by lateral bridgeless filling using sand and gravel soil. Its profile is non-uniform, so that a core has been provided in the central part, where 70 to 80% of head is lost. The dam drainage consists of composite concrete blocks. Figure 8.12b shows the granulometric curves of the dam soil. The quarries from which sandy gravel was taken had lenses of clay, which was hydraulically transported into the dam core. The apparatus for measurements of pore pressure was used to estimate the consolidation of the core and the dam stability; it was deployed in the dam core. Laboratory tests have exposed a linear relationship between the coefficient of pore pressure and the coeficient of side pressure, which made possible determination of dam stability. Performance of dam has supported the design findings and has shown full operational reliability of the dam.

Geotechnical parameters of fill soil in various parts of the dam body are determined in the course of design process for first and second class of soil by a test filling procedure. Computational methods for determination of grain size distribution of the filled soil are also available (Grishin 1979).

Determination of the dam stability during filling is possible by computations, in which a rectilinear profile of the slip surface is assumed (Grishin 1979) under additional assumption that the lateral prisms be not interconnected. The above method may be accompanied by traditional computations using circular cylindric slip surfaces (Chapter 6), which is also used for assessment of the slope stability of hydraulically filled dam under operation. Dam stability under filling can be controlled not only by the rate of construction but also by grain distribution of soil. If coarser soil is filled into the lateral prisms, then the angle of repose becomes higher, which prevents sliding. The strength and filtering properties of soil founda-

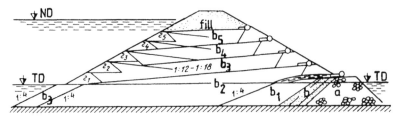

Figure 8.11. One-side hydraulic fill for dam with screen.

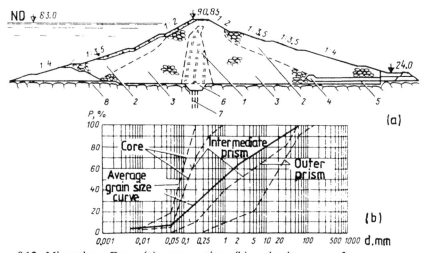

Figure 8.12. Mingechaur Dam. (a) cross-section; (b) grain size curves for separate zones of dam; (1) loam core; (2) external prism (20% of gravel, 40% of pebbles and 40% of sand); (3) transition zone (75% of sand and 25% of gravel and pebbles); (4) drainage; (5) drain outlet; (6) loam cut-off; (7) grout curtain; (8) rock.

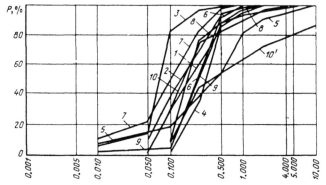

Figure 8.13. Average grain-size distribution of hydraulic-fill dam soil. (1)–(3) dykes of Rybinsk scheme (Volga, Sheksnin, No.46, respectively); (4) 22nd CPSU Assembly Dam on the Volga River; (5) Uglich; (6) Lenin Dam on the Volga River; (7) Tsimlansk; (8) Kakhovsk; (9) Gorkiy; (10) (10') Dubossar, central and lateral fills, respectively.

tion of the hydraulic fill dam should also be assessed. If weak clayey soil is found at the foot of dam, then the stability of slope computations should also account for possible unconsolidated condition of soil (Chapter 2).

Homogeneous hydraulic-fill dams are constructed from uniform sandy soil with high coefficients of uniformity ($\eta = 3$–14) and small contents (up to 10%) of fine fractions below 0.05 mm. Fractions are not separated in this case, and even in the opposite cases the spreading is very narrow. Under two-side filling, a central zone of finer material (compared with the remaining part of dam) is generated. One-side filling can also be used.

In the latter mode of filling, finer sand particles settle on the upstream side while coarser parcicles are deposited on the tailwater side. Sorting does not occur in practice of subaqueous filling, so that homogeneous dam is in the limit loose condition. The tailwater slope is generated in the course of filling, while the upstream slope under free settling of uniform sandy soil becomes very shallow ($m = 10$–11 for medium sand and $m = 30$–40 for fine sand); the so-called beach slopes are created. The necessary steepness of a slope can be secured by the use of seepage levees.

Slope stability of homogeneous dams is assessed by usual methods as in this case there is no core which could be considered a heavy liquid. Special attention must be drawn to seepage forces as the filtering water can exit directly on the slope and bring about abrasion.

In order to prevent washing of slopes or another infringement on its stability one uses special construction drainage, including forced drawdown of water by the use of well-point facilities. In this case the seepage field is changed, and soil becomes more consolidated. The consolidation effect can also be reached by other methods, in particular dynamic action on loose soil (Ivanov 1985).

Homogeneous hydraulic-fill dams are widely used in the Soviet Union; examples are hydroelectric power plants on the Volga and Dneper cascades. The grain composition of hydraulic-fill dams is depicted in Figure 8.13.

Tsimlansk hydraulic-engineering scheme also includes dams (Fig.8.14) which were hydraulically filled, and consists of fine and medium sand. The procedure was basically two-side filling with central settling basin. The lower part of the river dam was constructed by unilateral filling. The average density of soil in the lateral prisms was 1500 to 1750 kg/m^3, with the coefficient of permeability $k = 0.01$–0.001 cm/s. Different drainage was applied to various sections of the dam — closed drains and composite channels on the shallow terrace (Fig.8.14a), and drainage prism and inclined drainage in the river part (Fig.8.14b). During operation the depression curve found from piezometric measurements was practically close to the design line or below it along all sections. Slope revements of the dam are also shown.

Soviet engineeres have gained considerable experience from the construction of hydraulic-fill dams in the winter at negative temperatures (winter filling). Successful

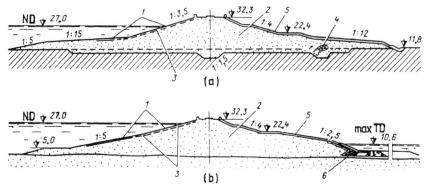

Figure 8.14. Hydraulic-fill dams of Tsimlansk scheme. (a) flood plain; (b) river; (1) armoured concrete slabs; (2) dam body sand; (3) three-layer sheet filter; (4) closed drainage; (5) rubble revetment; (6) drainage prism and sloping drainage.

winter filling requires continuous works at high rates. Dredgers and pumping stations (boosters) must be thermally insulated, and so must be primary pipelines in a number of cases. Open water areas (polynias) should be provided for suction dredgers. Field areas should be cleared from ice crust in thicknesses above 0.25 m; hot water must be added to slurry at temperatures below -15°C; special shelters and gates must be installed on slurry lines at lower locations so that water can be discharged during longer standby periods.

Hydraulic-fill dams have following advantages versus rockfill dams: high rates of construction, possible construction without the evacuation of water (dumping of soil into water); simple transportation and absence of heavy machinery for moving and consolidation of soil; lower labour consumption; generally lower cost per 1 m³. At the same time they have the following shortcomings: higher requirements put on quarry soil as to hydraulic transportation; higher consumption of electric energy for dredgers, boosters etc; considerable consumption of metals due to wear of pipelines, dredgers etc.

Quality control is necessary for any type of dam. As much as for rockfill dams, consolidation of soil (Section 1.7 and Chapter 3) is tested, along with its variation in time, and together with the grain size distribution, both in the quarry and in different parts of dam; slurry consistency is also one of the factors to be tested. Density of field soil can be controlled remotely with electronic devices.

8.4 EARTH-ROCKFILL DAMS

Design requirements. Earth-rockfill dams are the most widely used high dams made of earth materials. This wide application is a result of construction in mountainous regions, under complex engineering, geological, topographic, climatic and seismic conditions. In the Soviet Union, Charvak Dam, 168 m high and 13.9 million m³ in volume, and Nurek Dam, 300 m high and 58 million m³ by volume are already constructed while Rogun Dam, 333 m high and 71.1 million m³ in volume remains under construction. The dams constructed overseas include Mica in Canada (240 m high, 32.2 million m³), Oroville (USA, 224 m, 61 million m³) and Bennet in Canada, (183 m, 43.6 million m³).

Dams with screen or core are the commonest design of earth-rockfill dams. An intermediate version of these two types is a dam with sloping core. Selection of type depends on climatic and hydrologic conditions of region, complexity of engineering and geological structure of the foundation, composition of the hydraulic engineering scheme, availability of quarry for construction material, methods of construction etc. Since the basic posibility of lowering cost of construction consists in improved technology, one should try to adapt the dam design to rational construction technologies.

At the present stage, the dam design is based on analog analysis. This means that the design can be chosen by analogy to the existing structures and be checked by computations thereafter. The selected design includes dimensions of structural components, that is upstream and tailwater slope, thickness of water impoundment unit etc. In the selection of an analog structure one should prefer the design of recent years, in particular for the dams which were provided with complex measuring apparatus, as the analysis of data from field measurements sheds light on reliability of the structure and guarantees a successful analogy.

High earth-rockfill dams require design with a complex task of covering many topics; foundation, stability of slope material for transitions and inverse filters, settlement of dam under construction and operation, pore pressure in water confinement units etc. One should remember that high dams depend very much on the behaviour of coarse grained material under considerable load. It has been noted (Chapter 1) that high specific pressures induce nonlinear stress-strain relationships, which must be taken into account in the determination of stresses and strains in dams (Chapter 4).

Moreover, the relationship between the limit shear resistance and normal stress on shear plane is nonlinear in a wide range of loading, as emphasized earlier.

The complexity of design models for soil deformation and strength stipulates the necessity of sophisticated investigations to substantiate the design of high dams. Considerable heads on dams require particular attention paid to the structure of its underground contour. In dam design one must try to utilize in the dam body the excavated material and the cheapest soil. Practically any type of soil is presently used for dam construction. The design of dam must accomodate this variety of materials so as to secure reliable operation of the structure.

In the design of the water-confining components one must take the optimum soil out of those available about the site. One should prefer the soil with continuous grain size distribution, so called skeletal soil, which incorporates coarse and fine grains (for instance 35–50% of fine grains with diameters below 5 mm and 50–65% of coarse grains above 5 mm, up to the maximum size of 150–200 mm).

Rational economic sizes of seepage prevention components are designed not only by technological parameters but also by economic requirements, which embody the cost of soil filling in water-confining components, transitions and thrust prisms of dams (see Chapter 9).

Design of rock and earth dams. Dams with central core are most widely constructed both in the Soviet Union and abroad, which is due to a number of their advantages versus dams with screens. Dams with central core have the upstream slopes steeper than the dams with screens, which is associated with the higher strength of the upstream prism, compared with the strength of water-confining component. Hence a dam with core is more economic than a dam with screen.

In the construction of the central core one may use clayey soil with lower strength properties, compared with a soil filled in a screen. A dam with core has an easier and more reliable transition between the water-confining component and banks, particularly in the case of steep sides. For dams with core one may relax the conditions of deformability of coarse grained soil placed in the upstream prisms, as compared with requirements inposed on the soils for dams with screens. This is due to the fact that the settlement of thrust prisms of a dam affects cracking in the core to a lesser degree than the settlement of dam body, generating cracks in a screen.

Cores of earth-rockfill dams may be classified as thin and heavy. Thin cores are those in which the ratio of core width at the bottom b to the height of dam H_{dam} is below unity. Heavy cores are the ones in which the above ratio is higher than one. The ratio b/H_{dam} varies in wide range, from 0.2 to 2.4.

An example of a dam with thin central core is provided by El-Infiernillo Dam, the height of which is 148 m, volume — 5.7 million m^3 and the ratio b/H_{dam} is 0.2. The design of this dam is depicted in Figure 7.10. The width of core at bottom is 30 m, and about the base the core is widened to 45 m to provide a better transition between the core and the bedrock. The clayey soil is consolidated to mean density of 1590 kg/m^3, at $w_L = 49\%$ and $w_P = 24\%$. The material of the core is made more plastic by filling the soil with moisture content about 3.7% above the optimum. The upstream prisms consist of two regions, of which the internal one consists of fine stone with a diameter of 4.5 cm and the outer one is composed of rock. The upstream revetment is made of large size stone. Surface grouting, connected to a cement grout curtain has been arranged at the interface of the core and foundation; it consists of metamorphic siliceous conglomerates.

The size of the core is limited primarily due to the necessity of cracking prevention in view of inhomogeneous settlement of various parts of the dam. Seepage stability of core is usually secured with a high safety margin as the Soviet standards permit a critical mean hydraulic gradient in water-confining components up to 12, which corresponds to $b/H_{dam} \geq 0.08$. In thin cores with high hydraulic gradients considerable attention must be paid to the quality and dimensions of transition zones and to the construction of inverse filters.

Dams with medium cores are most widely used nowadays; their ratio d/H_{dam} is about 0.5 at mean $I_m=2$. An example is provided by Charvak Dam on the Chirchik River (Fig. 7.3). The seismicity of the construction site is 7 points. The dam is situated in a trough-shaped canyon, the bed and banks of which consist of Paleozoic limestones, separated by a system of structural cracks and tectonic faults filled with clayey material. Alluvial deposits up to 10 m thick are encountered in river beds. The design stipulated removal of the alluvium and cracked rock, surface grouting and construction of grout curtain to a depth of 68–70 m, conducted from a grouting gallery.

The core of Charvak Dam is made of loessial loam consisting primarily of fine grains; therefore transition zones (inverse filters) were necessary. They consist of two layers of sand and gravel aggregate, the first layer with a maximum diameter of 20–30 mm and the second layer with diameters up to 150–200 mm. The grain size distribution curves of the soil in Charvak Dam are depicted in Figure 8.15.

A reliable performance of the interface of dam core and dam foundation is secured if the core is made wider, as already noted for El-Infiernillo Dam; surface grouting and concrete plugs, in which the core is supported, are also adequate measures. The design of a concrete plug is shown for the cross-section of Charvak Dam (Fig.7.3).

A better connection of soil and foundation (bedrock or concrete plug) is provided if the grain size distribution of the lower region of the core is selected under the following conditions: the lower limit of the grain size distribution for the interface should secure a lower permeability of soil, equal to or smaller than permeability of the soil of core proper, while the upper limit is determined by the prevention of contact suffosion, as a function of rock cracking. As a rule, the soil at the interface has a water content by 2–4% above the optimum value. A recommended composition of the material od dam core and the interface of bedrock is shown in Figure 8.16 (Borovoy et al. 1973).

The earth-rockfill Nurek Dam on the Vakhsh River (Fig.I.1), 300 m high and 58 million m³ in volume, is the largest earth dam ever constructed in the world (Borovets 1961). Its central core has the ratio $b/H_{dam}=0.5$, and incorporates double-layer filters and pebble upstream prisms. The closing stage involved reinforcement of the upper part of the dam with the so-called antiseismic belts (Section 5.6). Unlike the core in Charvak Dam, the one used in Nurek Dam is made of safedobian sandy loam, which is a skeletal material (Fig.8.17). The original design anticipated the use of loessial loam from Langar deposit, which basically consists of clay and silt particles.

The following factors substantiate the preference given to soils with continuous grain size distributions (skeletal soil) for the use in bodies of water-confining components. First of all, skeletal soil permits high textural density of filling (1900–2200 kg/m³) and sufficient impermeability ($k=10^{-4}$–10^{-9} cm/s) provided the water content is selected adequately. Secondly, the soil possesses the capacity of crack healing; cracks can arise due to non-uniform settlement of dam components.

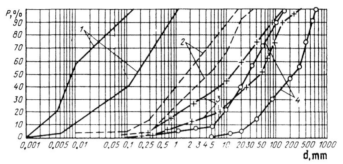

Figure 8.15. Grain-size curves of Charvak Dam soils. (1) core loam; (2) first filter layer; (3) second filter layer; (4) stone revetment.

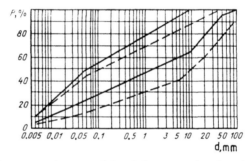

Figure 8.16. Grain-size curves for materials of dam core (- - -) and transition to bedrock (—). The coefficient of permeability is $A \cdot 10^{-4}$ cm/s for core soil and $10^{-5} - 10^{-7}$ for the transition.

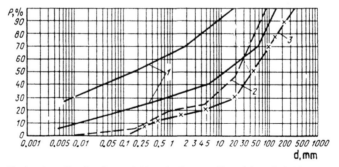

Figure 8.17. Grain-size distribution of Nurek Dam soils. (1) safedobian sandy loam; (2) transition layer; (3) thrust prisms of gravel and pebbles.

Upstream prisms of Nurek Dam consist of gravel and pebble, in contrast to the Charvak Dam made of stone fill. For dams higher than 150 m it is presently recommended that gravel and pebbles be used for upstream prisms, as they are less susceptible to deformations, compared with rubble fills. The considerable deformability of rubble fills is attributed to the failure at point contacts under high pressures.

Dams with inclined cores are also constructed. It is considered that an inclined core reacts better with the tailwater prism, and that cracks on the upstream side are less likely (Rasskazov & Sysoyev 1982). Examples of dams with inclined cores are provided by Mica Dam (Canada), 240 m high, Figure 8.19, Oroville Dam (224 m), Cougar Dam (158 m, Figure 8.19), Furnas Dam in Brasil (127 m), Miboro Dam in Japan (130 m) and others.

The world's highest earth-rockfill dam with inclined core, Rogun Dam on the Vakhsh River, 335 m high , is now under construction in the USSR (Osadchiy & Bakhtiyarov 1975). The experience of the design and construction of Nurek Dam has been utilized in the design of Rogun Dam. The core of Rogun Dam is designed of enriched pebble and loam mixture, while the transition filters and upstream prisms are to be made of cobbles and stone overburden (Fig.8.20).

The construction site is characterized by complex geologic and tectonic conditions — large Ionakhsh fault, filled with a salt stratum, is very close to the site, and secondary faults are also present. A number of measures aimed at protecting the salt stratum from leaching are foreseen, and the tectonic activity of the faults has also been analysed. The design seismicity of the site is 9 points.

Earth-rockfill dams with screen are usually lower. Their upstream slope is shallower than in dams with core (from 1:2 to 1:3). The thickness of the screen is usually smaller than the core thickness, and the ratio b/H_{dam} (in which b = thickness of screen along normal line) varies from 0.1 to 0.5. Since screens are more sensitive to deformations and are therefore more susceptible to cracking than cores, stringent requirements are imposed on soil compaction. The screens on the tailwater sides are protected with inverse filters while those on the upstream side have filters and cover layers. The cover layer can be very thick, whereupon the dam with screen differs only slightly from a dam with an inclined core.

The transition between screen and dam foundation is provided by analogy to core design, that is with a concrete plug or a cut-off, direct filling on grouted rock etc. Screen may also be connected to the upstream puddle blanket, which is constructed to elongate the seepage path.

One of the highest dams with screen was constructed in the USA (Brownley, 122 m, Fig.8.21). The thin screen is made of loam with plasticity index of 8.5%, optimum water content of 19.6%, and with 90% of fractions below 0.074 mm.

The strength properties of the loam are: $\varphi = 17°$, $c = 0.18$–0.24 MPa, $k_{mean} = 2 \cdot 10^{-5}$ cm/s. The transitions between the screen and the upstream rock prism consist of three layers of different sizes of material. Strongly permeable alluvial deposits, cut through by two trenches down to the basalt rock (with thin laminae of tuff) occur

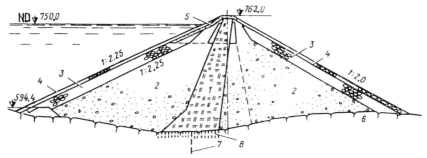

Figure 8.18. Mica Dam (Canada). (1) moraine core (30–40% of fractions ≪ 0.07 mm); (2) thrust prisms of gravel and pebbles; (3) overfill of gravel or stone; (4) boulder or large stone revetment; (5) zone of compacted gravel, pebbles and boulders; (6) cracking granitogneiss with interlayers of clayey shales; (7) seepage prevention curtain, down to 90 m; (8) area grouting.

at the dam foundation down to the depth of 30–33 m. A screen connected with the grout curtain is arranged in the upstream trench. A fairly economic profile, with upstream slope of 1:2 and downstream slope of 1:1.4, has been provided.

Viluy Dam with a 74.5 m high screen (Fig.8.22a) was constructed on the Viluy River in the USSR, where permafrost persists. Strong diabases overlain by Quaternary deposits up to 3 m thick are found at the foundation of the dam. The screen of the dam consists of cobbles and rotten stone in loam from quarries. The transition zones and the filter are made of disintegrated material, while the upstream prism consists of unsorted rubble. The transition between the screen and the foundation is provided by a concrete slab and a cut-off, which incorporates a gallery; a double-row grout curtain is arranged about the foundation of the cut-off.

Materials for rock and earth dams. As mentioned above, the earth-rockfill dams can incorporate various soil materials, but location of these materials in dam must be selected adequately in the design so as to make proper use of the deformation, strength and seepage properties of soil in right constituents of the structure.

Cores and screens of rock and earth dams can include loose and cohesive soil, from sand to fat clay. Permeability of soil determines the size of water-confining components. If clay is used for cores and screens, the latter can be very thin, with small ratio of b/H_{dam}, ranging from 0.1 to 0.5. Overall dimensions of a dam increase if sand is used. Sand poses problems due to its high permeability, while the use of clay is difficult because of filling and compaction problems. Therefore sandy loam and loam are most suitable for water-confining units as they possess low permeability and good plasticity (rolling capacity).

Skeletal materials, which are less deformable than clayey soil and have better suffosion prevention properties are preferred for water-confining components of dams. Such soils include cohesive soil with coarse additions, cobbles, gravel and pebble and above 50% of fine fractions (diameter below 2 mm).

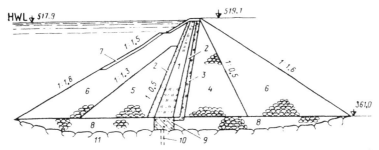

Figure 8.19. Cougar Dam (USA). (1) clayey core, with screenings of fractions $d > 15$ mm; (2) first filter layer of unsorted sandy gravel; (3) second filter layer of pebbles; (4) rubble of strong rock (d_{max}=4.6 cm); (5) stone with up to 25% of weathered blocks; (6) rubble of strong rock (d_{max}=61 cm); (7) large size stone; (8) strong basalt fill in old river channel; (9) concrete plug; (10) seepage prevention curtain; (11) weathered and cracked basalts and tuffs.

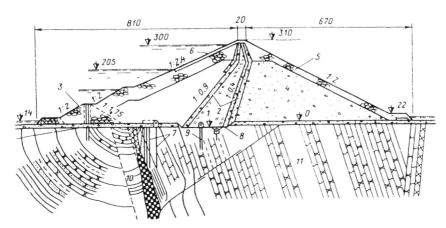

Figure 8.20. Rogun Dam (design drawing). (1) core; (2) transitions (filters); (3) upper construction cofferdam with injection core; (4) thrust prisms; (5) stone revetment; (6) reinforced-concrete revetment; (7) leaching prevention measures; (8) area grouting and concrete sputtering at core base; (9) deep grout curtain; (10) rock salt stratum; (11) interlayers of sandstones and aleurolites.

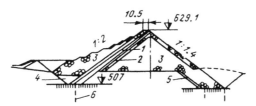

Figure 8.21. Brownley Dam (USA). (1) screen; (2) transition zones; (3) thrust prisms; (4) (5) upstream and downstream trenches; (6) curtain.

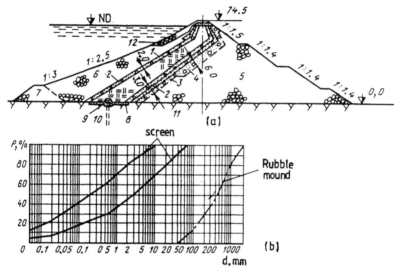

Figure 8.22. Viluy Dam No.1. (a) dam cross-section; (b) grain size distribution of screen and rubble; (1) screen of druss-cobble loam; (2) (3) transition zones and sandy gravel filter ($d=0$–40 mm; $D=0$–150 mm); (4) fine-stone adjustment layer; (5) rubble revetment of regular stone (diabase); (6) overburden of regular stone; (7) toe fill; (8) concrete slab; (9) gallery; (10) seepage prevention curtain; (11) diabases; (12) large-size stone cover.

Special preprocessed mixtures of enriched soil are often used for water-confining components of earth-rockfill dams, so as to improve rolling and reduce deformation of soil. For instance, the core of Oroville Dam has been made of rock waste with sorted clayey silty sand soil having the maximum fraction size of 228 mm, with technologically unacceptable high contents of grains above 4.7 mm. The material was therefore sieved prior to filling into the core of the dam, the fractions above 76 mm were removed, and the material obtained had less than 50% of grains above 4.7 mm. The coarse material sorted out was used for the construction of transitions. In Cougar Dam, a highly non-uniform (as to grain size distribution) material was used with a large amount of stone above 15 cm, so that the quarry product had to be processed prior to utilization.

The construction technology of earth-rockfill dams determines the reliability of their operation; the quality of filling controls to a considerable degree the performance of dam structure.

8.5 EARTH DAMS WITH NONSOIL SEEPAGE PREVENTION COMPONENTS (ROCKFILL DAMS)

Nonsoil seepage prevention constituents are used if suitable soil for water confining components is unavailable at site or if this is dictated by construction technology. These types of dams have been constructed since long, but it is recently that asphalt-concrete, metallic, reinforced-concrete or polymer seepage prevention components, including those for high dams, become widespread.

Asphalt-concrete screens and diaphragms are used extensively in the western practice of dam construction and are drawing more and more attention in the Soviet Union (for instance, Boguchan Dam with asphalt-concrete diaphragm now under construction).

Asphalt concrete is made of a mixture of bitumen, powder filler, sand and gravel. Depending on the proportions of these components, asphalt-concrete possesses different deformability, strength and density. In an average asphalt concrete, the content of bitumen varies from 7 to 15%, powder filler from 10 to 25%, sand from 20 to 50% and gravel from 30 to 50%.

Dams with asphalt-concrete screens have fairly steep slopes, depending on the material of dam body; this in particular holds for dams made of rubble. The slope factor (cotangent) reaches 0.7 in a number of cases. The hazard of slip failure or heave is quite essential for steep slopes, as proved by failure of 72-m El-Grib Dam (Algeria).

Screens are made in single or double layers, or a special subgrade of cobbles or rubble, not thinner than 0.3 m, is applied. The material of a dam body beneath the asphalt-concrete screen should be compacted to avoid substantial deformation, which would bring about failure of the screen. In double-layer screens, a certain transition drainage of porous asphalt-concrete is placed between two layers. In most cases there is no outer protection of the asphalt-concrete screen.

Loading due to freezing ice upon oscillation of the upstream water level is most hazardous to screens. In hot-climate regions, the creep of asphalt-concrete on the outer side is prevented by placement of concrete or reinforced-concrete slabs having a thickness of 8–15 cm. The design of asphalt-concrete screens in a number of dams is shown in Figure 8.23.

Figure 8.24 shows different versions of transition between an asphalt-concrete screen and a concrete slab at the foundation, along with a drainage gallery intended for collection and removal of water from the layer of porous asphalt-concrete designed for the drainage of the screen.

Asphalt-concrete screens are constructed with special machinery for filling and compaction of bituminous mixture directly on dam slopes. Selection of the mixture for the asphalt-concrete and the technology of construction are described in detail by Kasatkin et al. (1970). The composition of asphalt-concrete is controlled by climatic conditions of site, as rheologic properties (creep, yield) of bituminous material largely depend on temperature. For high mean yearly temperatures on site one must use more rigid mixtures with lower contents of bitumen.

Strain (shear modulus G_o) and strength (φ and c) properties of materials are primarily controlled by the proportions of bitumen and coarse filler, and may be found from the nomographs in Figure 8.25.

Asphalt-concrete diaphragms are sometimes preferred to screens as they are simpler and not susceptible to external effects. Reliable operation of diaphragms depends on the absence of bending and good fitness to deformation of the dam body.

The thickness of a diaphragm depends on the height of dam, and usually increases from dam crest towards dam foot. From the present practice it follows that the thickness of a diaphragm at dam foot should be 2–3% of the dam height. The minimum thickness at dam crest is 0.25–0.3 m.

Good transition between diaphragm and dam body made of coarse soil is secured by construction of transition zones made of pebbles or sand and pebble material, which prevent removal of asphalt-concrete into pores of upstream prisms. In deformation design of asphalt-concrete diaphragms (and screens) one must take into account that they are permeable and subject to total hydrostatic pressure of water on the upstream side. Deformations of diaphragms or screens are controlled by the flexibility of dam body and can be computed by the same design methods and schemes which are used for earth dams with water-confining components of earth materials (finite element method etc). In simplified computations the screen or diaphragm can be treated as a beam on elastic foundation.

Earth dams with metallic seepage prevention components. They have a number of advantages over dams with soil seepage prevention units such as possible construction in regions without traditional construction materials for soil cores, screens and inverse filters; replacement of troublesome high-quality filling of cores and screens, or transition zones, by simpler technology of metal assembly; reduced requirements on quality of construction in the dam body; absence of seasonal works in the construction of seepage prevention components under conditions of extremely low temperatures in the Far North or high humidity during rainy seasons of tropical and subtropical zones; possible passage of flood over structure under construction etc. Metallic screens and diaphragms are very strong and watertight.

More than 60 dams of the above type have been constructed in the world. The progress in this area depends on the developments in the skill of welding and corrosion protection. One must note that the construction of metalllic screens is generally more expensive than the construction of diaphragms. However, screens are better accessible for inspection and repairs. At the same time, the construction of screens requires a high-quality overburden, which should be little deformable in the course of reservoir filling and dam loading over the metallic screen; this brings about protection from considerable deformations and stresses in the screen. The range of soil materials utilized in the construction of diaphragms becomes much wider, from rubble, cobbles and gravel to fine sand. The material of screens and diaphragms is a special steel with good corrosion resistance properties. Ordinary steel requires special measures to satisfy the requirements of corrosion resistance.

Salazar Dam with metallic screen (Fig.8.26) was constructed in Portugal in 1941. The height of the dam is 63 m, for 192-m length at crest, the upstream slope of 1:1.25 and the tailwater slope of 1:1.4. The dam incorporates a metallic screen having a thickness from 8 mm at the base to 6.3 mm at the crest. The screen consists of a metallic sheet 7.5×2.5 m on a concrete subgrade to which it is anchored. Flexibility of the screen is secured by special compensators arranged along vertical joints, every 7.5 m, and at the foundation. Corrosion is reduced by bitumen filling

Site	Design of asphalt-concrete screen	Composition of screen
Kerngrund Dam (GDR)	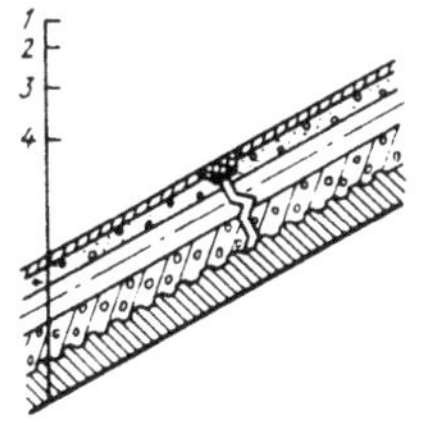	(1) textile asphalt-concrete; (2) concrete; (3) slab with drainage openings; (4) lean cement mortar
Upper Por"abka-"Zar (Poland)	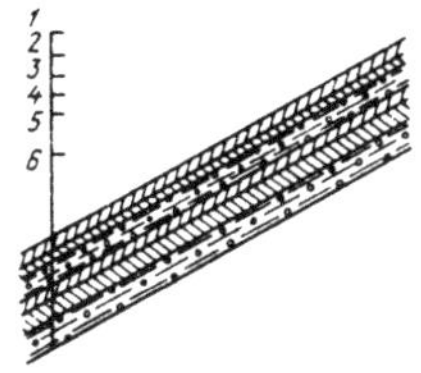	(1) (2) two 4.5-cm impervious layers; (3) 10-cm drainage layer; (4) impervious asphalt concrete; (5) underlying asphalt-concrete layer; (6) 15-cm drainage layer
Upper Cerny Vag	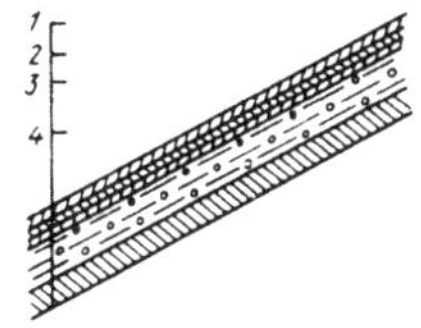	(1) two 4-cm layers of firm asphalt concrete; (2) porous asphalt concrete, 8 cm; (3) 40-cm drainage layer; (4) 15-cm adjustment layer
Radojna Dam at Bistrica (Yugoslavia)	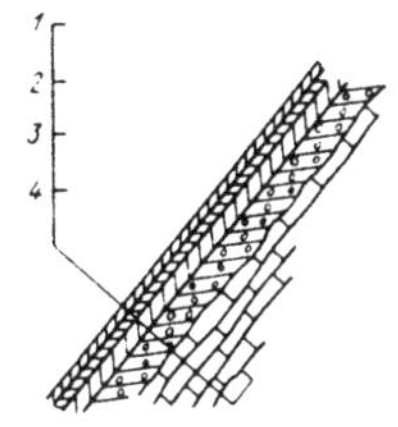	(1) two layers of firm asphalt concrete, 9 cm; (2) 15-cm porous concrete; (3) 30–60-cm concrete cushion; (4) dry masonry
Dobczyce Dam (Poland)	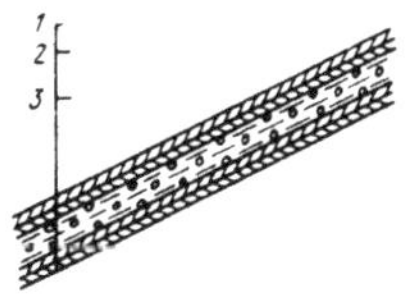	(1) two 4-cm layers of firm asphalt concrete; (2) drainage layer; (3) two 4-cm layers of firm asphalt concrete
Valia de Pest (Rumania)	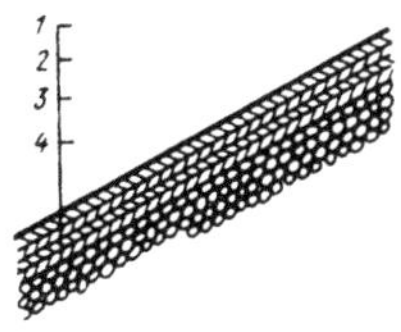	(1) bitumen lining; (2) two layers of firm asphalt concrete; (3) two layers of porous asphalt concrete; (4) cobble layer.

Figure 8.23. Foreign design of asphalt-concrete screens (Glebov 1985).

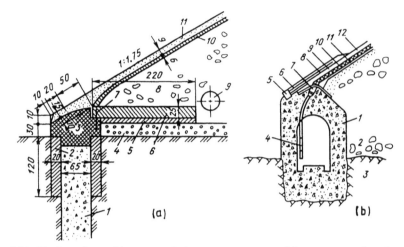

Figure 8.24. Design of transitions at asphalt-concrete screen. (a) concrete wall at foundation; (1) concrete wall; (2) guiding wall; (3) concrete thrust block; (4) adjustment fill; (5) concrete subgrade; (6) concrete slab; (7) metallic cut-off; (8) drainage layer; (9) drainage pipe; (10) bituminous drainage layer; (11) asphalt concrete; (b) drainage gallery; (1) concrete gallery; (2) stone revetment; (3) rock; (4) drainage pipe; (5) concrete beam; (6) two layers of copper foil; (7) 1-m sheet of copper foil at joints in concrete; (8) 4-cm asphalt-concrete cover layer; (9) two 4-cm asphalt-concrete layers; (10) 10-cm drainage layer of asphalt; (11) 4-cm asphalt-concrete layer; (12) bituminous adjustment layer.

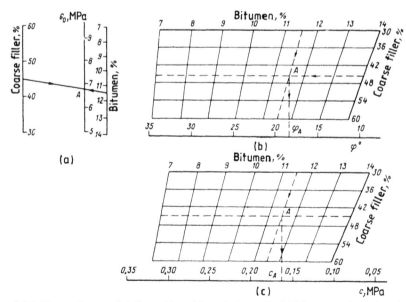

Figure 8.25. Dependence of deformation (a) and strength (b)(c) properties of asphalt concrete on contents of bitumen and coarse filler at 20^{0}C, after Chukin.

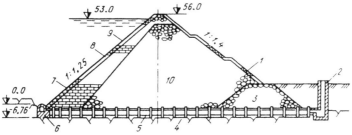

Figure 8.26. Salazar Dam (Portugal). Coarse stone revetment; (2) pump station; (3) thrust berm; (4) bedrock; (5) drainage gallery; (6) concrete cut-off; (7) masonry with mortar; (8) concrete subgrade; (9) metallic screen; (10) rubble mound.

of the screen on both sides. At the dam foundation the screen is connected with a concrete cut-off.

Serebransk Dam (GES-2) in the USSR, with a metallic diaphragm (Fig.8.27) is 64 m high, 1820 long (at crest), its slopes being 1 : 2. The thrust prisms are made of sand and gravel. The steel piled diaphragm is made of channel profile ShK 1, 10 mm thick, and is connected with the concrete cut-off protruding by 2–2.5 m into bedrock; a grout (cementation) curtain reaching the depth of 15 m is arranged below the cut-off. In the part of the bedrock overlain by sand and gravel (cross-section $C–C$) the diaphragm is directly connected by bolts with the piled curtain stretching downward to a depth of 28 m. The diaphragm is freely connected to the cut-off — its lower part may move at no constraints in the horizontal direction (over a bituminous mattress), with prevents from generation of high cutting forces in the diaphragm upon impoundment of reservoir.

Figures 8.28 and 8.29 illustrate the connection of dam diaphragm and bedrock. Examples of dams with metallic seepage prevention components are given by Anonymous (1976b).

Concrete and reinforced concrete screens and diaphragms. They are used extensively as seepage prevention components of earth dams. Dams with screens are most widespread as dams with reinforced-concrete diaphgrams are less reliable because the diaphragm may fail upon operation.

Concrete and reinforced-concrete screens are accessible for inspection and repairs. They may be of three major types: rigid monolithic, vulnerable to settlement deformation; sliding semi-rigid, subject to lity of deformation under settlement; and flexible, which may accommodate considerable deformations in the dam body.

Rigid screens are cut by thermal joints and used for low dams founded in bedrock. Semi-rigid sliding screens are cut every 4–5 m into sections by horizontal and vertical thermal-settlement joints (Fig.8.30). Wider spacing can be employed in the upper parts of a dam, while lower spacing is required at the foundation. Joints must secure sufficient mobility and tightness of screen under deformations. The

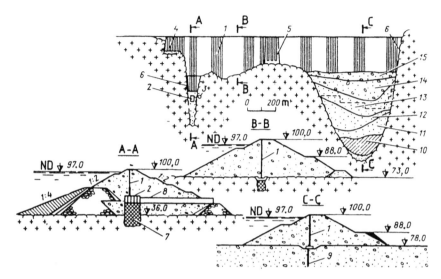

Figure 8.27. Serebransk Dam (GES-2). (1) metallic diaphragm; (2) grouting hole; (3) pipeline for passage of construction discharge; (4) spillway; (5) intake; (6) transition walls; (7) grout curtain; (8) transportation gallery; (9) sheet piling; (10) loam; (11) sandy loam; (12) medium sand; (13) coarse sand; (14) gravelly sand; (15) gravel.

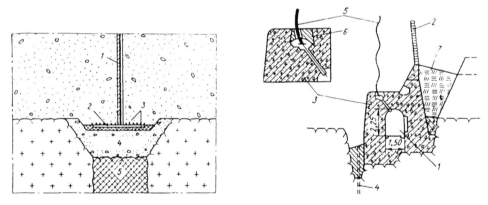

Figure 8.28. Transition between metallic diaphragm and bedrock of Serebransk Dam (left). (1) diaphragm; (2) thrust sheet; (3) bituminous mats; (4) cut-off; (5) grout curtain.

Figure 8.29. Transition between metallic diaphragm and bedrock of Berev Dam (right). (1) gallery; (2) filter; (3) concrete cut-off; (4) grout curtain; (5) diaphragm; (6) asphalt; (7) plastic clay.

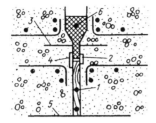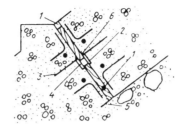

Figure 8.30. Transition joints of screen slabs. (1) plank; (2) plug; (3) metal platen; (4) asphalt; (5) tar oil; (6) plastic.

design of joints can be quite diversified, from metallic U-plates and bituminous cut-offs to beams etc. Screens of this type are arranged on a subgrade made of adequate material, filled with bitumen or covered with a bituminous material, and anchored in the dam body, so that sliding of reinforced-concrete slabs over the slope is prevented. Particular attention must be paid to the connection of screen and the foundation; a cut-off is usually arranged at the foundation. The design of the transition must secure free deformation of the screen.

Flexible (layered) screens consist of reinforced-concrete slabs in sizes up to 10 m, arranged in several layers with overlapping joints (Fig.8.30). Bituminous interlayers are arranged between the layers of slabs. The latter are anchored in their subgrade. Flexible screens are well adapted for considerable deformation of dam body, including seismic effects.

Dams with reinforced-concrete screens above 100 m high have already been constructed abroad — at Paradela (Portugal; 110 m), Salt Springs (USA; 101 m) etc. The reinforced-concrete screens in low and medium-size dams are however most widespread. High quality concrete (not below grade 200), with a permeability better than V8 must be used for the screens. The reinforcement ratio of the slabs varies from 0.5 to 1%. The thickness of the screen in its upper part is 20–30 cm and increases towards the foundation, reaching roughly 1% of dam height. If a special subgrade is arranged beneath the screen, the thickness of the latter can be made variable over height, as implied by the empirical formula $\delta = a + b \cdot H$, in which a = minimum thickness of screen at crest, and b = coefficient varying from 0.03 to 0.006, depending on quality of subgrade; H = head at given point, in metres.

Shown in Figure 8.32 is Urtotokoy Dam (USSR) with reinforced concrete double-layer screen made of 7.5×7.5 m slabs. Wool material impregnated with bitumen is placed between the layers. The layer under the screen consists of 10-cm concrete subgrade arranged on a 3-m gravel layer. Practically impermeable feldspar and coarse porphyries make up the dam foundation. Coarse stone is placed as masonry behind the cut-off at the dam foundation.

Dams with reinforced-concrete diaphragms are rather rare nowadays. As mentioned above, diaphragms are subject to deformations upon filling of reservoir, mostly due to deformations of the tailwater prism. Moreover, one encounters dif-

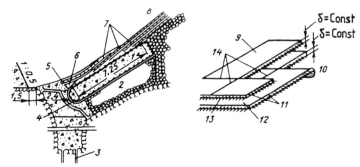

Figure 8.31. Layered screen. (1) reinforced-concrete thrust slab; (2) fine round stone; (3) grouting holes up to 16.5 m deep; (4) asphalt layer; (5) three asphalt layers with canvas interlayers; (6) asphalt concrete; (7) three slabs up to 10 cm thick; (8) layered screen; (9) upper slab; (10) lower slab; (11) continuous weld; (12) surface covered by a thin asphalt layer; (13) joints with copper platens; (14) asbestos-asphalt joints.

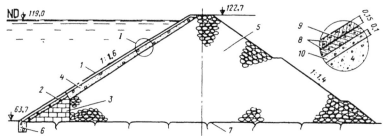

Figure 8.32. Urtotokoy Dam with reinforced-concrete (RC) screen. (1) flexible 0.4-m RC screen; (2) rigid 0.5-m RC screen; (3) dry masonry of coarse stone; (4) gravel subgrade; (5) rubble mound; (6) concrete cut-off; (7) porphyry; (8) 0.4–0.5-m screen; (9) sackcloth; (10) 10-cm concrete subgrade.

ficulties during construction due to the fact that the dumping of dam prisms must be coordinated with the construction of a diaphragm, and also due to complex compaction requirements for prisms at their interface with the diaphragm etc.

Shown in Figure 8.33 is the design of 143-m high Ayldon Dam in Australia. As a result of the sliding of the upstream slope, the diaphragm was deflected by 1.42 m and failed. The dam profile after repair is shown in the drawing. The thickness of the diaphragm at crest is 0.6 m, versus 1.5 m at the dam foundation. The diaphragm is armoured in two rows of bars, 12 mm in diameter. A better transition between the dam and the foundation is provided due to the fact that the diaphragm cuts into clayey deposits and is connected to the bedrock.

Screens and diaphragms of plastics and polymer materials. These materials are prospective construction materials for seepage prevention components of earth dams such as screens, diaphragms and upstream puddle blankets. In earth dams

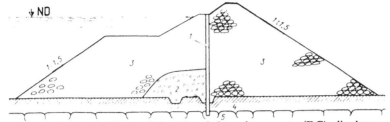

Figure 8.33. Ayldon Dam (Australia) with reinforced-concrete (RC) diaphragm. (1) RC diaphragm; (2) clay; (3) rubble mound; (4) clayey soil; (5) bedrock.

which confine reservoirs with highly toxic wastewater, screens and diaphragm are connected with an upstream puddle blanket, which sometimes covers completely the bottom of a reservoir and prevents percolation of harmful solutions into ground. Since polymer materials possess a relatively low durability, screens and diaphragms of the above type can be applied in temporary structures, for instance cofferdams. The durability and lifetime of polymers measured in decades can be reached by the use of a special stabilizing agents (anoxidants). In the construction of screens of polymer materials one may inspect the units and repair them, if necessary. At the same time this technology is more complex due to problems in the protection from atmospheric and solar effects at high levels, in the zone from dam crest to the dead volume datum. The failure of a screen on the upstream side is prevented by construction of a special protective layer. Another special layer to prevent mechanical abrasion is arranged beneath the screen inside the dam body. The thickness of the protective layer should be at least 0.5 m. In a number of dams (for instance, Dobcina and Landstein in Czechoslovakia), a polyethylen screen as a 0.1 mm foil is arranged between special reinforced-concrete profile slabs. Since the coefficient of friction for soil on polymer is 0.1–0.4, one recommends the slopes below screen not steeper than 1 : 3 (Guidelines 1983). The grouting works are simpler in the case of a plastic or polymer diaphragm.

Plastic foils, polyethylene foils or PVC and other materials for counterseepage components of dams have usually thicknesses from 0.2 to 2 mm. The thickness of the foil depends on the acting head and the grain size distribution of the soil in contact with the foil. Under unfavourable conditions, for instance if the underlying soil is coarse and its grains are sharp-edged, one recommends certain protective interlayers between the diaphragm and soil; rubberoid, glass cloth, porolone and other materials may be used for this purpose.

The transition between the foil and the flanks and foundation of a dam is of utmost importance. The contact seepage between foil elements and foundation is prevented if special measures are foreseen such as cut-offs filled with plastics or concrete. Special compensation compounds are needed for the screen-cutoff transition in deformable dam bodies. Figure 8.34 shows the usage of foil.

The screen thickness at various dam levels is a primary design factor for seepage

Site with screen	Design of polymer-foil screen	Composition of screen
River lining (GDR)	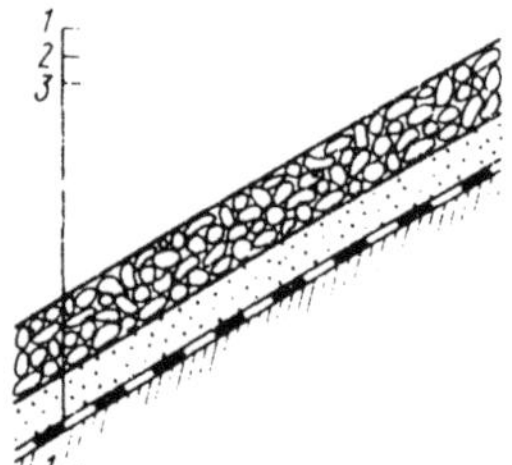	(1) 30-cm cobbles; (2)25-cm sand; (3) 1.1–1.5-mm foil
Dam slopes (GDR)	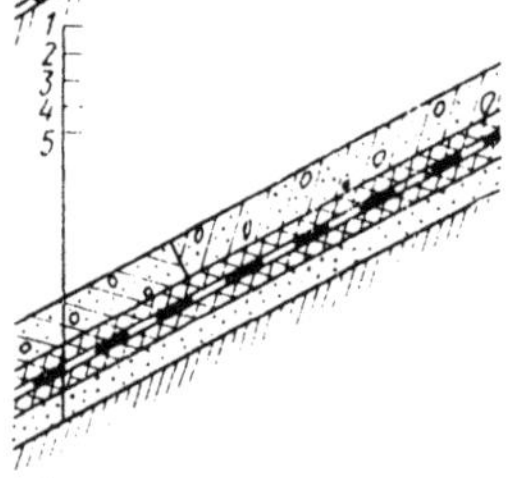	(1) RC slabs: (2) 2-cm polystyrene foam; (3) 1.5-mm foil; (4) 2-cm poly-styrene foam; (5) base layer
Dobcina Dam (CSSR)	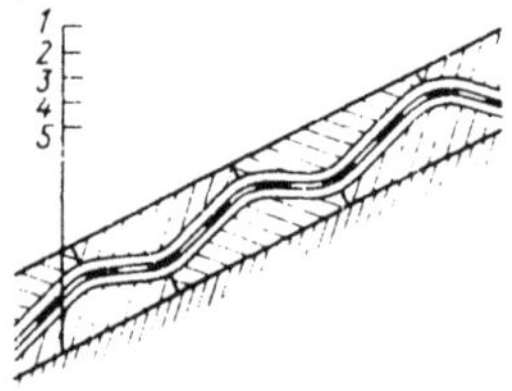	(1) corrugated RC slabs; (2) interlayer; (3) 0.9–0.11-mm foil; (4) interlayer; (5) corrugated slabs
Caplina Dam	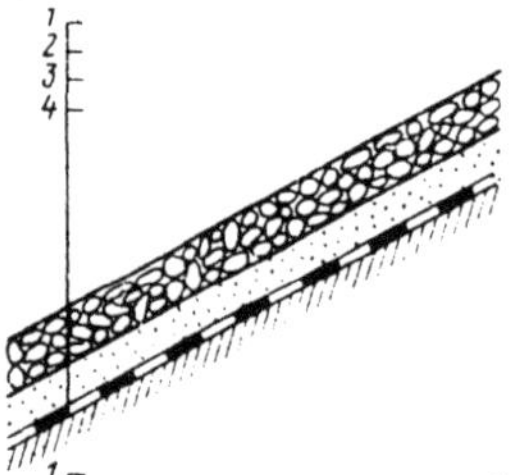	(1) 20-cm gravel; (2) 20-cm sand; (3) 0.8–1.5-mm foil; (4) consolidated base
Settling basin Moldavian SSR	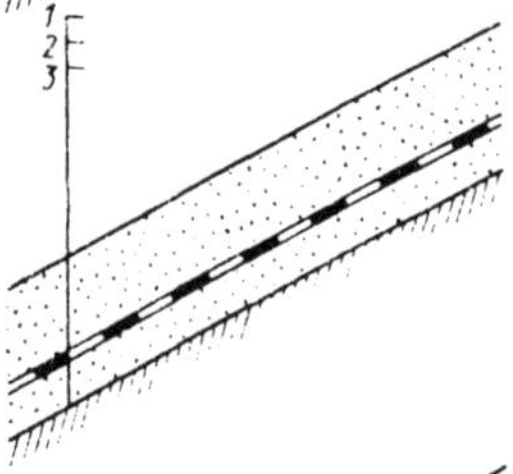	(1) 50-cm sand; (2) 0.2-mm foil 30-cm sand
Ash disposal Magadan plant (USSR)	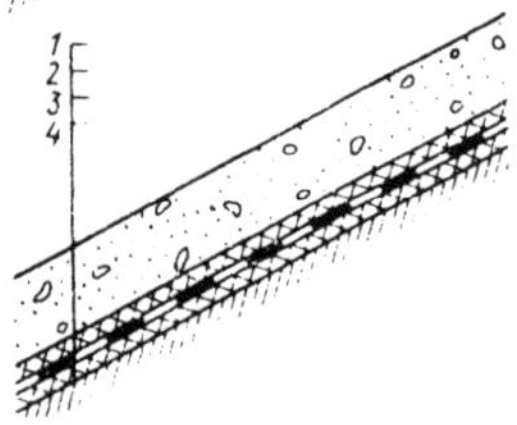	(1) sand and gravel; (2) rubberoid; (3) 0.2-mm foil; (4) rubberoid

Figure 8.34. Design of seepage prevention foil screens.

confining foil. For heads above 10 m and underlying soil grains above 2 mm, the foil thickness of screen or diaphragm must be thoroughly computed.

The film thickness under condition of its continuity (non-failure) reads

$$\delta = 0.1 d_{gr} \frac{q}{k_{int}} \qquad (8.11)$$

in which

d_{gr} = minimum diameter of the coarsest fraction behind foil,
q = load on foil,
k_{int} = effectiveness factor for additional protective interlayers; taken as 2–5 for rubberoid, 2–3 for glass cloth, 2–10 mm for porolon, 1.5–0.2 mm for polyethylene foil, and 1 in the absence of interlayers.

The thickness of the foil under conditions of tensile stress due to hydrostatic pressure reads

$$\delta = 0.135 d_{gr} q \sqrt{\frac{E}{[\sigma]^3}} \qquad (8.12)$$

in which

q = hydrostatic pressure, MPa,
E = modulus of elasticity of foil (120 MPa),
$[\sigma]$ = permissible tensile stress (1 MPa).

If computations by Equation 8.12 yield $\delta > \frac{1}{3} d_{gr}$ then one has

$$\delta = 0.586 d_{gr} \sqrt{\frac{q}{[\sigma]}} \qquad (8.13)$$

In the case $\delta < \frac{1}{3} d_{gr}$ one assumes $\delta = \frac{1}{3} d_{gr}$. The design thickness of foil is rounded up to the nearest standard thickness conforming to the Soviet standard (GOST).

A seepage prevention foil screen was used for 61-m Terzaghi Dam in Canada (Fig.8.35). Upon construction of a clayey screen it was covered with a 0.76-mm PVC foil in the lower one-third of the central part of the dam so as to prevent possible seepage in cracks of the screen due to deformation of the dam body. As seen in the drawing, the dam body is inhomogeneous due to diversified grain size distribution of its soils and differences in permeability.

A foil diaphragm has also been employed in the Soviet Union, in the upper part of Atbashin Dam (Fig.I.2). The diaphragm is made of 0.6-mm foil subject to a head of 41 m. It is connected with a grout diaphragm, 22 m deep and 20 m wide, incorporating 7 rows of holes at 3.5-m spacing. The polyethylene foil, 0.6 mm thick, is connected with the grout diaphragm via a grouting gallery. At the connection the foil is contained between elements anchored in the ceiling of the gallery; the foil itself is protected with rubber interlayers. At the sides of the reservoir the diaphragm is attached to concrete slabs. In the vertical direction, the diaphragm is protected with cloth and sand layers.

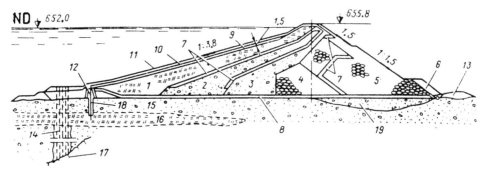

Figure 8.35. Terzaghi Dam (Canada) with PVC foil screen. (1) boulder clay with stones, $k=5\cdot10^{-5}$ cm/s; (2) coarse grained material with a large amount of silty particles, $k=5\cdot10^{-4}$ cm/s; (3) coarse grained material with a large amount of sand, $k=5\cdot10^{-3}$ cm/s; (4) coarse grained material, $k=5\cdot10^{-2}$ cm/s; (5) rubble mound; (6) dumped material; (7) transition zones; (8) filter; (9) clay layer; (10) 0.76-mm PVC foil; (11) cover layer of clay, sand and stone; (12) old dam; (13) tailwater cofferdam; (14) seepage prevention curtain; (15) sandy gravel deposits; (16) clay layer; (17) rock; (18) sheet piling; (19) adjustment backfill.

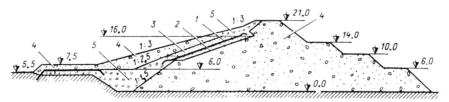

Figure 8.36. Ust'-Khantay cofferdam with PVC screen. (1) 0.2-mm PVC foil screen; (2) 0.2–0.3-m layer of fine sand; (3) 1-m layer of fine sand; (4) gravelly sand; (5) moraine soil.

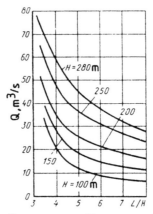

Figure 8.37. Seepage discharge Q versus head H and ratio L/H (Korchevskiy 1962).

A polyethylene screen has been constructed in the cofferdam of the Ust- Khantay hydroelectric power plant (Fig.8.36).

8.6 EARTH DAMS CONSTRUCTED BY CONTROLLED BLASTING

Construction of earth dams requires large quantities of rock and soil which in turn involves time-consuming transportation and consolidation of earth materials. Construction of high earth dams, tens of millions of cubic metres in volume, also requires special machinery for earth moving operations.

The method of controlled blasting is a prospective technology for construction of earth dams. It makes possible construction of reliable structures with sufficient seepage strength and general stability, and at the same time it secures a considerable reduction of the duration of construction works and elimination of transporting and soil rolling machinery.

In the technological and economic analysis of dam construction by controlled blasting one must take into account the cost of 1 m^3 of soil in dam, together with some features which do not exist in other cases, for instance preparation of foundation. Earlier operation of the structure, compared with traditional methods, is also worthwhile to emphasize.

The method of controlled blasting is widely used in the USSR, in the construction of cofferdams and earth dams, snow sheltering structures, and earth dams for water resources and energy systems.

In the design of dams constructed by the method of controlled blasting one must consider a variety of problems which arise in the design of ordinary dams. One must determine the grain size distribution of dam soil, its sorting and distribution, together with seepage conditions in dam body; the latter incorporate depression curve, seepage flow rate, seepage strength of soil, and also slope stability due to static loading and seismic effects. The specific method discussed herein requires special technologies for rock processing. In general, the design of dams constructed by controlled blasting differs from the traditional design.

Depending on the destination of a dam constructed by controlled blasting one distinguishes homogeneous and inhomogeneous dams, with water confining components in the form of screens or cores (Korchevskiy 1962).

Considerable attention in the design of homogeneous dams constructed by blasting must be given to geotechnical and suffosive characteristics of quarry material, which in turn depend on engineering and geologic properties of the site, geomechanical characteristics of rock, and the technology of blasting.

Seepage through the body of a homogeneous exploded dam is considerable. The reduction of groundwater flow rates may be reached if the dam profile is wide, so that the ratio of dam width at foot L to the acting head H is kept in prescribed proportions. Figure 8.37 illustrates the dependence of groundwater discharge on the head and shallowness of dam profile.

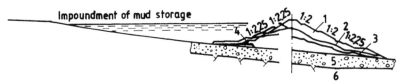

Figure 8.38. Cross-section of Medeo Dam. (1) rubble mound dumped after gorge fall; (2) heap contour due to LHS blasting; (3) heap contour due to RHS blasting; (4) loam screen; (5) alluvium; (6) granites.

Groundwater flow rate does not affect appreciably smaller structures for control of mud flows and local water resources. However, in power-oriented dams and reservoirs one must use special measures to reduce the groundwater rate so as to keep the head losses down to minimum. Colmatation of dam body by removed material is possible during operation, so groundwater flow rate decreases. Seepage computations for a homogeneous dam can be conducted by the methods described in Chapter 2.

Dams constructed by blasting and having water-confining components are similar to common dams with cores and screens. In construction of an exploded dam with core one first fills the upstream and tailwater prisms. The space between the latter is filled with a less permeable soil; one may use the blasting method applied to a soil volume prepared earlier, or an ordinary method of layered filling and compaction of soil. In the latter case the technology will be a combined one.

The water-confining component implemented as a core can be constructed in a body of a homogeneous dam by the use of injected clay-cement grout or other agents.

If a strongly permeable rock occurs in a dam foundation, the construction of the latter may be accomplished in two stages. In the first stage, the lower part of the dam is made with usual machinery and becomes an overburden, in which one often arranges a concrete slab. Bedrock grouting is implemented at this level. In the second stage the remaining part of the body is constructed by blasting.

A dam with a screen constructed by blasting can be implemented by two methods:

1. Screen is constructed by blasting simultaneously with the main body of dam (upstream prism).

2. Upstream prism is constructed first and screen (soil or special reinforced-concrete slabs, polymer foil, asphalt concrete etc) is constructed from the upstream prism thereafter. In the second case one must adjust the upstream slope of the prism prior to construction of the screen; transition zones and filters are also required. After the screen is constructed it must be covered or protected by other methods, if necessary. Construction of screens is more troublesome.

Examples of dams constructed by blasting. Medeo dam, above 100 m high, about 2.5 million m³ in volume, was constructed by controlled blasting in the years 1966–1967

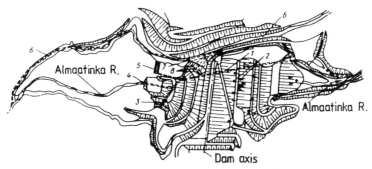

Figure 8.39. Plan view of Medeo Dam. (1) spillaway tunnel; (2) seepage exit on tailwater side; (3) temporary right-bank intake; (4) temporary discharge in construction conduit; (5) left-bank intake; (6) mud flow discharge; (7) temporary bank-based pumps; (8) temporary pontoon-based pumps.

to protect the city of Alma Ata from torrential currents generated in the catchment area of the Malaya Almaatinka River. The dam was constructed by two explosions on the left and right bank of the river, of which the right-bank explosion was first in October 1966. About 1.7 million m³ of soil was put in the dam at this stage. The second explosion, on the left bank was carried out in April 1967; more than 0.8 million m³ of soil was placed in this way. The mean consumption of explosives per 1 m³ of soil was 2.6 kg. The first explosion gave birth to a 60–62-m prism. On the upstream side the slope was 1:2, while that on the tailwater side was stepped, with slope tangent from 1:6 to 1:3. Upon second explosion the height of dam reached 90 m. Next 12–13 m of dam crest were dumped by ordinary filling with the use of rubble produced by blasting. A clay loam screen was arranged on the upstream slope. The dam survived a very intensive turbidity current generated as the upstream lake failed: 4 million cubic metres of sediment were accumulated in the settling basin of Medeo Dam.

Figure 8.38 shows the dam cross-section while Figure 8.39 illustrates the plan view of Medeo Dam. Aside from the mud-flow dam, the scheme incorporated a tunnel outlet for passage of the Almaatinka water; the latter has been situated on the left bank of the river, while the left bank mud-flow outlet for removal of excessive volumes has been constructed upstream of a dam. The tunnel outlet was completed prior to blasting. The mud-flow runoff filled the upstream area in July 1973, clogged the outlets and forced emergency pumping of water over the dam crest.

Another dam, 60 m high, was constructed by blasting on River Vakhsh downstream of the Nurek scheme, within a larger Baypazin hydraulic engineering system. The latter has been intended for irrigation of Yavan and Avikan valleys in Tadzhikistan. The blasting was executed in March 1969 on the right bank. The total mass of explosives was 1900 tons. The layout of charges for the dam construction is illustrated in Figure 8.40. About 1.6 million m³ of rock (limestone) was generated by blasting, of which 0.76 million m³ was incorporated in the dam body. Since a good transition from the dam body to the foundation was required, an upstream puddle blanket of special loam was made in river bed by the blasting method. After construction of the dam body, a screen (partial filling in water) was constructed from the upstream slope, connected later with the upstream puddle blanket. The profile of rock heaps upon blasting and the cross-section of Baypazin Dam are shown in Figure 8.41.

The blasting works were difficult because the charges arranged 300–350 m away from the left-bank outlet, and the right-bank outlet tunnel, 5.3 m in diameter, designed to draw irrigation waters into Yavan Valley, had to be protected from explosion hazards.

The Middle Asia department of Gidroproyekt considered a number of designs for Kambaratin hydroelectric power schemes on River Naryn. A high dam (265 m) was planned for Karabakh; it was to be a homogeneous rubble dam of Kambaratin scheme number one (Fig.8.42). The region has been favourable for blasting works because of considerable elongation of the river, V-shaped canyon, and high flanks above normal impoundment level, so that series of controlled blasting could be successful. The design has foreseen about 112 million m³ of soil to be placed in the dam body by blasting, and the total mass of explosives was 250 000–300 000 tons; the cost of 1 m³ of fill materials in dam body was estimated about 0.7–0.8 Rouble. The construction is planned in the wake of the construction of the main scheme, embodying a 1 600-MW hydroelectric power plant.

Geotechnical characteristics of soil in blasting-constructed dams. One must know geomechanical and seepage characteristics of soil placed in a dam if the properties of dam operation are to be estimated. Strength, deformation and seepage properties of soil in dam body depend on grain size distribution of the material generated by blasting and the density of the dumped material.

Grain size distribution may vary in wide ranges — it depends on the strength and cracking ratio of rock, and partly on the technology of blasting (mass of unit charge, sequence of explosions etc). More accurate prediction of soil properties requires test blastings which can link the site conditions to design characteristics of soil in dam.

Test blastings were carried out on River Burlikiya in Kirghizstan, so as to substantiate the design of Kambaratin Dam; subsequently, a dam 50 m in height and 330 m wide (along river) was constructed. The fill volume was 300 000 m³. The grain size distribution of the fill material generated for a number of dams is illustrated in Figure 8.43. It may be seen that the grain size curves do not differ too much.

Fairly high values of density can be reached if the coefficient of uniformity η is above 20. Good textural density of filling and low permeability are reached if fine fractions are in considerable quantities. Figure 8.44 illustrates the dependence of

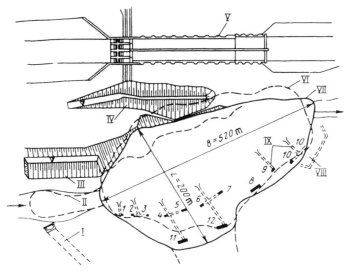

Figure 8.40. Layout of explosives upon construction of Baypazin Dam. (1)–(10) auxiliary charges; (11)–(12) primary charges; (I) 7.5-km conduit; (II) upstream puddle blanket; (III) (IV) protection dikes; (V) chute; (VI) design contour; (VII) real contour; (VIII) test pit; (IX) pits leading to charges.

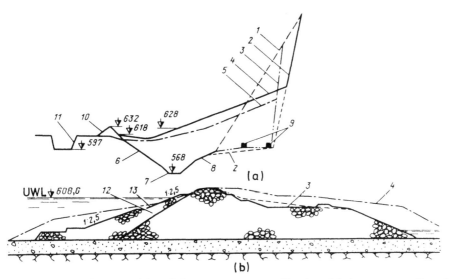

Figure 8.41. Baypazin Dam. (a) rock heap contour upon blasting; (b) dam cross-section; (1) right flank prior to blasting; (2) real detachment line; (3) design detachment line; (4) real heap contour; (5) design heap contour; (6) left bank; (7) Vakhsh River channel; (8) right bank; (9) explosive charges; (10) protection dike; (11) discharge; (12) screen; (13) rubble revetment — 0.8 m cover layer and 0.2 m filter layer.

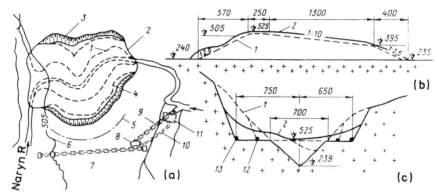

Figure 8.42. Layout of structures in Kambaratin No.1 scheme at Karabakh section. (a) plan view; (b) longitudinal dam section; (c) profile cross-section; (1) design dam contour; (2) rock heap contour; (3) slopes of blasting hole; (4) charge chambers; (5) zone of critical permissible seismic oscillations; (6) reservoir; (7) discharge conduit; (8) underground hydroplant; (9) take-off tunnels; (10) transportation-cable tunnel; (11) 500-kV supply area; (12) auxiliary charges; (13) primary charges.

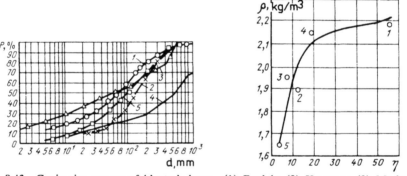

Figure 8.43. Grain-size curves of blasted dams. (1) Burlyk; (2) Kurpsay; (3) Medeo; (4) Baypazin; (5) Papan.

density on the coefficient of uniformity (a), and permeability versus the contents of fine and coarse fractions (b, c).

One must note that the grain size distribution results from two processes — rock disintegration upon blasting and granulation of material upon dynamic compaction under dam construction. The density of soil in dam body may be designed if one knows the processes of dynamic compaction investigated in special compression apparatus, in which soil is compacted by falling weight. The tests provide characteristics of dynamic compaction and variation of grain size distribution. More reliable data is provided from test blastings. Field tests are performed to verify design assumptions for dams constructed by blasting and to make more accurate and reliable the values taken in the design for the density and permeability of soil.

It is of interest to analyse the results of tests on geotechnical properties of soil for the test dam of Burlikiya (Fig.8.45). The geotechnical characteristics were found from a number of holes arranged in the dam body (Reyfman et al 1982). The deepest hole (No. 3) was bored from the dam crest. The graphs show that contents of fine soil and soil density increase with depth, which is obviously related to the dynamic action of compressive stresses. The textural density of fill is usually very high. Medeo Dam had soil density of 2170 kg/m^3 at a depth of 20 m; it did not vary deeper while that at the depth of 6 m from crest was 1950 kg/m^3. A higher density of fill (2200 kg/m^3) was reached at Baypazin Dam, which was due to homogeneous grain size distribution.

Computations of charge size and blasting technology. The layout and technology of blasting works must be determined for adequate construction of dam. The necessary quantity of explosives is found for given design parameters of dam, reliability of components of a scheme and other structures in the region.

Explosions and earth moving operations are usually divided into 'removal' and 'disposal' classes. The difference consists in the fact thet 'removal' takes place over areas with horizontal or weakly inclined surface, while 'disposal' is referred to the sections inclined to horizontal at 30° and more. In both cases there occurs disintegration of rock and its displacement beyond limits of the explosion funnel (Burshtein & Grebenshchikov 1972, Pokrovskiy 1974). Depending on the type of blasting, the quantity of explosives may be determined. The directionality of explosions is controlled by layout of charges and the magnitude of change and the sequence of blasting.

The charge mass (in kg) for removal and disposal is proportional to the third power of the so-called line of lowest resistance (LLR) that is the distance from the charge to the ground surface

$$Q = K_0 W^3 \tag{8.14}$$

in which
W = LLR length.

Boreskov found

$$K_o = K(0.4 + 0.6n^3)$$

in which
K = unit design consumption of explosives (kg/m^3), usually determined by tests,
n = explosion effectiveness equal to the ratio of the funnel semi-opening W_n to W (Fig.8.46) taken from 1 to 2.

Equation 8.14 provides good results for small depths of charge deployment. For higher depths (removal blasting) one uses a formula derived by Pokrovskiy:

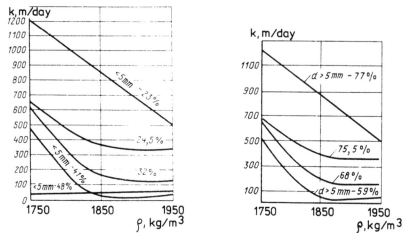

Figure 8.44. Dependence of soil density ρ of blasted dam on rock coefficient of uniformity η (a), coefficient of permeability k versus contents of fine grains $P_{d<5}$ (b) and rockfill coefficient of permeability versus $P_{d>5}$ (c). (1) Medeo; (2) Burlyk; (3) Papan; (4) Baypazin; (5) Kurpsay.

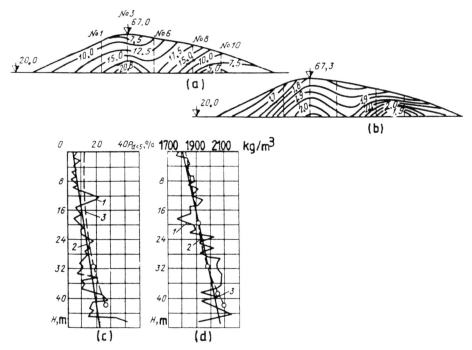

Figure 8.45. Distribution of geotechnical properties of soil in the body of blasted dam on River Burlykiya. (a) percentage of fractions below 5 mm; (b) density, t/m³; (c) variation over dam height of fraction below 5 mm; (d) variation over dam height of density, t/m³; (1) test curve; (2) linear approximation; (3) prediction curve.

$$Q = KW^3(\frac{1 + n^2}{2})^2(1 + 0.02W) \qquad (8.15)$$

For deep charges the effectiveness of blasting decreases.

In disposal blasting, one seeks a low specific consumption of explosives which is linked to the configuration of the centre of gravity. The following formula derived by Burshtein may be used for disposal blasting

$$Q = qW^3n^2K_r \qquad (8.16)$$

in which

q = specific consumption of explosives for loosening of soil (kg/m³), which is equal to $K/3$,
$K_r = [0.75(1/n+n)^2(1+0.02W)]^{\cos\alpha}$,
α = slope of free surface in grads.

Equation 8.16 holds true for a single charge. The interaction of simultaneously acting charges is accounted for in Equation 8.16 by the coefficient K_r varying from 0.4 to 2. The spacing of concentrated charges can be determined from the formula

$$a = 0.5W(n + 1) \qquad (8.17)$$

The construction of the heap profile generated by blasting may be done by approximate methods. The simplest way consists in discrimination of the profile of rock from which the material was detached in the cross-section passing through the centre of the charge and incorporating the LLR line. The detachment countour is determined from the detachment radius on the uphill side (Fig.8.46b):

$$r_1 = W\sqrt{1 + n^2} \qquad (8.18)$$

while that on the downhill side reads:

$$r_2 = cW \qquad (8.19)$$

in which

$c = 1.7$–3.5, depending on geology and topography of site and design parameters of blasting.

The heap area is found as follows

$$S_H = \mu S_o$$

in which

S_o = detachment area.

For rock one has $\mu = 1.25$ for central charges and $\mu = 1.1$ for lateral ones. For semirock blasting μ decreases by 0.1.

The flight of soil is $L = 4nW$, and the shape of fill is regarded as triangular. The maximum heap height of fill is $h = 2S_H/L$.

The schematic fill due to a few charges is illustrated in Figure 8.46d. Upon graphical construction the profile is corrected not to surpass the natural angle of repose.

The specific consumption of explosives depends on the cracking strength of rock. The standard consumption q_e of 6ZhV ammonite varies from 0.3 to 1.4 kg/m^3, for weak and cracked rock to strong rock, respectively. For disintegrated rock, the consumption reads $q = q_e(500/d)^{0.4}$, in which d = size of acceptable fraction in mm. For bulk blasting one has $q_e = 0.00027\gamma$, in which γ = unit weight of soil in kN/m^3.

Charges are usually arranged in two rows, of which the first (auxiliary) is blasted earlier to provide the initial controlled shape for the primary charge. The depth of charge is recommended as 0.7 to 0.9 of the heap body, so as to facilitate more uniform disintegration of rock. Several tiers of charges are advisable on steep slopes of canyons. The charges are exploded consecutively, starting from the lowest tier.

Explosives should be computed by analogy to the existing field counterparts and by reference to tests. Large charges must be checked on seismic hazards posed to surface as well as underground objects. The check can employ the relationship between celerity v in soil and earthquake intensity J (in points), cf. Table 5.1. The celerity in soil reads

$$v = \lambda(Q^{1/3}/R)^{\nu}$$

in which
   R = distance from charge centre, m
   λ, ν = coefficients; for Baypazin Dam they are 315 and 1.8 while those for Medeo Dam are 420 and 1.73,
   Q = mass of charge, kg.

Rock is damaged by blastings; this fact must be remembered in the design of works so as to avoid weakening of rock about the hydraulic engineering scheme, in particular below the water datum, where lateral seepage might arise and special preventive measures may prove necessary.

8.7 DAMS CONSTRUCTED IN THE FAR NORTH

Characteristics of northern climatic zones. Northern regions (the Soviet Far North, and the adjacent Siberia and Far East) are characterized by severe climatic conditions with low winter temperatures, which on the average vary from -15 to 20°C in the European North and from -40 to 50°C in the northern regions of Eastern Siberia. The summer is short with low mean temperatures but, because of the continental type of climate, the absolute temperatures are high, about 22–36°C in the European North and northwest Siberia or 18–38°C in the lower Yenisey, eastern Siberia and Yakutia. Similar temperatures are encountered in Chukotka and Madagan Province.

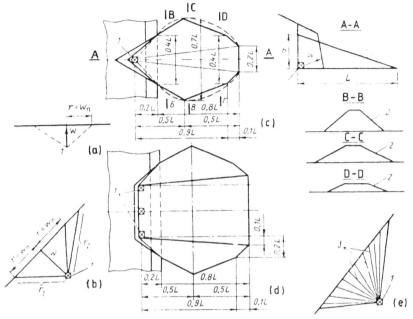

Figure 8.46. Types of blasting and schemes of dumping. (a) outburst blasting; (b) discharge blasting; (c) heap due to single charge; (d) heap due to three charges; (e) discharge blast with fan-type arrangement of charges; (1) charge; (2) free heap slope; (3) fan-type layout of charges.

Engineering and geological conditions of the above regions are very diversified — European North and west Siberian North are characterized by considerable strata of Quaternary deposits, ranging in thickness from 100 to 120 m. In a number of areas this thickness is lower, and somewhere native rock outcrops. In the northern part of the Yenisey River, thick strata of native rock prevail, which in the Yenisey valley are covered by a stratum of Quarternary deposits up to 200 m thick. Eastern Siberia displays a variety of geological conditions with predominant native rock in western Yakutia, sedimentary deposits in the Leno-Viluy Depression, and thick alluvia in the depression along the northern Arctic Sea coastline.

Geocryologic (frost-soil) conditions of the above regions take the form of seasonal and permanent permafrost. Seasonal permafrost is characterized by negative temperatures in the winter, while permanent permafrost shows negative temperatures in a long run. The soil in the range of positive temperatures is referred to as thawing soil.

Permafrost in the Soviet Union spreads above 70°N, but considerable areas in Yakutia and basins of the Lena and Viluy, 63–67°N also occur. The thickness of permafrost varies from 3.5 to above 500 m. Permafrost may lye directly below seasonal permafrost and thaw soil, or may be contained within the latter.

The depth of seasonal and permanent frost is diversified. In the North, the thawing layer is usually about 0.2–0.3 m thick, while in the South it can reach 3–5 m. Thawing and freezing cycles cause changes in the strength and deformation properties of soil, which is primarily due to the presence of frozen water in soil pores (Tsytovich 1973, Vyalov 1959). The course of freezing and thawing also largely affects the seepage properties.

Depending on frost conditions, groundwater may be grouped in three classes: superfrost, interfrost and subfrost. The first kind of groundwater occurs in Quaternary deposits and is determined by the seasonally thawing soil. The water-confining stratum for this structure consists of permafrost. The second type of groundwater belongs to underground lakes inside permafrost, while the third kind incorporates the groundwater below permafrost.

Hydrological conditions of rivers in the northern regions are characterized by huge spring floods and heavy ice conditions, small flow rates during low water periods in winter, generation of brash ice, which all together complicate the passage of construction flow rates, particularly if heavy ice conditions occur.

Negative mean yearly temperatures, ubiquitous permafrost in bedrock, and the hydrological conditions of rivers flowing into the Arctic Sea determine the peculiar features of the construction and design of earth dams in the Far North and the adjacent territories of Siberia and Far East. An experience gained in the design, construction and operation of earth dams under severe northern conditions is very substantial, both in the Soviet Union and abroad. Earth dams in these environments are the most economic structures and may be constructed of both thawing and freezing soil (Biyanov 1983).

Temperature conditions in dam body and foundation. In the design of earth dams for severe climatic conditions, particular attention should be drawn to the stability of slopes, seepage, consolidation of clayey soil, settlement of dam body etc. The stress-strain problems for dam and its foundation, slope stability, infiltration of water from reservoir etc. should be analysed in terms of temperature variations in the dam foundation and body during construction and operation. Particular temperature conditions provide a basis for selection of dam type, which may be classified in the following groups: dams with permafrost curtains (non-seepage) and dams without permafrost curtains (thawing or permeable).

Type of dam and selection of its design may be decided upon if one knows thermal conditions and the state of foundation, which require forecast of temperatures and moisture contents in dams and foundations. A variety of factors and their interactions in the processes of heat transfer in dams and their foundations must be explored. The processes of heat and mass transfer in earth-rockfill dams are very complex because the dam foundation and body are complex multi-phase media.

For fine dispersed soils of dam body and for soil structure in dam foundations, the thermal conditions are primarily determined by the following factors: heat conductivity of soil and bedrock, heat content, heat transfer by groundwater, melting

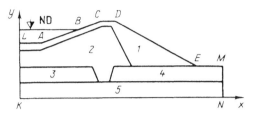

Figure 8.47. Design dam section. (1) rubble mound; (2) soft soil; (3) pebbles and gravel; (4) wet clay and loam; (5) bedrock.

of ice, and freezing of water. In coarse soils of upstream prisms, an essential factor in the heat transfer between the dam and the ambient environment consists in the convective heat and mass transfer due to the motion of air in soil pores. The convection between pores is determined not only by the thermal processes but also water content (mass transfer) in the dam, which brings about generation of ice in the dam body. Hence the thermal conditions of dam made of coarse soil are complex due to the transfer of heat by convection with air and water vapour in soil pores, heat exchange between the coarse soil and ice, air and water vapour moving in pores, phase transformations of water vapour, liberation or absorption of heat upon phase transformations of water vapour, and heat contents of fill material.

Hence periodic (seasonal, monthly etc) variations of ambient temperture, initial temperature conditions of foundations, heat transfer due to the motion of ground-water from reservoir during construction and operation of a hydraulic engineering scheme, convective transfer of heat etc. (including special measures bringing about changes in thermal conditions of dam foundation discussed below) give rise to variations of thermal conditions in the dam itself and its foundation both upon construction and thereafter.

In the simplest case, without inclusion of heat and mass transfer, temperature conditions of dam foundation can be determined by solution of the so-called Stephan problem, which is reduced to the following quasi-linear differential equation

$$C(T)\frac{\partial T}{\partial t} = \frac{\partial}{\partial x}[\lambda_x(T)\frac{\partial T}{\partial x}] + \frac{\partial}{\partial y}[\lambda_y(T)\frac{\partial T}{\partial y}] + \frac{\partial}{\partial z}[\lambda_z(T)\frac{\partial T}{\partial z}] \qquad (8.20)$$

in which
    T = temperature,
    t = time,
    $C(T)$ = effective heat content of soil depending on temperature,
    $\lambda(T)$ = heat conductivity of soil, also depending on temperature.

In the simplest case one has $C(T) = C$ = volumetric heat content of soil and $\lambda_x(T) = \lambda_y(T) = \lambda_z(T) = \lambda$ = constant heat conductivity (both in time and space). Equation 8.20 holds true for both thawing zone of soil (in which $C = C_T$ and $\lambda = \lambda_T$) and freezing soil (in which $C = C_F$ and $\lambda = \lambda_F$).

Accordingly, for the assumptions made, Equation 8.20 can be written down as follows

$$C_{T,F}\frac{\partial T_{T,F}}{\partial t} = \lambda_{T,F}\nabla^2 T_{T,F} \quad \text{or} \quad \frac{\partial T_{T,F}}{\partial t} = \alpha_{T,F}\nabla^2 T_{T,F} \tag{8.21}$$

in which
  a = soil temperature conductivity.

The indices 'T' and 'F' refer to thawing and freezing soil, respectively.

Solution to heat condictivity equations may be obtained for known initial and boundary conditions. For the scheme illustrated in Figure 8.47, upon assumption that the distribution of temperatures in dam and its foundation is known at the end of dam construction, one may take the following initial and boundary conditions, as derived from field data:

a) initial condition

$$T(x,y,z) = \varphi(x,y) \neq 0 \tag{8.22}$$

b) conjugation condition at the interface of inhomogeneous soils (thawing or freezing) — heat continuity equation in both cases of

b1) identical heat conductivity in ith and $(i+1)$th layers of thawing soil or jth and $(j+1)$th layers of freezing soils

$$(T_T)_i = (T_T)_{i+1} \quad \text{and} \quad (T_F)_j = (T_F)_{j+1} \tag{8.23}$$

b2) for different heat conductivity in the adjacent layers:

$$(\lambda_T)_i \left(\frac{\partial T_T}{\partial n}\right)_i = (\lambda_T)_{i+1} \left(\frac{\partial T_T}{\partial n}\right)_{i+1}$$

$$(\lambda_F)_i \left(\frac{\partial T_F}{\partial n}\right)_j = (\lambda_F)_{j+1} \left(\frac{\partial T_F}{\partial n}\right)_{j+1} \tag{8.24}$$

in which
  n = normal line of the interface;

c) heat exchange conditions for a solid body with air, along $LABCDEM$

$$T_H = f(t), \quad \frac{\partial T(x,y,t)}{\partial n} = -\frac{\alpha_c}{\lambda}[T(x,y,t) - T_H] \tag{8.25}$$

in which
  T_H = temperature of external air about the interface,
  α_c = coefficient of convective heat exchange between solid body and air,
  λ = heat conductivity of materials at the interface with air.

The coefficient α_c can be taken as a function of the mean wind speed v_w as $\alpha_c = 1.2v_w^{0.5}(6 + 6.2/v_w^2)$; in approximation it may be taken as 23.26 Watt/(m²·°C);

d) at the boundary LAB after filling of reservoir the soil temperture is equal to temperature of water, which is assumed to be known

$$T(x, y, t) = T_B(x, y, t) \tag{8.26}$$

e) on vertical edges KL and MN one assumes the absence of heat flux:

$$\frac{\partial T}{\partial x} = 0 \tag{8.27}$$

f) at the lower boundary KN one also takes heat flux in the vertical direction

$$\frac{\partial T}{\partial y} = 0 \tag{8.28}$$

g) condition at the interface of phases

$$T_F = T_T = T_{Fr}; \quad \lambda_M \frac{\partial T_F}{\partial n} - \lambda_T \frac{\partial T_T}{\partial n} = \pm \rho \kappa W \frac{d\xi}{dt} \tag{8.29}$$

in which

T_{Fr} = freezing temperature of water in soil,
$\rho; w$ = density and water content of soil, respectively,
κ = latent heat of phase transformation,
$\frac{d\xi}{dt}$ = velocity of propagation of the interface of phase in the course
of freezing (+) and thawing (-).

The equation of heat conductivity with initial and boundary conditions is solved numerically by methods of finite differences or finite elements. These methods are intensively exercised at VNIIG, VNII VODGEO, MISI, GISI and other research and teaching institutions. Solutions to a number of problems are given by Deresevich (1961).

The mathematical formulation of the problems becomes more complex if the heat and mass transfer of groundwater in the dam foundation and body is considered, in particular if the heat and mass transfer and the generation of ice due to the motion of air in the tailwater prism is added.

If the problem is solved, one determines the motion of the zero isotherm, the ultimate configuration of which becomes quasi-stable in time and, in the case of dam operation, depends only on the seasonal variation of temperatures of the ambient air, so that the problem is practically reduced to the steady state. The steady layout of isotherms, when Equation 8.20 is reduced to the Laplace equation $\frac{\partial T}{\partial t} = 0$ may be found by the method of electrohydrodynamic analogy because the equation of heat conductivity in steady state is analogous to the equation for steady motion of groundwater. A practical method for determination of zero isotherm by electrohydrodynamic analogy is described by Grishin (1979).

An important selection of thermophysical characteristics of soil and rock must be done in the solution of heat conductivity problems applied to earth dams and foundations, much as in solution of any problems of soil mechanics. In accordance with *Guidelines* (1976), for soils of dam body one may take average characteristics given in Table 1.2.

Hence one of the primary elements in the design of earth dams is the necessity of temperature analysis for the system of dam foundation, this temperature controlling the seepage condition, nature of settlements and displacements, stability of dam slopes and banks etc. Settlement computations for foundations and earth structures subject to thawing are very particular. These computations differ from usual stress-strain computations for dams and foundations discussed in Chapter 4 for common (unfrozen) soils. In the case of thawing and freezing one must take into account deformation characteristics of soil, and particularly the modulus of deformation.

Gorelik & Nuller (1983) consider the variation of the modulus of deformation for freezing and thawing soil in terms of elastic model. In one-dimensional case the reduction of the modulus corresponds to the elastic element which initially consists of two parallel elastic springs. After thawing of soil its compressibility increases, which is simulated by removal of one spring, and additional deformation is generated due to this force. Increasing modulus of deformation is simulated by addition of another stressless spring to the spring which was originally compressed under the applied load. This additional spring is set in operation only if load is changing.

In three-dimensional stress state the reduction of the moduli of elasticity is described by the following law:

$$\sigma_{ij} = E e_{ij} \delta_{ij} + 2G\epsilon_{ij}$$

while for increasing moduli Hooke's law will hold true only for increments

$$\tilde{\sigma}_{ij} = E \Delta e \delta_{ij} + 2G e_{ij}$$

in which

$$\tilde{\sigma}_{ij} = \sigma_{ij} - \sigma_{0ij}; \quad \tilde{e}_{ij} = e_{ij} - e_{0ij}.$$

The notation e_{0ij} and σ_{0ij} corresponds to the original moduli of elasticity E_0 and G_0, while the symbols e_{ij} and σ_{ij} denote their limit values.

The solution of the stress-strain problem is reduced to solution of an auxiliary problem of the classical theory of elasticity with additional volumetric and surface forces. The model proposed by Gorelik & Nuller (1983) has been extended to the case of any increase and decrease in load.

One must note that the formulation of the problem and its numerical solution is more complex if stagewise construction is considered, along with the temperature of dumped soil, filling of reservoir, variation of water temperature over depth of the reservoir etc.

Depending on temperature conditions, types of dams are primarily determined by the condition of soil in the water-confining unit (Biyanov 1983).

Unfrozen dams are characterized by thawed soil of the water-confining unit. Frozen dams have the water-confining unit in frozen condition. Selection of dam type by thermal conditions is determined by a set-up of engineering, geologic and geocryologic conditions of site, in particular deformability and strength of foundation soil upon thawing (during construction in permafrost regions).

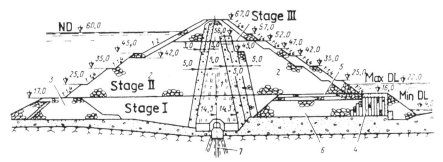

Figure 8.48. Ust-Khantay channel dam. (1) moraine loam core; (2) rubble mound; (3) upstream cofferdam; (4) cribwork; (5) irregular blocks up to 10 tons; (6) downstream rockfill of first construction stage; (7) grout curtain.

If cracked rock and semi-rock strata are filled with ice or loose alluvia and rock waste, the thawing of which is accompanied by considerable deformations and changes in the permeability within a dam foundation, one recommends unfrozen dams which are well adaptable to non-uniform deformations due to settlement. Dams of this type are even more adequate for the case in which strength and deformation properties do not change as the result of thawing of foundation (rock and semi-rock, dense non-rock soil).

If changes in temperature conditions of foundation made of thick ice-saturated clayey, sandy, gravel or mud deposits (particularly if the quantities of ice are considerable) bring about essential settlement, thermal karst processes and substantial reduction in soil strength, then one must construct dams of frozen type.

Unfrozen dams constructed in the Far North are designed similarly to the dams constructed in other regions — they can be homogeneous, inhomogeneous, with screen or core, with water-confining non-soil units etc. If a permeable soil occurs in the foundation of small thickness, one uses special structures such as upstream blankets, cut-offs etc, or piles and grout (cementation) curtains for permeable soil in thick strata.

If sufficient amounts of impermeable soil, loam or clay are available on site, until recently it has been most economic to consider earth-rockfill or rockfill dams as designs with a screen or core. Fills (thrust prisms) of such dams were constructed of rock, gravel or pebble materials (Viluy, Kolym, Ust-Khantay Dams etc). Another alternative is presently offered by dams with asphalt and grout (injection) diaphragms.

Particular attention in the construction of unfrozen dams is drawn to preparation of foundations, especially at the interface shared with the foundation of seepage prevention facilities. In the construction of Viluy Dam with a screen, gravel, pebble and boulder deposits, 3 m thick, were not removed from the foundation of the thrust prism. If soil with a lot of ice occurs in the foundation of a dam, this ice being a potential compressibility hazard due to thawing, then it is necessary to take special measures, viz. preconstruction soil thawing i.e. hydraulic thawing by the use of well points into which hot water is pumped, electric thawing for cohesive soil, and the seepage-drainage technique for strongly permeable soil. In the latter case, thawing is forced by water percolating from ground surface.

An example of an unfrozen dam with a water-confining unit made as a core is provided by Ust-Khantay Dam, 67 m high and 300 m long at crest (Fig.8.48). Considering different design versions of this dam (Kuperman et al. 1987), preference was given to a dam with core, as core consolidation and filling of possible cracks are facilitated by a considerable load due to core weight. Moreover, deformation of rubble affects the core less than a screen. In order to reduce core strength hazards due to considerable settlement of rubblemound one designed a thick three-layer transition. Suffosion stability of core soil was also an important design factor. The eepage through the core was investigated for various versions of its freezing during operation. The three-layer transition on the tailwater side incorporates a two-layer filter (a layer of sorted sand, 0–10 mm, and another layer of natural sand-gravel mixture) and a third transition layer of sorted rubble, 10–80 mm. The thickness of the filter is 5 m, or 3 m in the upper part. On the upstream side the core is protected with a single filter layer made of sand and gravel. The upstream and tail-water cofferdams, 17 and 16 m high, are included into the dam cross-section. The upstream cofferdam is made of soil and has a moraine screen, while the tailwater cofferdam is cribwork type.

Another design has been chosen for the right-bank Ust-Khantay Dam, having the maximum height of 33 m and 2.5 km long (Fig.8.49); this was due to complex engineering and geologic conditions. A particular feature is the presence of horizontal drainage with layered inverse filter embedded in the foundation behind the core of the dam. The drainage enhances consolidation of foundation and prevents suffosion. In order to accelerate the consolidation of the foundation beneath the tailwater prism, a vertical drainage was provided in the form of 10–15 m deep holes, 30 cm in diameter; the scheme was eventually given up as the foundation was consolidated upon construction.

The highest earth dam constructed under severe climatic conditions is the earth-rockfill Kolyma Dam, 130 m high (Fig.8.50). The mean yearly temperature in this region is -12°C, the minimum temperatures going down to -60°C, and the maximum ones in the summer reaching 36°C. The dam foundation consists of permafrost rock. The dam core is made of rock waste and solifluction loam and sandy loam, while the transitions are made of natural sand and gravel or fine boulders. The thrust prisms consist of rubble mound. A concrete cut-off with a grouting hole is built into the core.

Frozen dams can incorporate facilities for artifical freezing of dam body and foundation, or may freeze naturally as cold air enters the dam on the tailwater side. Frozen dams differ from unfrozen ones by the absence of drainage facilities.

A frozen curtain should be removed far away from the source of heat, and a frozen screen must be quite heavy.

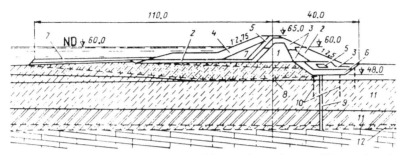

Figure 8.49. Right-bank Ust-Khantay Dam. (1) moraine soil core; (2) gravel and pebbles; (3) sand and gravel; (4) rubble mound; (5) 10–80-mm gravel fractions; (6) sand; (7) rock (limestone); (8) permafrost foundation soil; (9) vertical unloading drainage; (10) consolidation drainage; (11) lacustrine-glacial deposits; (12) glacial deposits.

Frozen dams are usually made with central core of thawing impermeable (or practically impermeable) soil, which is consolidated and frozen. The freezing procedures incorporate an air- or liquid-freezing system, using for instance kerosene, freon or brine. The liquid system consists of refrigerating columns in which 40% of solution of $CaCl_2$ or any other of the above liquids circulates at a temperature of -15 to -25°. The column usually consists of pipes, outer and inner. The outer pipe delivers the liquid which is removed in the inner pipe. A cylinder of frozen soil is generated about the column. The distance between the columns should provide a satisfactory mutual contact of the cylinders. The thickness of the so generated frozen curtain should be sufficient to minimize the thawing process during summer (2–4 months), because the circulation of the refrigerating liquid is arrested in the summer.

More progressive technologies use air as a cooling agent. In this case many shortcomings of the liquid system are eliminated such as the necessity of keeping accurate concentration of brine, frequent replacements, intensive corrosion of pipes, leakage of brine, complexity of generation, high cost of other liquids etc.

In the winter, the air in the cooling system is sucked by a fan at the outer pipes of the column and enters the air collector between the pipes upon cooling of the soil.

Dam on Irelakh River, 20 m high, with an air-cooled system (Fig.8.51) was constructed in 1964. The dam incorporates rolled thawed loam, overfilled with sand, stone and thermal insulation layer on slopes. The foundation consists of marl clay, dolomites, and marl. The alluvium on the native rock contains ice lenses up to 40 cm thick. The air system consists of 327 columns 8.5 to 25 m deep, with the spacing of 1.5 m. The system was decomposed into seven independent sections, and the diameter of the piles was 140 mm. The discharge of the air supplied to the fan in each section was 13 000 m³/hour. In 1966 the thickness of the frozen curtain was 10 m. The cooling system did not work during summer. The primary

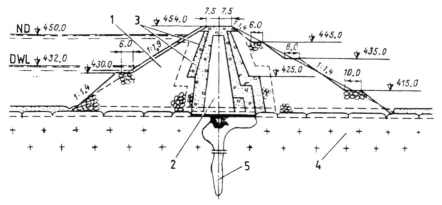

Figure 8.50. Earth-rockfill Kolyma Dam. (1) rubble mound; (2) core; (3) sand-gravel filter; (4) granite; (5) grout curtain.

disadvantage of the system, clogging of pipes with ice due to condensation and freezing of air moisture, was evident from that period of operation.

Computations of the time of air system freezing and some other temperature computations are based on the equation of heat conductivity (see for instance Grishin 1979).

In the design of earth dams one must avoid overhanging of frozen soil above a thaw zone.

In impervious dams (with frozen curtains) it is recommended that about 50% of the area of each design cross section be frozen in order to make the design reliable. In the summer, in order to prevent warming of dam on the tailwater side and at crest, one must provide timber shelters or ice galleries covered with a layer of peat, as done at Norylsk on Lake Dolgoye.

Considerable attention must be paid to field investigations under operation of earth dams in the Far North. Aside form usual observations, one should also study the dynamics of temperature fields in the dam foundation. Since the prediction of temperature conditions in dams and foundations is often based on many assumptions, and the thermophysical characteristics are usually taken with low accuracy, the results of field investigations may also be used, aside from their primary destination, in the assessment of the effectiveness of design methods employed, not to mention the increase in the accuracy of thermophysical estimates for soil. In the latter case one must solve the so-called identification problem for soil properties.

Field investigations show that the initial temperatures of both foundation and the dumped soil play a significant role in the temperature field of the structure only in the initial period of operation. The primary factors affecting changes in temperature conditions are due to seasonal variation of ambient air temperature and water temperature in reservoir, groundwater flow and heat transfer in rockfill.

The results of field investigations on temperature conditions at Viluy Dam (Olo-

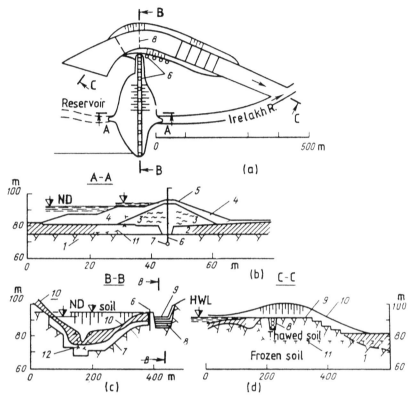

Figure 8.51. Irelakh hydraulic engineering scheme. (a) plan view of scheme; (b) dam cross-section; (c) dam axial section; (d) spillway axial section; (1) bedrock — consolidated clay, cracked dolomites, marls and limestones; (2) mud, peat and loam with ice contents up to 60%; (3) dam loam core; (4) sand fill with slope revetment of stone; (5) moss and and peat for thermal insulation of core under dam crest; (6) rows of holes for dam core freezing and generation of frozen soil between dam and spillway; (7) downstream side of freezing holes; (8) horizontal freezing pipes under spillway in dam section; (9) concrete protection of bed, flanks and stepped spillway; (10) natural ground level; (11) interface of frozen and thawed soil during dam operation; (12) natural interface of frozen and thawed soil under river bed.

vin & Medvedev 1980) are very interesting. The dam was constructed all over the year. The material was frozen during the winter filling of loam into the dam screen, and frost accumulated during the entire period. The screen was also cooled down on both sides. During the summer filling, soil temperature was above zero, 12–16°C. Prior to impoundment of the reservoir, the screen incorporated interlayers of frozen and thawed soils. After the reservoir was filled, in 5 to 6 years, the frozen zones of screen became thawed, the temperatures were in mutual balance and stabilized about the level of mean temperature of the reservoir water (roughly 4°C). The downstream fill was kept at -25°C due to winter filling and convective heat exchange with cool air.

In view of the generation of ice in the fill body due to condensation and freezing of water vapour, the role of convection in the generation of thermal conditions becomes minor. The frozen rock at dam foundation gradually thawed off, and the left-bank slope of the dam, exposed to the south, underwent more intensive thawing. In addition, the passage of spring and summer floods in the construction trench also affected the course of thawing.

8.8 SELECTION OF EARTH DAM TYPE

Selection of dam type in the case of earth materials is an important issue in the design of hydraulic engineering schemes, and many technological and economic factors are involved. They are listed below.

Climatic conditions in many situations dictate the selection of dam type and soil filled in dam components. Under conditions of the Far North homogeneous dams should be given priority, together with dams with screens and cores, whereupon the latter design is recommended if frozen type dams are chosen.

Homogeneous dams of 'lean soil' are recommended because of convenient organization of works, as filling of clayey soil at negative temperatures is difficult and dangerous in view of the continuity of water-confining elements. Under severe climatic conditions one should try to construct dams with metallic and asphalt-concrete diaphragms. In regions with long rainy seasons, which make difficult the filling of clayey soil, it is purposeful to construct dams with screen, for filling of dam body with coarse grained materials may lead in the construction schedule.

Topographic conditions of site often predetermine the type of dam. In hardly accessible mountainous regions, the construction of dam by explosions should be suggested as a rule for one may choose the cross-section suitable for this type of dam, and additionally more complex topographic conditions make difficult the transportation of construction materials.

Hydrological features of rivers and streams, for instance in mountainous regions, in particular the presence of extreme floods, require high cofferdams, which should be incorporated in the design of a dam proper. Dimensions of cofferdams should be suited to dimensions of the construction conduits designed for passage of construction discharges, and to the conditions of their operation.

Geological conditions of a hydraulic engineering scheme are also important for the selection of dam type. Under plain river conditions, for permeable dam foundations, the design should take into account special seepage prevention elements which extend the seepage path at the foundation. Shallow wide-profile dams should be given preference together with dams with extended upstream puddle blankets, piles in foundation etc. In mountainous construction on bedrock, the underground contour of dam should also incorporate a grout curtain of various depth and configuration, embodying the flank transitions so as the bypassing groundwater is kept down to minimum. A curtain is often combined with drainage which relieves the groundwater flow in bedrock and rock bodies. The presence of bedrock opens the opportunity of the construction of high and very high dams from soil materials which transfer heavy loads to a reliable foundation.

In many cases the type of dam is determined by the presence of soil available for filling. One should thoroughly investigate quarries for availability of soil about the site of a hydraulic engineering scheme; this must be done in the stage of dam design. In the construction in plain regions, a hydraulic-fill dam is most economic. In this case the homogeneous dam made of different sand sizes or sand-gravel soil is the most widespread type, while dams with cores generated as a result of the fractioning of soil sizes during hydraulic filling are less frequent. In the montainous construction, inhomogeneous dams with core or screen are common. Cohesive soil is used for the confining elements, and the so-called skeletal (i.e. heterogeneous) soil is given priority due to its nonsuffosivity. Application of clay for seepage prevention elements is purposeful at times, as the growth of filters and transition zones may be suitably chosen in this case. Gravel, gravel with pebbles, and rubble are recommended for dam fills; these materials are less compressible than stone and rubble. This factor is of particular importance in the construction of high dams since considerable pressures at lower parts of dam bring about failures at the contact of rubble blocks. A tendency towards the use of semi-rock soil with low permeability, subject to fast weathering, can be noted in the recent years.

Destination of dam and its height affect the selection of dam type and design. Dams for power schemes should have higher impermeability as losses of water due to seepage bring about lower effectiveness of energy conversion. Special dams, for instance in tailing reservoirs, including those for highly toxic wastes, should be designed with practically impermeable water-confining elements to prevent the transport of contaminants to groundwater. These water-confining elements should incorporate special measures to prevent the penetration of wastewater from the upstream side; for instance upstream puddle blankets of cohesive soil or PVC foils arranged at the bottom of reservoir. Counterseepage elements may utilize clay up to very fat clay. Mud-flow dams can be permeable and can be constructed by controlled blasting. Fractioning of soil upon filling in this case has no practical implications.

Dams in seismic regions should be primarily constructed with rock and earth as the use of fine soil enhances the hazard of liquefaction. The danger of tectonic failures should suggest the design providing a reliable operation of a dam upon considerable differential displacements at bedrock and foundations.

Layout of construction site is also important in view of the industrial network and development of transportation links. Proximity of railway and waterways is an advantage with respect to the transportation of construction materials for the nonsoil water confining elements, for instance metallic and asphalt-concrete screens and diaphragms etc.

Demographic features may also affect the selection of dam type in view of availability of skilled manpower.

The quality of personnel, flawless logistics, and technical background also play role in the selection of dam type. The Soviet experience in hydraulic engineering shows that the selection of dam type is often dictated by conditions of construction works. A simple design facilitating the procedures of filling and compaction of soil in dam body is given priority.

The present stage in the dam construction permits selection of any kind of soil. Effectiveness of dam in this case largely depends on the design. Formal approach to design procedures, underestimation of engineering and geologic investigations at site, or simplified attitude towards dam construction may have serious consequences, including reshaping of the dam design during the construction, which obviously would make the whole project more expensive and longer.

Hence the design of earth dams requires highly professional skills, a broad overview of engineering specialties and a kind of designer's intuition. Recent methods basing on the analysis of various construction and technological factors, the so-called factor analysis, can be employed successfully to substantiate the design of dams.

CHAPTER 9

Optimum design of earth dams by factor analysis

9.1 CONCEPT OF DAM DESIGN FACTORS

It has been noted that an earth dam is a complex system which depends on many elements, henceforth referred to as factors. These factors are both technological and design type. Design factors include the inclination of dam slopes, the width of core at foundation, the angle of core etc, while technological factors incorporate the density or compactness of core, thrust prisms, transitions, and the type of construction by horizontal and inclined layers (see Chapter 4), type of materials used for various dam components etc. Quantitative factors (textural density of soil in various dam components) and qualitative ones (type of dam, material from various quarries etc) are distinguished.

The factors can be deterministic and stochastic. We will discuss the deterministic factors.

Type of dam should be decided upon in various design stages, with respect to material (concrete or earth) followed by a decision on design (gravity, buttress or arched dam for concrete etc).

The set of all dams (or any other engineering structures) may be split up into classes (Pronina 1983) $A_1, A_2 \ldots A_n$.. Each object of class a_{ij} in n-dimensional space of properties X_j, in which $A_1 = \{a_{11}, a_{12}, \ldots\}$, $A_2 = \{a_{21}, a_{22}, \ldots\}$ etc is a vector $X_j = (x_1, x_2, \ldots, x_n)$. The coordinates of the vector X_j characterize properties of the object. Figure 9.1 illustrates the decomposition of a two-dimensional space of properties into the classes A_1, A_2, A_3. With respect to dams one may have: A_1=class of concrete dams, A_2=class of earth dams, A_3=class of timber dams. The restrictions imposed on the properties in each class are outlined in Figure 9.1. For class A_1 (shaded area) the following limits hold: $a \leq x_1 \leq b$; $c \leq x_2 \leq d$; $f_1(x_1, x_2) \geq 0$, $f_2(x_1, x_2) \geq 0$, $f_3(x_1, x_2) \geq 0$.

The design procedure for a hydraulic engineering scheme can be categorized in the following stages:

1. Selection of the river section in a system of cross-sections, for primary construction.

2. Selection of the composition of hydraulic engineering structures.

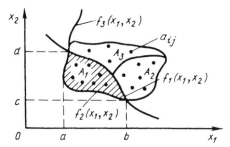

Figure 9.1. Classes of properties in two-dimensional space.

3. Selection of dam class for a given hydraulic engineering scheme.

4. Selection of dam design with inclusion of technical construction factors chosen from among a given class of dams.

Solution of the problem in each design stage has some general properties. Depending on given conditions, the stages can be generalized or more specific.

In the following we will discuss the selection of an optimum design for dams within the class of earth dams, that is we will dwell on the last stage of the design of a hydraulic engineering scheme.

9.2 FACTOR ANALYSIS IN DESIGN OF EARTH DAMS; METHOD OF COMPLETE FACTOR EXPERIMENT

Factor analysis in dam design. The problem of optimum design begins with selection of an architectural scheme for a dam design — earth-rockfill dam with core (central or inclined), homogeneous earth dam etc. Once the architectural scheme is chosen one determines the range of variation for parameters (factors) in this design.

The variable parameters (design elements) will be referred to as factors. Fairly many factors may be identified in earth dams. The number of factors considered determines the dimensionality of the factor space n. One should try to have as few dimensions as possible, preferably three to six. The optimum dimensionality is $n=4$–5, the reasons being discussed below. Reduction of the dimensionality of the factor space (the latter is defined as factors together with the boundaries of their variability) is reached through a more complete utilization of *a priori* information withdrawn from similarly designed structures. The *a priori* information is very important in optimization of dam design.

For instance, one designs an earth-rockfill dam of medium height in moderate climate. The following factors may be assumed $\bar{x} = m_1 =$ inclination of upstream slope; $\bar{x}_2 = m_2 =$ inclination of tailwater slope; $\bar{x}_3 = \frac{x}{H} =$ ratio of core width at bottom (for rectilinear boundaries of core) to height; $\bar{x}_4 = \alpha =$ inclination of core axis to vertical; $\bar{x}_5 = \gamma_{dry.c} =$ specific weight of dry soil of core; $\bar{x}_6 = \gamma_{dry.fill} =$ unit weight of dry soil in fill (thrust prisms); $\bar{x}_7 = IL$ or $HL =$ type of dam construction (by inclined layers, see Figure 4.8b, or horizontal layers along the entire width, see

Figure 4.8a); $\bar{x}_8 = \gamma_{dry.tr}$ = unit weight of soil in transition zones and similar areas.

The whole list of variable factors can be confined to $\bar{x}_1, \bar{x}_2, \bar{x}_3, \bar{x}_4, \bar{x}_5$, as one chooses moderate climate and the density index for coarse grained soil can reach $I_D \geq 0.9$ (obviously, if compaction machinery does exist); $\bar{x}_6$ can be assumed constant although its variability under severe climatic conditions might be substantial, so that the adequacy of the compaction of fills (thrust prisms) must be checked on. The factor $\bar{x}_7$ may also be variable as medium height dams (or even higher dams) are usually constructed by horizontal layers across the entire profile. In the construction of very high dams, as shown below, horizontal layers across the entire width are sometimes recommendable, and variation of $\bar{x}_7$ might affect the design of dam. The factor $\bar{x}_8$ can also be variable if the soil of core is hardly deformable, and high compactness of fill in transition zones is not difficult to achieve; $\bar{x}_8$ may be considered in strongly deformable soil of core and little deformable soil of thrust prisms (for instance gravel and pebbles rolled up to $I_D \geq 0.9$). In this case one may select a value of $\bar{x}_8$ that will ensure the cracking stability of core, for a given set of $\bar{x}_1, \bar{x}_2, \bar{x}_3$.

Hence from eight factors of analysis one may decide on five only. Further reduction of the number of factors can be reached by combining $\bar{x}_1$ and $\bar{x}_2$ into one factor $\bar{x}_1^*$, the variation of which brings about simultaneous variation of m_1 and m_2. Alternatively, variation of $\bar{x}_4$ may also be given up if α is constant and substantiated by geologic investigations (for instance, the presence of tectonic cracks requires fixed configuration of core (screen) in between cracks).

In some cases the optimum problem may be more specific — for fixed values of m_1 and m_2 one may find the most adequate values of $\bar{x}_3$, $\bar{x}_4$, $\bar{x}_5$ to ensure the cracking stability of dam core. The design problem in this case is to select variable factors from among those for specific construction conditions. The variation of the factors provides an optimum configuration for given conditions within a selected 'architectural scheme'.

If the variable factors for a given scheme do not provide a satisfactory optimum solution, then one may change the factor space or even the architectural scheme of the structure under design.

Once the variable factors are found one must find the functional relationship between these factors for:

1. Slope stability (upstream and downstream slopes).
2. Cracking stability in a seepage prevention facility, in vertical and horizontal areas.
3. Slope stability for a particular combination of load, for instance under seismic effects.

Each variation of selected factors provides numerical values for characterictics of slope stability, cracking etc. Each numerical value is referred to as system response, or simply — response. Responses provide functional relationships which shed light on the operation of the structure and provide information on the quantitative effect of factors or their combinations in various criteria of dam effectiveness. If in addition each variable factor is attributed a certain capital investment cost, then the technical criteria may also be regarded as economic ones.

Before we pass to the search for the optimum design let us consider criterial relationships.

Selection of variable factors is one of the important stages of optimization. The following stage consists in determination of boundaries for each variable factors, that is the determination of the factor space, the notion which unifies the factors and the ranges of their variability.

The ranges of variability for construction factors are found from the *a priori* information. For instance, inclination of the upstream slope for an earth-rockfill dam made of gravel and pebbles, with a thin central core on bedrock in seismically active regions (9 points) can be taken from 1:2 to 1:5 (higher value of m_1 corresponds to inclined core). The angle of core axis to vertical can vary from 0 to 20–25°, a higher value corresponding to dams with screens, that is for another architectural scheme of dam.

The factor space can obviously be expanded by inclusion of dams with screens, but an extreme increase in factor space dimensionality may have detrimental consequences, as will be discussed later.

The relative width of core $\frac{D}{H}$ can be taken from 0.2 to 0.25, up to 0.5, as the widths of central or slightly inclined cores exist in these ranges. Contemporary dams sometimes have cores expanded at their foundation, for instance Charvak Dam. In this case one may consider the highest width above the expanded section. If a wide core is anticipated at the beginning (for instance, a loam quarry is nearby the site and a quarry of coarse soil is far away), then the factor space may be changed ($0.5 \leq \frac{B}{H} \leq 1.0$).

Boundaries for variable factors of technological character, $\gamma_{dry.c}$ and $\gamma_{dry.fill}$, are determined by tests; recommendations given in Section 1.7 may be used if the tests are unavailable. Boundaries for the density of clayey soil depend on the degree of filling and moisture control of clayey soil, together with possible rollings. Analysis of possible technological versions of soil filling provides boundaries for the variation of compactness due to filling. Boundaries for the density of loose soil are determined by analogy — certain possible schemes for soil filling are considered and the adequate value of γ_{dry} is chosen. Similarly, one determines boundaries for the variation of other factors. The parameters φ, c and k should not be varied as they depend entirely on the density and moisture content of a given soil.

The basic rule in selection of factors reads that factors should be mutually independent and independently controlled. All factors cited above obey this rule.

For simplification of solution one must standardize the factors, which can be reduced to operation of centering and scaling (Pronina 1983).

Centering is the translation of the origin of coordinates to the centre of the factor space. For each factor the centre will have the following coordinate

$$\bar{x}_{0i} = 0.5(\bar{x}_{i.max} + \bar{x}_{i.min}) \qquad (9.1)$$

Scaling is a c-fold compression or expansion of the centered values of a factor

$$c = \frac{1}{0.5(\overline{x}_{i\,max} - \overline{x}_{i\,min})} = \frac{1}{\Delta\overline{x}_i} \tag{9.2}$$

in which

$\Delta\overline{x}_i$ = semi-interval of variation for i-th factor.

Standardization of the factor reads

$$x_i = \frac{\overline{x}_i - \overline{x}_{0i}}{\Delta\overline{x}_i} \tag{9.3}$$

while by returning to natural variables one has

$$\overline{x}_i = x_i\Delta\overline{x}_i + \overline{x}_{0i} \tag{9.4}$$

At the boundaries the factors x_i assume the following values (Eq.9.3): $x_{min} = -1$ = lower level, $x_{imax} = +1$ = upper level. At the origin of coordinates of the factor space one has $x_{oi} = 0$, and the corresponding natural value $\overline{x}_{oi}$ is referred to as the basic level of variable factor.

One should remember that the value of $\overline{x}_{i.max}$ or $\overline{x}_{i.min}$ may be assigned to any of the boundaries.

The transformation of natural variables (factors) $\overline{x}_i$ into standardized x_i by Equation 5.3 is linear. The standardization brings about dimensionless factors.

Upon transition to standardized factors the factor space in all cases assumes the form of a cube in n-dimensional space with the side length of 2. Details of factors and their properties are provided by Adler (1976).

Method of complete factor experiment. The most convenient type of response function inside the factor space is the following polynomial

$$\begin{aligned} y_i &= b_0 + b_1x_1 + b_2x_2 + \cdots + b_nx_n + b_{1\,2}x_1x_2 + \\ &\quad + b_{1\,3}x_1x_3 + \cdots + b_{1\,n}x_1x_n + \cdots + b_{n-1}x_{n-1}x_n + \\ &\quad + b_{1,2,3}x_1x_2x_3 + \cdots + b_{n-2,n-1}x_{n-2}x_{n-1}x_n + \cdots \\ &\quad + b_{1...n}x_1...x_n + b_{1\,1}x_1^2 + b_{2\,2}x_2^2 + \cdots + b_{n\,n}x_n^2 + \cdots \end{aligned} \tag{9.5}$$

One usually tries to describe the response function by a quasi-linear polynomial, in which all terms beginning from $b_{11}x_1^2$ are abondoned. In the theory of experiment planning (Adler et al. 1976, Zedginidze 1976), such polynomials are considered in terms of complete factor experiment. If in addition none of the interaction terms ($b_{1,...,n}x_1...x_n$) intervenes then one refers to fractional factor experiment.

The extreme case of a polynomial corresponding to fractional factor experiment is a linear polynomial, in which no interaction term exists. If the terms $b_{nn}x_n^2$ are introduced into the polynomial then the latter is of second order. Polynomials of higher order are impossible to utilize in problems of optimum design, and it is then better to reduce dimensionality of the factor space.

From Equation 9.5 it follows that the number of polynomial terms corresponding to the method of complete factor experiment reads $N=2^n$ (in which $n=$number of variable factors). For $n=2$ one has

$$y_i = b_0 + b_1 x_1 + b_2 x_2 + b_{1\ 2} x_1 x_2 \tag{9.6}$$

while for $n=3$ one gets

$$y_i = y_i(n = 2) + b_{1\ 3} x_1 x_3 + b_{2\ 3} x_2 x_3 + b_{1\ 2\ 3} x_1 x_2 x_3 \tag{9.7}$$

and so forth.

The coefficients at the factors b_0, b_1 ... are unknown as the standardized factors have variable values.

For independent determination of all unknown coefficients, the matrix of design planning (plan of dam design) should comply with some requirements — the number of various computations with variable factors N should be 2^n, that is the number of unknown coefficients; the sum of elements in matrix columns, with exclusion of the column x_0, where all elements are 1, should be zero $\Sigma x_i = 0$; the product of any two columns should be zero (orthogonality of the matrix; $\Sigma x_i x_j = 0$ at $i \neq j$). The factors are made variable at two levels only: upper (+1) and lower (-1), and therefore the standardization factor is taken either +1 or -1. Unities are skipped in the planning matrix and the signs remain only (Table 9.1).

Take n factors and assume that the response (for instance, the safety factors for the upstream slope or tailwater slope of dam, $k_{s.m_1}$ or $k_{s.m_2}$, respectively) and the factors are interrelated linearly:

$$y = b_0 x_0 + b_1 x_1 + b_2 x_2 + \cdots + b_n x_n \tag{9.8}$$

Write out for this case the matrices $[X]$, $\{Y\}$ and $\{B\}$ (Zedginidze 1976); in which $[X]$ = matrix of factors at nodes of factor space (planning matrix). The number of rows N in the matrix $[X]$ (number of design schemes) equals $n+1=$number of unknown coefficients with inclusion of b_0; $\{Y\}$ = vector-column of responses, i.e. values $k_{s.m_1}$ for various design schemes determined by rows in the design matrix (Table 9.1); $\{B\}=$vector-column of the unknown coefficient in Equation 9.8.

In this case the input system of linear equations in the matrix form will read

$$\{Y\} = [X]\{B\} \tag{9.9}$$

Equation 9.9 should be solved for $\{B\}$:

$$\{B\} = \left([X]^T[X]\right)^{-1} [X]^T \{Y\} \tag{9.10}$$

in which
$[X]^T$ = transposed matrix, or

$$
\begin{matrix} b_0 \\ b_1 \\ \cdot \\ \cdot \\ = \\ \cdot \\ \cdot \\ b_n \end{matrix}
\left(
\begin{bmatrix}
x_{0\ 1} & x_{0\ 2} & \cdots & x_{0\ N} \\
x_{1\ 1} & x_{1\ 2} & \cdots & x_{1\ N} \\
\cdots & \cdots & \cdots & \cdots \\
\cdots & \cdots & \cdots & \cdots \\
x_{n\ 1} & x_{n\ 2} & \cdots & x_{n\ N}
\end{bmatrix}
\begin{bmatrix}
x_{0\ 1} & x_{1\ 1} & \cdots & x_{n\ 1} \\
x_{0\ 2} & x_{1\ 2} & \cdots & x_{n\ 2} \\
\cdots & \cdots & \cdots & \cdots \\
\cdots & \cdots & \cdots & \cdots \\
x_{0\ N} & x_{1\ N} & \cdots & x_{n\ N}
\end{bmatrix}
\right)^{-1}
$$

$$
\times
\begin{bmatrix}
x_{0\ 1} & x_{0\ 2} & \cdots & x_{0\ N} \\
x_{1\ 1} & x_{1\ 2} & \cdots & x_{1\ N} \\
\cdots & \cdots & \cdots & \cdots \\
\cdots & \cdots & \cdots & \cdots \\
x_{n\ 1} & x_{n\ 2} & \cdots & x_{n\ N}
\end{bmatrix}
\cdot
\begin{Bmatrix}
y_1 \\ y_2 \\ \cdot \\ \cdot \\ \cdot \\ \cdot \\ y_N
\end{Bmatrix}
\tag{9.11}
$$

Since the elements in the planning matrix X_{ij} assume values $+1$ or -1 and the matrix

$$
X_1 X_j = \begin{cases} 0 \text{ for } i \neq 1 \\ 1 \text{ for } i = 1 \end{cases}
$$

is orthogonal, then one has

$$
[X]^T[X] =
\begin{bmatrix}
n & 0 & 0 & \ldots & 0 \\
0 & N & 0 & \ldots & 0 \\
\ldots & \ldots & \ldots & \ldots & \ldots \\
\ldots & \ldots & \ldots & \ldots & \ldots \\
0 & 0 & 0 & \ldots & N
\end{bmatrix}
$$

One also has

$$
([X]^T[X])^{-1} =
\begin{bmatrix}
1/n & 0 & 0 & \ldots & 0 \\
0 & 1/n & 0 & \ldots & 0 \\
\ldots & \ldots & \ldots & \ldots & \ldots \\
\ldots & \ldots & \ldots & \ldots & \ldots \\
0 & 0 & 0 & \ldots & 1/N
\end{bmatrix}
; [X]^T\{Y\} =
\begin{bmatrix}
\sum y_m x_0 \\
\sum y_m x_1 \\
\cdot \\
\cdot \\
\cdot \\
\sum y_m x_n
\end{bmatrix}
$$

From Equation 9.10, with the above transformations, one obtains

$$
b_i = \frac{\sum\limits_{j=1} y_i x_{ij}}{N}
\tag{9.12}
$$

in which
$$ i = 0, 1, 2, \ldots, n. $$

Hence the above properties of the planning matrix make it possible to find the

Table 9.1. Design plan and responses.

Sch. No	x_0	Planning x_1	x_2	...	x_n	x_1x_2	x_1x_3	...	x_1x_n
1	+	+	+	...	+	+	+	...	+
2	+	−	+	...	+	−	−	...	−
3	+	+	−	...	+	−	+	...	+
4	+	−	−	...	+	+	−	...	−
5	+	+	+	...	+	+	−	...	+
6	+	−	+	...	+	−	−	...	−
N-2	+	−	+	...	−	−	+	...	+
N-1	+	+	−	...	−	−	−	...	−
$N=2^n$	+	−	−	...	−	+	+	...	+

$\Sigma x_1=0; \ \Sigma x_2=0; \ ... \ \Sigma x_n=0; \ \Sigma x_1x_2=0; \ \Sigma x_1x_3=0; \ ... \ \Sigma x_1x_n=0$

No	...	$x_{n-1}x_n$	$x_1x_2x_2$	...	$x_{n-2}x_{n-1}x_n$	...	$x_1...x_n$	$y_1...$	$d_1...D \ E$
1	...	+	+	...	+	...	+		
2	...	+	−	...	+	...	−		
3	...	+	−	...	+	...	−		
4	...	+	+	...	+	...	+		
5	...	+	−	...	+	...	−		
6	...	+	+	...	+	...	−		
N-2	...	−	+	...	−	...	−		
N-1	...	−	−	...	−	...	−		
$N=2^n$	...	−	−	...	−	...	+		

$\Sigma x_{n-1}x_n = 0; \ \Sigma x_1x_2x_3 = 0; \ ... \Sigma x_{n-2}x_{n-1}x_n = 0; \ ... \Sigma x_1...x_n = 0$

Table 9.2. Properties of materials.

Level	E_0^*,MPa	n	Γ_0	M	G_0^*, MPa	U_0, MJ	b	γ^{**}, kN/m^3
+1	54.76	0.5	0.015	0.0	3.8	0.06	20	24.9
-1	12.9	0.68	0.017	0.02	2.87	0.05	40	24

(*) E_0 and G_0 are determined at $\sigma=0.01$ MPa. (**) Core soil is deemed saturated.

Table 9.3. Design plan for earth dam $N=2^3$.

Scheme No.	x_0	Planning core x_1	x_2	x_3	x_1x_2	x_1x_3	x_2x_3	$x_1x_2x_3$	$k_{s.0.y}$
1	+	-	-	+	+	-	-	+	1.21
2	+	-	+	-	-	+	-	+	1.27
3	+	+	-	-	-	-	+	+	1.79
4	+	+	+	+	+	+	+	+	1.70
5	+	-	-	-	+	+	+	-	1.27
6	+	-	+	+	-	-	+	-	1.48
7	+	+	-	+	-	+	-	-	1.94
8	+	+	+	-	+	-	-	-	1.58

coefficients b_i in Equation 9.8, independent of each other by Equation 9.12, that is any coefficient equals the sum of the product of elements in the vector columns of responses, for instance, values of $k_{s.m}$ for various variable versions of dam design with respective technological elements (Table 9.1) with respect to x_i, divided by the number of design versions N. In the case of Equation 9.7 the matrix $[X]$ can be expanded by inclusion of the interaction columns which possess all above properties. In the vector column $\{B\}$ one encounters the elements corresponding to interaction effects. Equation 9.12 in this case is suitable for determination of the coefficients intervening in Equation 9.7. This expansion of Equation 9.2 is equivalent to replacement of the interaction effects by new linear terms. These equations are referred to as linear by parameters.

In general, one usually obtains several responses as a result of solution of each problem utilizing the planning matrix — the stability index for upstream slope y_{j1}, the stability index for downstream slope y_{j2}, the safety index with respect to cracking in horizontal areas y_{j3}, the safety index for vertical areas y_{j4} etc, so for each response one finds the coefficients $b_{im} = \frac{\sum_{j=1} y_{im} x_{ij}}{N}$ and obtains k functions of Equation 9.7 type relationships reflecting the effect of the variable factor on one or another response.

The next stage consists in verification of adequacy of the obtained functions within the factor space. The best verification is the one about the centre of the factor space, where all factors are at the primary level ($x_i = 0$) and all terms of Equation 9.5 are zero, aside from the free term $b_0 (y_m = b_{0m})$. One usually takes the design scheme for a dam in the case $x_i = 0$ and standard computations are carried out. If the responses y_m and b_m differ by more than 5% (any other criterion is also possible), then the condition of adequacy can be regarded as satisfied, that is one may assume that the approximation by Equation 9.7 is sufficiently accurate. The above guidelines on the maximum possible reduction of the factor space dimensionality have been provided for the adequacy condition formulated as Equation 9.7. If y_m differs from b_{0m} by more than 5%, then the adequacy is not satisfied and one must either reduce the dimensionality of the factor space, and repeat the whole procedure again, or add quadratic terms to Equation 9.6 or Equation 9.7, that is pass to Equation 9.5. Additional computations are required if, for determination of the coefficients of the new polynomial, a new matrix must be formed (of second order) and the earlier matrix will be a part of it. Plans for second-order procedures are discussed in special references, for instance Zedginidze (1976). Statistical methods may also be applied (Adler et al. 1976) to verify the adequacy.

The above procedure was employed in the selection of rational design for a very high (300 m) dam (Rasskazov & Dzhkha 1979). Taken as x_1 was soil for the dam core. Since it originated from different quarries, two different types of material were considered: small deformability (upper level +1) and a greater deformability (level -1). Since no-one of the materials could have continuous properties corresponding to x_1 between +1 and -1, then x_1 must be considered a qualitative factor, as it can assume two values only of +1 and -1. The properties of the materials in terms of the energetics model (see Sections 1.3 and 1.4) are summarized in Table 9.2.

The second factor $\bar{x}_2$ was taken as the angle α of the core axis to vertical. The upper level of $\bar{x}_2$ was taken as a sloping core ($\alpha=24°$), while the lower limit corresponded to the upright layout ($\alpha=0°$). In all cases the core width was taken as $H/2$.

The third independent variable $\bar{x}_3$ was considered as a scheme of works (Fig.4.8). The upper limit (+1) was taken as the construction of a dam by horizontal layers, while the lower one (-1) corresponded to inclined layers. Intermediate values of this factor are associated with a part of dam (percentage of dam height) constructed by horizontal layers. The basic level corresponded to the lower 50% of dam height constructed by horizontal layers across the entire dam width. Other possible independent factors were not considered.

The response function incorporated the dependence of the safety factor for overall stability $k_{s.o.y}$ on the variable factors.

The layout of computations (planning matrix) was arranged as shown in Table 9.3.

The planning core in Table 9.3 must be identified. It is the three columns that reflect the variability of the three factors chosen above. The remaining columns are derivatives of the planning core. The column x_o always entirely consists of +1, and does not enter the plan. By direct check one may make sure that the conditions of orthogonality and zeroing of the sum of elements in each column are satisfied. The number of rows in the matrix is $2^3=8$.

Each row of the matrix represents a design version of dam. For instance, the third version describes a core soil of low deformability ($x_1=+1$), with a central core ($x_2=-1$), constructed by inclined layers ($x_3=-1$). As $k_{s.o.y}=1.79$ was obtained, the possible failure surface is profound, and embodies two slopes (which substantiates the term 'safety factor by overall stability').

Basing on Equation 9.12 one obtains the following coefficients of the polynomial

$$y = k_{s.0.y} = 1.54 + 0.21x_1 - 0.01x_2 + 0.04x_3 - 0.1x_1x_2 + \\ +0.03x_1x_3 + 0.02x_2x_3 - 0.02x_1x_2x_3 \tag{9.13}$$

Since the coefficient at the first factor x_1 is higher by one order of magnitude than the remaining coefficients, this factor (material of core) is of primary importance in comparison with the other factors intervening in this problem. Second in row is the coefficient at the interaction term of x_1 and x_2. The remaining terms of Equation 9.13 are secondary and may be neglected if Equation 9.13 is simplified.

Equation 9.13 is shown graphically in Figure 9.2. The ordinate axis shows the angle of core inclination to vertical while the abscissae axis, the third factor, indicates the fraction of dam height above foundation $x\%$ constructed by horizontal layers. One family of isolines depicts core soil of low deformability (skeletal soil shown is solid lines), while the other family illustrates the variation for the more deformable soil (dashed line).

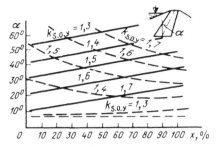

Figure 9.2. Isolines of overall stability factor $k_{s.o.y}$ for 300-m dam.

If the above results are confined to the distance above the abscissae axis up to the line AB (the extrapolation area being above it), one may note that $k_{s.o.y}$ is reduced by roughly 15% if the angle of core axis increases from 0 to 40°.

The data obtained for a 300-m dam (solid line in Figure 9.2) indicates that higher deformability of fills (thrust prisms), compared with dam core, stipulates construction of dams with a thin central core. On the other hand, if the soil of thrust prisms is much less deformable than the core soil (dashed line in Figure 9.2; the most frequent case), then a dam with inclined core is more advisable.

It is interesting to see that the safety factor increases in both cases if dam is constructed by horizontal layers across its entire width.

Equation 9.13 and Figure 9.2 facilitate selection of dam design for specific conditions in early design stages. It is also possible to estimate quantitatively the effect of various design schemes, selection of material during construction, and other factors that control $k_{s.o.y}$.

The above factor analysis is only a part of the general optimization problem for dam design. Examples of more complete utilization of factor analysis are given by Rasskazov (1985) and Reifman et al. (1982).

9.3 METHOD OF FRACTIONAL FACTOR ANALYSIS

An in-depth analysis of the working capacity of an earth dam might require variability of eight factors (or even more). In such cases, a complete factor analysis would require $N=2^8=256$ versions of dams, which would involve the analysis and construction of quasi-linear polynomials with 96 terms. This involves some difficulties. Reduction of the number of versions is possible in terms of the method of fractional factor analysis.

Examination of response functions has shown that the lowest effect is usually attributed to the interaction terms of the highest orders — in the case of three factors this is $x_1 x_2 x_3$. Indeed, the coefficient b_{123} in Equation 9.13 has the lowest value (together with the double interaction term $x_2 x_3$.) Accordingly, the interaction of high orders can be replaced by additional factors. In this case, the polynomial accounting for the variation of three factors in Equation 9.7 can assume the following form upon substitution of the triple interaction term by x_4:

$$y_0 = b_0 + b_1 x_1 + b_2 x_2 + b_3 x_3 + b_4 x_4 + b_{12} x_{12} + b_{13} x_{13} + b_{23} x_{23} \qquad (9.14)$$

In this case one refers to the problem as a '1/2 replica'. If double interaction terms had been further replaced by the fifth factor x_5, it could have been referred to as '1/4 replica' etc. The number of dam versions using fractional replicas will be

$$N = 2^{n-q} \qquad (9.15)$$

in which

q = number of interaction terms replaced by additional factors,
n = number of variable factors.

The ultimate case appears if all interaction terms are replaced by additional factors and one uses linear polynomial of Equation 9.8 type; this transition is usually possible within a fairly small range of variability as otherwise the adequate representation is not ensured. Each transition from complete to fractional analysis requires thorough consideration of all circumstances, preferably based on *a priori* information. In other words, one must substantiate the substitution of the interaction terms by additional factors and the general reduction of computational work.

Examples of fractional factor analysis, together with selection of variable factors and representation of computations, are given in Section 9.5.

Each response corresponds to a quasi-linear polynomial consisting of eight terms. In view of complexity of this polynomial one usually replaces it by simpler forms neglecting secondary factors and interactions, which is illustrated by low coefficients at these terms (as all terms are standardized). The simplified polynomial can be given as a nomograph. One proceeds by analogy to other polynomials reflecting the response of output to the other factors. It is however difficult to recommend a structure which would simultaneously respond to requirements of all outputs, as the importance of various factors is different on various levels, and conflicting results are often obtained.

A solution of the identification problem for classes of technically acceptable structures is presented in Section 9.4.

9.4 SEARCH FOR OPTIMUM STRUCTURE

The presence of various response functions of Equation 9.7 type poses the question of an optimum structure satisfying all response requirements, their limitations inclusive. Each response has it own criterion. For instance, the safety factors for the upstream and downstream slopes under a normal combination of load are given by the Soviet standard SNiP II-50-74 as $k_{s.up.} = k_{s.dwn.} \geq 1.25$, and those corresponding to particular combinations of loads are $k_{s.up.} = k_{s.dwn.} \geq 1.1$ (for first class dams). By analogy one has $k_{s.crack} \geq 1.1$, that is one may assume that the safety factor for seepage prevention design as to cracking corresponds to the safety of slope stability under a particular combination of loads. Moreover, the selected structure should cost as little as possible:

$$E = f(x_1, x_2, ..., x_n) \to E_{min}$$

Hence one has a few criteria which should be satisfied by the structure at the same time.

In the case of two criteria and two factors, the search for optimum design should give geometric interpretation on a plane (Pronina 1983). If two functions (response function and cost function) are shown as isolines, the optimum should be sought at the common points of their isolines. The set of these points is referred to as the Pareto set and a separate optimum point taken out of the Pareto set is called the Pareto point (to commemorate the name of Italian mathematician).

For two criteria of optimum $F_1(x_1, x_2)$ and $F_2(x_1, x_2)$, the equations of matched optimum are:

$$dF_1(x_1, x_2) = \frac{\partial F_1}{\partial x_1}dx_1 + \frac{\partial F_1}{\partial x_2}dx_2 = 0$$

$$dF_2(x_1, x_2) = \frac{\partial F_2}{\partial x_1}dx_1 + \frac{\partial F_2}{\partial x_2}dx_2 = 0$$

(9.16)

They represent a homogeneous system of two linear equations with respect to dx_1 and dx_2. This system has a nontrivial solution if the determinant is zero:

$$\begin{vmatrix} \frac{\partial F_1}{\partial x_1} & \frac{\partial F_1}{\partial x_2} \\ \frac{\partial F_2}{\partial x_1} & \frac{\partial F_2}{\partial x_2} \end{vmatrix} = 0$$

(9.17)

If the factor space is two-dimensional, then Equation 9.17 and its respective restrictive criteria yield generally two nonlinear equations with two unknows, the solution of which provides a unique optimum point. If the limitations correspond to acceptablity of the set of points from among $F_1(x_1, x_2) \geq C_1$ or $F_2(x_1, x_2) \geq C_2$, then the Pareto set is identified. An example of such limitation may be seen as $k_{s.up.} \geq 1.25$ etc.

If the number of variable factors is above three, for two criteria, the geometric interpretation of the matched optimum is impossible. In this case the Pareto set is sought by formulation and solution of a system of ordinary nonlinear equations (in which the determinant of Equation 9.17 type is transposed)

$$\begin{vmatrix} \frac{\partial F_1}{\partial x_1} & \frac{\partial F_2}{\partial x_1} \\ \frac{\partial F_1}{\partial x_2} & \frac{\partial F_2}{\partial x_2} \end{vmatrix} = 0; \quad \begin{vmatrix} \frac{\partial F_1}{\partial x_1} & \frac{\partial F_2}{\partial x_1} \\ \frac{\partial F_1}{\partial x_3} & \frac{\partial F_2}{\partial x_3} \end{vmatrix} = 0$$

$$\cdots\cdots\cdots\cdots\cdots\cdots$$

$$\cdots\cdots\cdots\cdots\cdots\cdots$$

$$\begin{vmatrix} \frac{\partial F_1}{\partial x_1} & \frac{\partial F_2}{\partial x_1} \\ \frac{\partial F_1}{\partial x_n} & \frac{\partial F_2}{\partial x_n} \end{vmatrix} = 0$$

(9.18)

The determinants in Equation 9.18 make it possible to obtain n-1 equations for n unknowns. With inclusion of the restrictions imposed on the function, $F_1(x_1, x_2, \ldots x_n)$ or $F_2(x_1, x_2, \ldots, x_n)$ one may find the set of optimum points (Pareto points) as simple lines. By moving on the line of matched optimum one may find the minimum criterial function, for instance $F_2(x_1, x_2, \ldots, x_n) = E(x_1, x_2, \ldots, x_n)$.

The $(n-1)$th equation obtained from the zero determinant condition in Equation 9.18 may be added to the criterial equation with the limitations of the type $F_1(x_1, x_2, \ldots, x_n) = C'$. In this case one generally obtains more than one system of n equations; this depends on how many values of C' are given *a priori*. By solving the ith system of equations one finds coordinates of the common points for the isolines C'_i and the one given by $F_2(x_1, x_2, \ldots, x_n) = C''$. This point is referred to as the local optimum point.

Putting coordinates of the obtained point into $F_1(x_1, x_2, \ldots, x_n)$ one obtains F_2 corresponding to this point. By varying C_i within the chosen interval one finds other points of the Pareto set and their respective F_2, from which one may try to idenfity the overall optimum. The limitations put on the factor space (for instance with inclusion of 20% extrapolation of the results of solution $1.2 \geq x_1 \geq -1.2$ and $1.2 \geq x_2 \geq -1.2$ etc.) one reduces the number of points in the search of optimum design.

The Pareto principle is not the only choice in the selection of optimum design. In other alternatives one seeks the optimum $E(x_1, x_2, \ldots, x_n)$ within the factor space, with limitations in the form of inequalities $D(x_1, x_2 \ldots x_n) \geq 0$. One may have several inequalities of this type (several technical criteria); examples are shown by Rekleitis et al. (1986).

If one has more than two criteria within the Pareto approach, the problem becomes more complex and one requires convolution of criteria. There are different principles of convolution (also those using the Pareto analysis) but we shall confine ourselves to the approach basing on a unique optimization parameter (Adler et al. 1976).

The convolution is implemented only among the technical responses of a system which provide one response $D(x_1, x_2, \ldots, x_n)$; there remains yet another response, an economic one; the matched optimum is sought ultimately. In the general form, each Pareto point provides a local optimum so that the overall optimum is sought by the minimum $E(x_1, x_2, \ldots, x_n)$ within a certain extended factor space (if necessary).

The convolution of technological criteria can be implemented with the Harrington desirability function. This function makes it possible to translate the natural response into dimensionless desirability scale (psychophysical scale), which secures a correspondence between physical and psychological parameters. Physical parameters are all possible responses characterizing the working capacity of a dam (safety factors for the stability of upstream and tailwater slope under static and seismic forces, and cracking in horizontal and vertical planes in seepage prevention facilities). Psychological parameters are referred to as purely subjective estimates of desirability for one or another response.

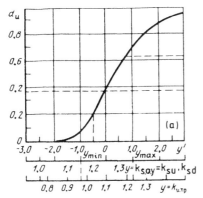
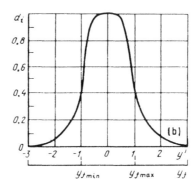

Figure 9.3. Desirability function. (a) properties bounded on one side; (b) properties bounded on two sides.

Partial responses are translated into dimensionless values denoted by d_u ($u=1,2$, $\ldots$, n) referred to as partial desirability; the scale stretches from 0 to 1. The quantity $d_u=0$ corresponds to absolutely unacceptable level of a given property, while $d_u=1$ provides the best value. The Harrington desirability function (Fig.9.3a) reads

$$d = exp[-exp(-y')] \qquad (9.19)$$

From Figure 9.3a it is seen that the desirability function may have two characteristic points: $d_j = \frac{1}{e} \cong 0.37$ and $d_j = e^{-e^{-1}} \cong 0.69$. The value of 0.37 corresponds to the boundary of acceptable responses ($k_{s.up.} = k_{s.dwn.} = 1.25$). This value corresponds to $y'=0$ on the abscissae axis (inflection of the curve). For response falling below the acceptable value its counterpart d_u decreases sharply. However, for the response growing above the value corresponding to $y' \cong 0.69$, the quantity d_u increases more and more slowly.

Transition from response to parameter d_u is provided by the relationship

$$y' = \frac{2y - (y_{max} + y_{min})}{y_{max} - y_{min}} \qquad (9.20)$$

in which

$y_{max}; y_{min}$ = conditional responses corresponding to the scale of y' from -1 to +1.

For instance, one may assume $y_{max}=1.35$ and $y_{min}=1.15$, which is convenient in the assessment of safety factors for normal conditions of dam operation.

Determine y' for the case with $k_s=1.25$:

$$y' = \frac{2 \cdot 1.25 - (1.35 + 1.15)}{1.35 - 1.15} = 0$$

and $d_1 \cong 0.37$, in agreement with Equation 9.19.

Figure 9.3b illustrates the desirability function for the properties bounded on two sides. This function may be utilized in analysis of concrete structures if $y_{j.min}$ corresponds to the tensile strength of concrete in the structure while $y_{j.max}$ describes the compressive strength of concrete. The desirability function in this case reads:

$$d = exp[-(y')^k] \tag{9.21}$$

in which

k = parameter controlling the scale of desirability function, most often taken as 2–2.5.

Upon translation of all partial responses into partial desirability functions by Equation 9.19 (Tables 9.1 and 9.4) one may construct a generalized index D, which is referred to as a generalized desirability function:

$$D = \sqrt[n]{\Pi_{y=1}^n d_u} \tag{9.22}$$

The primary peculiarity of Equation 9.22 consists in difficult overlapping of a small range of one response by a larger range of another response. If one of d_u goes to zero, the entire system also assumes a small value D.

The generalized desirability function is sensitive to small partial desirabilities. Each design scheme can correspond to the parameter D_j by general rules analogous to those for partial functions; hence one may construct the function $D = f(x_1, x_2, \ldots, x_n)$. A matched optimum exists between the two optimization criteria $D(x_1, x_2 \ldots, x_n)$ and $E(x_1, x_2, \ldots, x_n)$. Solution is also possible for three and more criteria. The Pareto principle will reduce by one the dimensionality of the criterial space.

9.5 APPLICATION OF FACTOR ANALYSIS IN OPTIMIZATION OF STRUCTURES

The method presented above can be fully illustrated for a selection of adequate design for a 70-m dam (Rasskazov & Orekhova 1985). The dam foundation consists of gravel and pebbles. The dam incorporates a seepage prevention facility made as a core or screen of loam, and fills (thrust prisms) constructed of gravel and pebbles. The distance from the centre of gravity of the dam to quarries is about 3 km. Tests on the loam have shown that consolidation by average rolling provides $\gamma_{dry} = 17.8$ kN/m³. Consolidation by heavy rolling does not improve the weight ($\gamma_{dry} = 18.2$ kN/m³, cf. Section 1.7).

In selection of the optimum design, four criteria were chosen: x_1 = core width in the range from 34 to 17 m; x_2 = angle of core axis from 0 to 58°; x_3 = unit weight of dry core soil, from 18.2 to 17.8 kN/m³; x_4 = slope of upstream and downstream sides from 2.75 and 2.5 to 2.0 and 1.8, respectively, denoted by m_1, and m_2. Simultaneous variation of the slopes m_1 and m_2 was required for reduction of factors. At the same time, it involves limitations due to the necessity of proportional simultaneous variation of both slopes upon transition to intermediate levels.

Basing on the present experience of the construction of earth-rockfill dams it was assumed that filling of gravel and pebbles in thrust prisms corresponds to $I_D=0.9$; with transition to the density as specified in Chapter 1.

The optimum design was sought within planning matrices (Table 9.4) constructed by the method of fractional factor analysis of 2^{4-1} type; i.e. semi-replicas were utilized. The factor x_4 was taken instead of the triple interaction term.

All computations were carried out in terms of the energetics soil model (Sections 1.3 and 1.4). Soil properties were determined in triaxial tests (they are not given here as they do not affect the method of dam optimization). Partial responses are given in Table 9.4, i.e. safety factors with respect to stability of the upstream ($k_{s.up.}$) and downstream ($k_{s.dwn.}$) slopes and cracking conditions in vertical and horizontal planes, $k_{s.crack.v}$ and $k_{s.crack.h}$; respectively. Partial responses in the Harrington formulation are attributed the coded values (Eq.9.19 and Fig.9.3a); generalized technological parameters of optimization were found by Equation 9.22 (Table 9.4). Coefficients of the polynomial $D = f(x_1, x_2, x_3, x_4, x_1x_2, x_1x_3, x_2x_3)$ were found by Equation 9.12 and thereby the effect of all factors on the generalized working capacity of the dam was determined (see Appendix of Table 9.4).

Table 9.4. Plan and results of dam computations. Matrix $N=2^{4-1}$.

Scheme x_0 No.		Plan of computations by factors							
		Cod. x_1	Abs. B,m^2	Cod. x_2	Abs. α^o	Cod. x_3	Abs. γkN/m^3	Cod. x_4	Abs. $m_1; m_2$
1	+	+	34	+	58	-	17.8	-	2.0;1.8
2	+	-	17	-	0	-	17.8	-	2.0;1.8
3	+	+	34	-	0	-	17.8	+	2.75;2.5
4	+	-	17	+	58	-	17.8	+	2.75;2.5
5	+	+	34	+	58	+	18.2	+	2.75;2.5
6	+	-	17	-	0	+	18.2	-	2.75;2.5
7	+	+	34	-	0	+	18.2	-	2.0;1.8
8	+	-	17	+	58	+	18.2	-	2.0;1.8

Interactions			Absolute and coded technical responses							
x_1x_2	x_1x_3	x_2x_3	$k_{s.up}$	$d_{k_{s.up}}$	$k_{s.dwn}$	$d_{k_{s.dwn}}$	$k_{cr.h.}$	$d_{k_{cr.h}}$	$k_{cr.v}$	$d_{k_{cr.v}}$
+	-	-	1.24	0.32	1.28	0.48	2.0	0.9	2.0	0.9
+	+	+	1.25	0.37	1.21	0.225	6.0	0.995	4.3	0.985
-	-	+	1.32	0.61	1.35	0.63	1.67	0.8	1.5	0.68
-	+	-	1.35	0.63	1.35	0.63	5.0	0.99	2.0	0.9
+	+	+	1.32	0.61	1.29	0.52	1.9	0.87	1.22	0.48
+	+	-	1.4	0.8	1.38	0.78	1.0	0.28	0.6	0.03
-	+	-	1.24	0.32	1.26	0.41	1.3	0.55	1.15	0.4
-	-	+	1.32	0.61	1.29	0.52	0.7	0.07	0.7	0.07

Table 9.4a.

Scheme No.	Gen. D_j	Economic responses of central section, roub/m E_{1j}	E_{2j}	E_{3j}	E_{4j}
1	0.594	30 075.8	57 240.8	73 971.8	101 136.8
2	0.534	13 186.1	38 779.3	16 510.6	42 093.8
3	0.676	17 810.7	51 539.6	23 212.1	56 941.0
4	0.771	20 289.5	61 828.9	22 813.5	64 352.9
5	0.603	38 410.1	77 682.3	82 306.1	121 578.3
6	0.269	21 114.8	63 313.0	24 429.3	66 627.5
7	0.412	13 448.8	37 068.6	18 850.3	42 470.1
8	0.198	14 561.7	43 427.7	17 085.7	45 951.7

Version	Dist. to quarry, km grav+pebb	loam	Soil cost, roub/m^3 grav+pebb	loose cl.loam	dense cl.loam
1	3	3	1.73	1.48	1.71
2	15	3	5.45	1.71	1.48
3	3	15	1.73	5.2	5.43
4	15	15	5.45	5.2	5.43

(Cod) coded value; (Abs) absolute value; (Gen) generalized index. $D = 0.507 + 0.064x_1 + 0.0344x_2 - 0.1366x_3 + 0.0726x_4 - 0.0071x_1x_2 + 0.0729x_1x_3 - 0.0044x_2x_3$; $E_1 = 21\ 113.4 + 3822.9x_1 + 4720.8x_2 + 770.4x_3 + 3292.8x_4 + 4585.76x_1x_2 + 222.68x_1x_3 - 118.78x_2x_3$;

$E_2 = 53860 + 2022.8x_1 + 6184.9x_2 + 1512.9x_3 + 9730.9x_4 + 5393.8x_1x_2 - 20.25x_1x_3 - 1002.8x_2x_3$; $E_3 = 34897 + 14688x_1 + 14147x_2 + 770x_3 + 3293x_4 + 14407x_1x_3 + 223x_1x_3 - 119x_2x_3$;

$E_4 = 67644 + 12887.5x_1 + 15610.9x_2 + 1512.9x_3 + 9730.9x_4 + 15215.0x_1x_2 - 20.24x_1x_3 - 1002.8x_2x_3$.

For convenient independent use, the function can be given in the form of nomographs (Fig.9.4). One may also provide graphical representation for partial responses (Rasskazov & Dzkha 1978, Rasskazov & Sysov 1982).

The primary factor is given by the cost response E_1=cost of 1 metre of dam length. It is determined on the condition that the quarry lies 3 km from dam. The respective specific costs of transportation at a distance of 3 km and filling of soil into dam body are given in Appendix of Table 9.4. By and large, the response function for the variable factors changing with the cost of 1 metre of dam length in the central cross-section was obtained: $E_1 = \varphi(x_1, x_2, x_3, x_4, x_1x_2, x_1x_3, x_2x_3)$. The functions E_2, E_3, E_4 were similarly constructed for analysis of various versions of quarry configuration with respect to the central cross-section of dam (Table 9.4); this is so because the cost of materials does not increase proportionally to the distance from quarry.

Upon substitution of $F_1 = D, F_2 = E_1$ in the three first determinants of Equation 9.18 one obtains three equations with four unknown factors. The first equation reads:

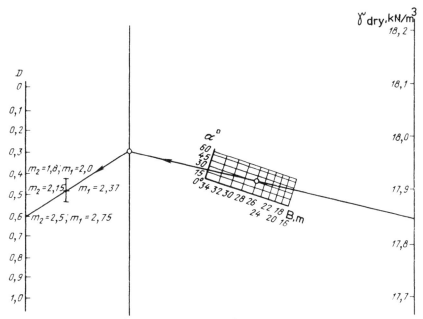

Figure 9.4. Nomograph for $D(\gamma_{dry}, B, \alpha, m_1, m_2)$.

$$\left.\begin{array}{l} 0.064 - 0.0071x_2 + 0.0729x_3; \quad -411 + 325x_2 - 81x_3 \\[2mm] 0.0344 - 0.0071x_1 - 0.0044x_3; \quad 485 + 352x_1 - 425x_3 \end{array}\right\} = 0 \qquad (9.23)$$

The system in Equation 9.23 is reduced to the cubic equation

$$0.104x_2^3 - 0.8x_2^2 + 1.29x_2 - 0.58 = 0$$

Solution of this equation yields three roots $x_2' = +0.984$, $x_2'' = +0.991$ and $x_3''' = +5.711$. With inclusion of the restrictions imposed on the factor space $-1 \leq x_2 \leq 1$ one obtains one root only (as x_2' and x_2'' coincide) — $x_2 = +0.99$. The remaining roots are $x_1 = +1.897$ and $x_3 = -0.781$.

Upon substitution of x_1, x_2 and x_3 into $D(x)$ and $E_1(x)$ and subsequent analysis of these functions one obtains $D = 0.5$ already for $x_4 \ll -1$, that is for very steep slopes of thrust prisms. The lower limit of variation of this factor is $x_4 = -1$; ($m_1 = 2$, $m_2 = 1.8$), and its reduction is undesirable; therefore one obtains $D = 0.578$ and $E = 14\ 222$ roubles/m at $x_4 = -1$.

The value $x_1 = 1.897$ is beyond the factor space ($-1 \leq x \leq 1$). By using the gradient method one finds coordinates of the optimum point as the closest to the given factor space and yet on its surface.

Parameters of the optimum dam design are given in Table 9.5 in terms of the distance from quarries to the central section of the dam. The solution is obtained for E_2, E_3, and E_4 by analogy to E_1 (see Table 9.4).

Hence, as a result of the solution, the optimum design for the 70-m dam on

Table 9.5. Parameters of optimum dam design.

Fig.9.5 ref.	Distance to quarry, km loam	grav.& pebb.	Standardized parameter	Natural parameter
a	3	3	$x_1 = 1.45$ $x_2 = -1.0$ $x_3 = -0.64$ $x_4 = -1.0$ $D = 0.52$ $E = 11\ 797$ r/m	$B = 38$ m $\alpha = 0°$ $\gamma_{dry} = 17.9$ kN/m^3 $m_1 = 2.0;\ m_2 = 1.8$
b	3	15	$x_1 = 2.13$ $x_2 = -0.35$ $x_3 = -0.957$ $x_4 = -1$ $D = 0.544$ $E = 33\ 013$ r/m	$B = 43.6$ m $\alpha = 19°$ $\gamma_{dry} = 17.8$ kN/m^3 $m_1 = 2.0;\ m_2 = 1.8$
c	15	3	$x_1 = -1.0$ $x_2 = -1.0$ $x_3 = -1.0$ $x_4 = -1.0$ $D = 0.534$ $E = 16\ 511$ r/m	$B = 17$ m $\alpha = 0°$ $\gamma_{dry} = 17.8$ kN/m^3 $m_1 = 2.0;\ m_2 = 1.8$
d	15	15	$x_1 = 1.871$ $x_2 = -1.0$ $x_3 = -0.575$ $x_4 = -1.0$ $D = 0.53$ $E = 37\ 157$ r/m	$B = 41.4$ m $\alpha = 0°$ $\gamma_{dry} = 17.9$ kN/m^3 $m_1 = 2.0;\ m_2 = 1.8$

(r/m) roubles per metre.

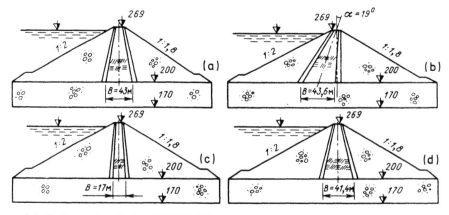

Figure 9.5. Optimum dam design (cf. Table 9.5).

compressible foundation, in four versions of quarry configuration and slope of thrust prisms has been found (Fig.9.5). The parameters obtained for the optimum design provide estimates for the differences between the actual dam and the optimum one if quarries change during construction.

The adequacy of the functions D and E is checked prior to ultimate selection of the optimum design; this means that an additional setup is considered, preferably at the centre of the factor space, and deviations of D from these given by relationships between D and E are examined.

Solution of the optimization problem may be based not only on an analysis of stress-strain conditions in dams but also on simpler and widely used methods. For instance, slope stability can be determined by the method of circular cylindric surfaces with inclusion or without seismic forces. Cracking stability may be analysed in terms of pore pressure effects in approximate stresses etc. However, in all these cases the above procedure provides a unique, most economic design in which all technological responses assume the required, or better, values. It is stressed that the most economical design must not incorporate the strength parameters (stability parameters) at their limits.

The existence of the functions $D(x_1, x_2, \ldots, x_n)$ and $E(x_1, x_2, \ldots, x_n)$ makes it possible to conduct continuous quality analysis of the effect of deviations from technological conditions on the construction and working capacity of a dam. Economy and construction go hand in hand in the optimum analysis of the distance from quarries.

The method of fractional factor analysis reduces the amount of computations and simplifies the selection of the optimum solution, which is a most welcome feature. One is advised to take not less than 1/2 replica, that is to introduce not more than one linear factor.

One must not be overwhelmed by the number of variable factors. The optimum number is 4, but one may increase the number of factors to 6 (by taking for instance 1/2 replicas). Replicas above 1/2 should not be taken, as the uniqueness of solution would be at stake.

One must stress once again that the factor space should be minimized in order to ensure adequate conditions.

If the optimum solution lies beyond the factor space, the root outside this space must be given the closest value at the boundary of the factor space, and the system of equations (simplified this time) must be solved, with new roots to be determined. This approach can be repeated for each factor, the optimum value of which lies beyond the factor space. In the case of the roots of solution outside the factor space one must remember that the sign of the root indicates the direction of the variation of the factor in the optimization of design. The value at the acceptable boundary of the factor space (for instance +1 or -1, depending on the sign of the root) must be taken in this case. The rank of the system is reduced, the system of nonlinear equations is solved again, the roots are analysed etc. This algorithm has been implemented in a numerical program worked out at the Moscow Institution of Civil Engineers.

References

Adamovich, A.N. 1980. *Soil reinforcement and counterseepage screens* (in Russian). Energiya, Moscow.

Adler, Yu. P., E.V.Markova. & Yu.V.Granovskiy 1976. *Planning of experiments in selection of optimum conditions* (in Russian). Nauka, Moscow.

Aiby, J.A. 1982. *Earthquakes* (in Russian). Nedra, Moscow.

Anonymous 1976a. *Dams and dykes from soil materials with asphalt-concrete and metallic counterseepage elements (A review)* (in Russian). VNIIG, Leningrad.

Anonymous 1976b. *Dams from soil materials with metallic counterseepage constituents* (in Russian). Informenergo, Moscow.

Anonymous 1982. Asphalt-concrete screens and diaphragms in earth dams (in Russian). *VNIIG Reviews, Ser. Gidroelektrostantsii* 4.

Anonymous 1984. Designing foundations of hydraulic engineering structures (in Russian). *Posobie SNiP-II-16-76*, Moscow.

Aravin, V.I. & S.N. Numerov 1955. *Seepage computations for hydraulic engineering structures* (in Russian). Gosstroyizdat, Moscow-Leningrad.

Argiris, J. 1968. *Developments in matrix design of structures* (in Russian). Stroyizdat, Moscow.

Bakhtiyarov, P.I., G.I.Kornyakov & V.F.Korchevskiy 1976. Construction by controlled blasting of hydraulic engineering schemes with rock dams, without screens (in Russian). *Gidrotekhn. Stroit.* 6: 10–13.

Balykov, B.I. et al. 1973. Some investigations of core consolidation in Charvak Dam (in Russian). *Izvest. VNIIG* 102: 177–181.

Belakov, A.A. 1983. Effect of arc forms on working capacity of large earth-rockfill dam (in Russian). *Gidrotekhn. Stroit.* 9: 19–22.

Belgorodskaya, G.N. 1969. *Seismic design of dams made of local materials, with inclusion of elastoplastic strains* (in Russian). 1: 102–105. Donish, Dushanbe.

Bezukhov, N.I. & O.V.Luzhin 1974. *Theory of elasticity and plasticity methods applied to engineering problems* (in Russian). Vysshaya Shkola, Moscow.

Biot, M.A. 1941. General theory of three-dimensional consolidation. *Jour. Appl. Phys.* 12(2): 155–165.

Biot, M.A. 1956. Theory of deformation of a porous viscoelastic anisotropic solid. *Journ. Appl. Phys.* 27(5): 459–467.

Biyanov, G.F. 1983. *Permafrost dams* (in Russian). Energoatomizdat, Moscow.

Bolt, B. 1981. *Earthquakes* (in Russian). Mir, Moscow.

Booker, J.R. & J.C.Small 1977. Finite-element analysis of immediate consolidation. *Intern. Journ. Solid and Struct.* 13(2): 137–149.

Borovets, S.A. 1961. Nurek hydroelectric plant on the Vahsh River (in Russian). *Gidrotekhn. Stroit.* 6: 3–8.

Borovoy, A.A. & V.I.Vutsel 1974. Design and construction problems of earth-rockfill dams (in Russian). *Gidrotekhn. Stroit.* 3: 51-55.

Borovoy, A.A., P.D.Yevdokimov & G.Kh.Pravednyy 1973. Water-confining elements of dams made of local materials (in Russian). *Gidrotekhn. Stroit.* 5: 4–8.

Botkin, A.I. 1940. Strength of granular and brittle materials (in Russian). *Izvestiya VNIIG* 26: 205–236.

Brzhoushek, M. 1981. Use of asphalt-concrete diaphragms in foreign hydraulic engineering (in Russian). *Energ. Stroit. za rub.* 29-31.

Budweg, F. 1973. Safety improvements taught by dam incidents and accidents in Brasil. *14th Intern. Congr. Large Dams* 1(Q52R73): 1254–1262.

Bugrov, A.K. 1975. Finite-element method in consolidation computations for water-saturated soils (in Russian). *Gidrotekhn. Stroit.* 7: 35–38.

Burenkova, V.V. & B.A.Mokran 1983. On reliability and suffosion resistance of soil in hydraulic engineering structures. *Energ. stroit.* 12: 70–73.

Burshteyn, M.F. & Yu.S.Grebenshchikov 1972. Fundamentals of complex design for dam construction by controlled explosions (in Russian). *Ob. nauch. trud. Gidroproy.* 37: 56–71.

Bushkanets, S.S., V.V.Berezoshvili & L.V.Gorelik 1973. Experimental validation of consolidation computations in closed systems. *Izvest. VNIIG* 102: 131–136.

Cambefort, A. 1971. *Soil grouting* (in Russian). Energiya, Moscow.

Chernous'ko, F.L. 1965. Method of local variations for numerical solution of variational problems (in Russian). *Zhur. vych. matem. mat. fiz.* 5(4): 749–754.

Chernous'ko, F.L. & N.V.Bonichuk 1973. *Variational problems in mechanics and control* (in Russian). Nauka, Moscow.

Chugayev, R.R. 1967. *Hydraulic-engineering earth structures* (in Russian). Energiya, Leningrad.

Churakov, A.I. 1976. *Special works in hydraulic engineering* (in Russian). Stroyizdat, Moscow.

Clough, K.W. & R.S.Woodward 1967. Analysis of embankment stresses and strains. *Proc. ASCE SM* 93(4): 529.

Connor, D.M. & C.Brebbia 1979. *Finite-element method in fluid dynamics* (in Russian). Sudostroyeniye, Leningrad.

Cough, R.U. 1979. *Finite-element method in solution of plane problems in the theory of elasticity* (in Russian). Stroyizdat, Moscow.

Deresevich, G. 1961. Mechanics of granular medium (in Russian). *Problems of mechanics* 3: 91–149. Izd. Inostr. Lit., Moscow.

Design guidelines on transitions of earth-rockfill dams (in Russian) 1971. VSN 47-71 VNIIG, Leningrad.

Design guidelines on consolidation of earth dams and foundations (in Russian) 1975. P 36-75 VNIIG, Leningrad.

Design guidelines for earth dams in northern climatic zones (in Russian) 1976a. P 48-76 VNIIG, Leningrad.

Design guidelines on seepage strength of earth dams (in Russian) 1976b. P 55-76 VNIIG, Leningrad.

Design guidelines on inverse filters for hydraulic engineering structures (in Russian) 1981. P 92-80 VNIIG, Leningrad.

Design guidelines on inverse filters in earth dams (in Russian) 1982. VNII VODGEO, Moscow.

Didukh, B.I. 1970. On solution of the nonlinear problem of consolidation in clayey-core rockfill dam (in Russian). *Construction on weak soil*: 181–186. Rizhskiy politekh. in-t, Riga.

Dolezhalova, M. 1967. Tensile testing of clayey soil (in Russian). *Trudy VODGEO* 18: 7–13.

Dolezhalova, M. 1969. Use of grid method in assessment of the effect of crack edge steepness on stresses and strains in the central core of earth-rockfill dams (in Russian). *VNII VODGEO* 5: 8-18.

Dolezhalova, M. 1983. Unusual strains in Jirkov Dam, and their mathematical modelling (in Russian)., *Seventh Danube-Europ. Confer. Soil Mech.*: 231-238. Kishinev.

Drucker, D. 1962. Stress-strain relationships for metals in plastic region: experimental data and basic notions (in Russian). *Rheology. Theory and Applications*: 127–158. Izd. Inostr. Lit. Moscow.

Eisenstein, Z. & S.Zaw 1977. Analysis of consolidation behavior of Mica dam. *Proc. ASCE GT* 103(8): 879–895.

Esta, Y.B. & M.Hajal 1973. Behaviour of sands before failure. *8th Intern. Conf. Found. Eng. Soil Mech.* 1(1): 111–116.

Fedorov, I.V. 1962. *Slope stability computations* (in Russian). Gosstroyizdat, Moscow.

Filchakov, P.F. 1960. *Theory of seepage in hydraulic engineering structures* (in Russian). Izd. AN SSSR, Moscow.

Finn Liam, V.D. & A.P.Troitskiy 1968. Stress-strain computations for earth dams slopes and foundations by FEM (in Russian). *Gidrot. stroit.* 6: 22–27.

Florin, V.A. 1939. Basic equations for rock dynamics (in Russian). *Izvestiya VNIIG* 25: 190–196.

Florin, V.A. 1948. *Theory of hardening of earth masses* (in Russian). Stroyizdat, Moscow.

Florin, V.A. 1953a. One-dimensional consolidation problem for a compressible creeping porous earth medium (in Russian). *Izv. AN SSSR OTN* 6.

Florin, V.A. 1953b. One-dimensional problem of hardening of earth medium with nonlinear ageing creep and structural failure (in Russian). *Izv. AN SSSR OTN* 9: 1229–1234.

Florin, V.A. 1961. *Fundamentals of soil mechanics* (in Russian). TPM Gosstroyizdat, Moscow.

Fradkin, B.V. 1973. FEM computations for gravity structures (in Russian). *Ob. nauchn. trud. Gidroproy.* 28: 29–36.

Freidental, A. & H.Geiriger 1962. *Mathematical theories of inelastic continuous medium* (in Russian). Fizmatgiz, Moscow.

Frenkel, Ya.I. 1944. On the theory of seismic and seismoelectric phenomena in wet soil (in Russian). *Izv. AN SSSR ser. geogr. geofiz.* VIII(4): 133–150.

Gersevanov, N.M. & D.Ye.Polshin 1948. *Theoretical fundamentals of soil mechanics* (in Russian). Gosstroyizdat, Moscow.

Glebov, V.D. 1985. Counterseepage measures in hydraulic-engineering earth structures (in Russian). *Gidrotekhn. Stroitel.* 1: 17–20.

Goldin, A.L. 1966. Consolidation computations for clayey core in high dam, with inclusion of viscous properties of soil skeleton (in Russian). *Izvestiya VNIIG* 80: 141–150.

Goldshteyn, M.N. 1977. *Mechanical properties of soil* (in Russian). Stroyizdat, Moscow.

Goldstheyn, M.N., S.G.Kushner & M.I.Shevchenko 1977. *Computation of settlement andstrength of foundations in buildings and structures* (in Russian): 16—19. Budivelnik, Kiev.

Gorelik, L.V. 1975. *Computation of consolidation of earth foundations and dams* (in Russian). Energiya, Leningrad.

Gorelik, L.V. & B.M.Nuller 1983. A method for assessment of strain in earth structures and dams, with inclusion of freezing and thawing. *Mechanics and physics of ice* (in Russian): 95—100. Nauka, Moscow.

Gorelik, L.V., B.M.Nuller & B.A.Shoykhet 1981. Design models for soil subject to freezing and thawing (in Russian). *Izvestiya VNIIG* 151: 66-71.

Grishin, M.M.(ed.) 1979. *Hydraulic engineering structures* (in Russian). Vysshaya Shkola, Moscow.

Guidelines on design and construction of counterseepage measures using PVC foil for artificial reservoirs 1983. SN-551-82 (in Russian). Stroyizdat, Moscow.

Gun, S.Ya. 1971. Determination of dam stress state by theory of elasticity methods (in Russian). *Trudy VODGEO* 30: 7-9.

Hydraulic engineering in permafrost and severe conditions (in Russian) 1979. *Proceed. Mater. konfer. sovesh. gidrotekh.* Energiya, Leningrad.

Ioselevich, V.A. 1967. Deformability laws for nonrock soil (in Russian). *Osnov. fundam.* 4: 3–7.

Ioselevich, V.A. & B.I.Didukh 1970. Use of the theory of plastic consolidation for description of soil deformability (in Russian). *Soil mechanics and construction on loessial foundations*: 125–133. Groznyj Chech. Ingush. kn. izd..

Ioselevich, V.A., L.N.Rasskazov & Yu.M.Sysoyev 1979. Growth of surface load upon plastic hardening of soil (in Russian). *Izv. AN SSSR Ser. Mekh. twerd. tela* 2: 155–161.

Ioselevich, V.A., V.V.Zuyev & G.A.Chakhaturi 1975. Effects of plastic hardening of nonrock soil (in Russian). *Nauch. tr. inst. mekh. MGU* 2: 96-112.

Istomina, V.S. 1957. *Seepage stability of soil* (in Russian). Stroyizdat, Moscow.

Istomina, V.S., V.V.Burenkova & G.V.Mirushova 1975. *Seepage stability of clayey soil* (in Russian). Stroyizdat, Moscow.

Ivanov, P.L. 1983. *Consolidation by explosion of slightly cohesive soil* (in Russian). Nedra, Moscow.

Ivanov, P.L. 1985. *Soil in foundations of hydraulic engineering structures* (in Russian). Vysshaya Shkola, Moscow.

Ivashchenko, I.N. & M.N.Zakharov 1973. Application of flow theory to soil (in Russian). *8th Intern. Congr. Soil Mech. Found. Eng.* 225-228. Stroyizdat, Moscow.

Ivlev, D.D. & G.I.Bykovtsev 1971. *Theory of hardening plastic body* (in Russian). Nauka, Moscow.

Kasatkin, Yu.N. 1981. Designing composition of transition layers in asphalt-concrete diaphgrams and earth dams (in Russian). *Gidrotekh. Stroit.* 6: 32–34.

Kasatkin, Yu.N, S.M.Popchenko & G.V.Borisov 1970. *Asphalt-concrete linings and screens in hydraulic engineering structures* (in Russian). Energiya, Leningrad.

Kerchman, V.I. 1974. Interface problem of the theory of consolidation in water-saturated medium (in Russian). *Izv. AN SSSR* 3: 102–109.

Khristianovich, S.A. 1940. Groundwater motion which does not obey Darcy's law (in Russian). *Prikl. matem. mekh. AN SSSR* IV: 33–52.

Kolar, V., J.Kratochvil & A.Leitner 1975. *Finite-element computations for plane and spatial structures* (in German). Springer Verlag.

Korchevskiy, V.F. 1962. Design of dams constructed by controlled explosion (in Russian). *Sbor. trud. Gidropr.* 88: 29–41.

Krasnikov, N.D. 1979. *Dynamic properties of soil and methods of their determination* (in Russian). Gosstroyizdat, Moscow.

Kryzhanovskiy, A.L. 1970. Equations relating stress components and soil strains in spatial stress condition (in Russian). *Voprosy mekh. grunt.*: 68–79. Checheno Ingush. knizh. izd.

Kudryavtsev, B.A., V.Z.Parton, Yu.A. Pyeskov & G.Z.Cherepanov 1970. Local plastic zone about extremity of gap (in Russian). *Mekh. tverd. tela* 1: 61–64.

Kulhawy, F.H. & J.M.Duncan 1972. Stresses and movements in Oroville Dam. *Proc. ASCE SM* 98(7): 653–665.

Kuperman, V.L., Yu.N.Myznikov & L.N.Topolov 1987. *Hydropower construction in the North* (in Russian). Energoatomizdat, Moscow.

Lagichev, Yu.P. 1982. *Use of finite-element method in design of earth dams* (in Russian). Univ. druzhby narodov, Moscow.

Latkher, V.M. & A.T.Li 1984. Seismic stability of Nurek Dam (in Russian). *Gidrot. stroit.* 12: 14–19.

Leonards, G.A. & J.Narain 1963. Flexibility of clay upon cracking of earth dam. *Proc. ASCE SM* 6(89): 43–59.

Lin, T.G. 1976. Physical thery of plasticity. *Probl. teor. plast.* 7: 7–68. Mir, Moscow.

Lombardo, V.N. 1973. Algorithm of numerical solution of dynamic and static plane problems in the theory of elasticity (in Russian). *Izvestiya VNIIG* 103: 152–163.

Lombardo, V.N. 1974. Correction of instrumental records of displacements and accelerations due to earthquakes (in Russian). *Izvestiya VNIIG* 105: 156–161.

Lombardo, V.N. 1978. Formulation of seismic data in seismic computations of gravity structures coworking with foundations (in Russian). *Izvestiya VNIIG* 103: 164–170.

Lombardo, V.N. 1983. Inclusion of elastic and inertia forces in determination of seismic loads on Kurpsay Dam (in Russian). *Gidrotekh. sooruzh.* 4: 16–23.

Lomize, G.M. 1963. Assessment of strength of clayey soil by laboratory data (in Russian). *Sb. trud. Gidropr.* 9: 1–23.

Lomize, G.M. & V.G.Fedorov 1975. Effect of initial condition of skeletal-clay soil on deformability and strength of soil (in Russian). *Gidr. Stroit.* 12: 19–28.

Lomize, G.M., A.M.Gutkin & N.V.Zhukov 1963. Investigations on rheological properties of plastic clay (in Russian). *Osnov. fundam. mekh. gruntov* 2: 1–4.

Lomize, G.M., N.I.Ivashchenko & M.N.Zakharov 1970. Deformability of clayey soil under complex loading (in Russian). *Osnovaniya* 6: 3–5.

MacCracken D. & U.Dorn 1977. *Numerical methods and FORTRAN programming* (in Russian). Mir, Moscow.

Malyshev, M.V. 1963. Effect of mean principal stress on soil strength and slip surface (in Russian). *Osnov. fund.* 1: 7–11.

Malyshev, M.V. 1964. Computation of pore pressure during construction in soil fills containing water and air in pores (in Russian). *Osnov. fund* 5: 5–7.

Malyshev, M.V. & E.D.Fradic 1968. Strength conditions in sandy soil. *Proceed. Hungar. Acad. Sc.* 1-4: 167–175.

Marsal, R.L. & L.R.Arellano 1965. *Observations on El-Infiernillo Dam upon construction...* (in Spanish). Mexico.

Maslov, N.N. 1982. *Fundamentals of engineering geology and soil mechanics* (in Russian). Vysshaya Shkola, Moscow.

Matsumoto, T. 1976. Finite-element analysis of immediate and consolidation strain based on effective stress principle. *Soil and Found.* 16(4): 23–34.

Melentov, V.A., N.P.Kolpashnikov & B.A.Volnin 1973. *Hydraulic-fill hydraulic engineering structures* (in Russian). Energiya, Moscow.

Meschyan, S.R. 1967. *Creep of clayey soil* (in Russian). Izd. AN ArmSSR, Yerevan.

Meschyan, S.R. 1985. *Experimental rheology of clayey soil* (in Russian). Nedra, Moscow.

Meschyan, S.R. & I.S.Moiseyev 1977. *Earth-rockfill dams* (in Russian). Energiya, Moscow.

Mikhailov, G.K. 1954. Simplified method of seepage computations for homogeneousanisotropic soil (in Russian). *Inzh. ob.* 19: 159–160.

Mostkov, M.A. 1954. *Hydraulics manual* (in Russian). Gosstroyizdat, Moscow.

Nedriga, V.P. 1967. Seepage in homogeneous hydraulic-fill dams on permeable foundation (in Russian). *Trudy VODGEO* 32.

Nedriga, V.P. 1983. *Hydraulic engineering structures. Design manual* (in Russian). Stroyizdat, Moscow.

Nedriga, V.P. & G.I.Pokrovskiy 1976. Method of seepage computations for large dams of local materials and horizontal water-confining stratum (in Russian). *Trudy VODGEO* 52: 37–40.

Nedriga, V.P. & G.I.Pokrovskiy 1977. Seepage computations for homogeneous dams constructed by controlled explosions (in Russian). *Trudy Gidropr.* 59.

Nelson-Skornyakov, F.B. 1949. *Seepage in homogeneous dam* (in Russian). Sovet. Nauka, Moscow.

Newmark, N. & E.Rozenbluet 1980. *Fundamentals of seismically stable construction* (in Russian). Stroyizdat, Moscow.

Nichiporovich, A.A. 1973. *Dams from local materials* (in Russian). Stroyizdat, Moscow.

Nichiporovich, A.A. & T.I.Tsybulnik 1963. Determination of pore pressure is slightly permeable soil of dam body upon consolidation (in Russian). *Voprosy proyekt*: 5–35. Stroyizdat, Moscow.

Nikolaevskiy, V.N. 1962. Dynamics of liquid-saturated consolidating porous media (in Russian). *Inzh. zhurn.* 3: 52–56.

Okamoto, Sh. 1980. *Seismic stability of engineering structures* (in Russian). Stroyizdat, Moscow.

Olovin, B.A. & B.A.Medvedov 1980. *Dynamics of temperature field in Viluy hydroelectric power plant* (in Russian). Nauka, Novosybirsk.

Osadchiy, L.G. & P.U.Bakhtiyarov 1975. Rogun scheme on Vakhsh River (in Russian). *Gidrot. Stroit.* 4: 10–13.

Panovko, Ya.G. 1971. *Introduction to theory of mechanical oscillations* (in Russian). Nauka, Moscow.

Pavlovskiy, N.N. 1956. *Groundwater flow* (in Russian). Izd. AN SSSR, Moscow.

Pokrovskiy, G.I. 1974. *Dam construction by controlled explosion* (in Russian). Nedra, Moscow.

Pol, B. 1975. Macroscopic criteria of plastic flow and brittle failure (in Russian). *Razrusheniye* 2: 336–520. Mir, Moscow.

Polubarinova-Kochina, P.Ya. 1977. *Theory of groundwater motion* (in Russian). Nauka, Moscow.

Pravednyy, G.Kh. 1966. *Design and selection of grain composition of filters in transitions of large dams* (in Russian). Energiya, Leningrad.

Pronina, G.E. 1983. *Mathematical analysis and optimum design of complex technological systems* (in Russian). VIA, Moscow.

Rabotnov, Yu.N. 1979. *Mechanics of deformable solid body* (in Russian). Nauka, Moscow.

Radchenko, V.G. & B.A.Zairova 1971. *Earth-rockfill and rockfill dams* (in Russian). Energiya, Leningrad.

Rasskazov, L.N. 1968. Experiments on shear resistance of coarse grained soil (in Russian). *Trudy VODGEO* 1: 92–97.

Rasskazov, L.N. 1969. Selection of soil characteristics in stress CAD computations for dams from local materials (in Russian). *Trudy VODGEO* 5: 115–125.

Rasskazov, L.N. 1973. Soil as material for dam body (in Russian). *Gidrot. stroit.* 8: 40–43.

Rasskazov, L.N. 1974. Soil strength condition (in Russian). *Trudy VODGEO* 44: 53–59.

Rasskazov, L.N. 1977. Construction scheme and stress-strain condition in earth dam with central core (in Russian). *Energ. stroit.* 2: 65–75.

Rasskazov, L.N. & A.A.Belakov 1982. Stress-stress computations for earth-rockfill dam (in Russian). *Gidrot. stroit.* 2: 16–22.

Rasskazov, L.N. & D.Dzhkha 1977. Deformability and strength of soil in computation of large earth dams (in Russian). *Gidrot. stroit.* 7: 31–36.

Rasskazov, L.N. & D.Dzhkha 1978. Selection of rational design of earth-rockfill dam (in Russian). *Energ. stroit.* 2: 60–67.

Rasskazov, L.N. & I.L.Orekhova 1982. Unique optimization parameter for optimum dam design (in Russian). *Energ. stroit.* 8: 42–44.

Rasskazov, L.N. & I.L.Orekhova 1985. Optimum design of earth dams (in Russian). *Gidrot. stroit.* 7: 32–37.

Rasskazov, L.N., O.A.Orlova & K.I.Orlov 1983. Investigation of dynamic properties of soil under complex stress conditions (in Russian). *Seventh Danube-Eur. Confer.* 1: 133–136. Kishinev.

Rasskazov, L.N. & Yu.N.Sysoyev 1982. Analysis of operation of large earth-rockfill dam on bedrock (in Russian). *Energ. stroit.* 3: 70–74.

Rasskazov, L.N. & M.V.Vitenberg 1972. Stress-strain condition of dams from local materials, and their stability (in Russian). *Trudy VODGEO* 34: 18–32.

Rasskazov, L.N. & M.N.Volokhova 1974. Stress-strain condition of dams from local materials with seismic effects (in Russian). *Trudy VODGEO* 44: 75–80.

Reiner, M. 1947. *Ten lectures on theoretical rheology* (in Russian). OGIZ Gostekhizdat, Moscow.

Rekleitis, G., A.Reivindran & K.Rexdell 1986. *Optimization in technology* (in Russian). Mir, Moscow.

Reyfman, L.S., V.F.Korchevskiy & G.N.Petrov 1982. Prediction methods for geotechnical and seepage properties of dams constructed by controlled explosion (in Russian). *Sb. nauch. trudov Gidropr.* 88: 41–55.

Rozanov, N.N. 1983. *Dams from earth materials* (in Russian). Stroyizdat, Moscow.

Rozin, L.A. 1978. *Variational formulation for elastic systems* (in Russian). Izd. LGU, Leningrad.

Sherard, J., R.J.Woodward, S.F.Gizienski & W.A.Clevenger 1967. *Earth and earth-rockfill dams*. New York.

Shevchenko, I.N. V.A.Savvina & A.I.Teytelbaum 1974. Determination of parameters of clayey soil for assessment of cracking in dam cores (in Russian). *Trudy VODGEO* 44: 86–89.

Shirinkulov, T. & A.Dasibiekov 1966. Solution of one-dimensional problem of consolidation for three-phase earth medium with inclusion of nonlinear creep (in Russian). *AN UzSSR ser. tekhn. nauk* 5: 27–32.

Sinitsyn, A.P. 1960. *Oscillations of triangular wedge and dams vibrations* (in Russian). VIA, Moscow.

Sneddon, I. 1955. *Fourier transforms* (in Russian). Izd. inostr. lit., Moscow.

SNiP P 16-76, Foundations of hydraulic engineering structures (in Russian) 1977. Stroyizdat, Moscow.

SNiP 2.06.04-82 Loads and effects on hydraulic engineering structures (ice-, wave- and ship-induced) (in Russian) 1983. Stroyizdat, Moscow.

SNiP 2.06.05.84 Earth dams (in Russian) 1984. Stroyizdat, Moscow.

Tan-Tuon-Kie 1957. Secondary time effects and consolidation of clays. *Acad. Sin./Inst. Civ. Eng. and Arch. Soil Mech. Lab.* 57(7): 1–17.

Taylor, P.W. & J.M.Parton 1966. Dynamic torsion tests of soil (in French). *Proc. 7th Intern. Congr. Soil Mech. Found. Eng.* 425–432.

Ter-Martirosyan, Z.G. 1973. One-dimensional problem of consolidation of multi-phase soil under variable load and head at boundary (in Russian). *Dokl. Mezhdunar. kongr*: 87-92. Stroyizdat, Moscow.

Terzaghi, K. 1933. *Soil mechanics for civil engineering* (in Russian). Gosstroyizdat, Moscow.

Teytelbaum, A.I., V.G.Melnik & V.A.Savvina 1975. *Cracking in cores and screens of earth-rockfill dams* (in Russian). Stroyizdat, Moscow.

Tikhonov, A.N. & A.A.Samarskiy 1966. *Equations of mathematical physics* (in Russian). Nauka, Moscow.

Tsytovich, N.A. 1973a. *Mechanics of frozen soil* (in Russian). Vysshaya shkola, Moscow.

Tsytovich, N.A. 1973b. *Soil mechanics* (in Russian). Vysshaya shkola, Moscow.

Vazov, V. & J.Forsyth 1963. *Difference solutions of partial differential equations* (in Russian). Izd. Inostr. Lit., Moscow.

Verigin, N.N. 1957. *Kinetics of leaching and removal of salts by groundwater flow: Dissolution and leaching of rock (in Russian)*. Gosstroyizdat, Moscow.

Verigin, N.N. 1960. Consolidation of water-saturated soil due to external load normal to boundary of halfspace (in Russian). *Sixth Intern. Congr. Soil Mech. Found. Engng*: 26–31. VNII VODGEO, Moscow.

Verigin, N.N. & A.Ye.Oradovskaya 1960. *Methodology of tests on salt dissolution in soil of hydraulic engineering foundations* (in Russian). VNII VODGEO, Moscow.

Volokhova, M.N. & L.N.Rasskazov 1974. Some seismic design problems for dams constructed from local materials (in Russian). *Trudy VODGEO* 44: 80–86.

Vovkushevskiy, A.K. & V.A.Zeyliger 1980. *Computer program for elasticity problem with unilateral bands and Poisson's equation by FEM for BESM-6* (in Russian). VNIIG, Leningrad.

Vyalov, S.S. 1959. *Rheological properties and bearing capacity of frozen soil* (in Russian). AN SSSR, Moscow.

Yokoo, Y. K.Yamagata & H.Nagaoka 1971. Finite-element method applied to Biot's consolidation theory. *Soil and Found.* 11(1): 16–27.

Yevdokimov, P.D., S.S.Bushkanets & T.F.Lipovetskaya 1977. Experience from tests on dam construction on weak soil (in Russian). Izvestiya VNIIG 117: 76-81.

Zaretskiy, Yu. K. 1967. *Theory of soil consolidation* (in Russian). Nauka, Moscow.

Zaretskiy, Yu. K. 1983. *Statics and dynamics of earth dams* (in Russian). Energoatomizdat, Moscow.

Zedginidze, I.G. 1976. *Planning of experiments in investigation of multi-component systems* (in Russian). Nauka, Moscow.

Zeldovich, Ya.Yu. & I.M.Yaglom 1982. *Higher mathematics for beginning physicists and technicians* (in Russian). Nauka, Moscow.

Zhilenkov, V.N. 1967. Seepage in fissured rock (in Russian). *Izvestiya VNIIG* 84: 86-93.

Zhurek, Ya. 1965. Deformability of coarse materials under high specific pressure (in Russian). *Trudy VODGEO* 11: 10–17.

Zienkiewicz, O. 1975. *Method of finite elements in technology* (in Russian). Mir, Moscow.